Advances in Experimental Medicine and Biology

Volume 1018

More information about this series at http://www.springer.com/series/5584

Qiliang Cai • Zhenghong Yuan • Ke Lan
Editors

Infectious Agents Associated Cancers: Epidemiology and Molecular Biology

Springer

Editors
Qiliang Cai
Key Laboratory of Medical Molecular Virology (Ministries of Education and Health), School of Basic Medical Science
Fudan University
Shanghai, People's Republic of China

Zhenghong Yuan
Key Laboratory of Medical Molecular Virology (Ministries of Education and Health), School of Basic Medical Science
Fudan University
Shanghai, People's Republic of China

Ke Lan
State Key Laboratory of Virology, College of Life Science
Wuhan University
Wuhan, People's Republic of China

ISSN 0065-2598 ISSN 2214-8019 (electronic)
Advances in Experimental Medicine and Biology
ISBN 978-981-10-5764-9 ISBN 978-981-10-5765-6 (eBook)
DOI 10.1007/978-981-10-5765-6

Library of Congress Control Number: 2017954956

Printed on acid-free paper

This Springer imprint is published by Springer Nature
The registered company is Springer Nature Singapore Pte Ltd.
The registered company address is: 152 Beach Road, #21-01/04 Gateway East, Singapore 189721, Singapore

Preface

Recent advances in the fields of molecular biology and epidemiology have led to significant development of studying revelations between infectious agents and cancer, and provide valuable insights into the molecular basis of carcinogenesis. Since the first oncogenic virus was discovered by Rous in 1911, many infectious agents including viruses, bacteria, and parasites are known to associate with about one-fifth of all human cancers worldwide. Their impact on global health is significant.

Despite that the link between cancer and microbe infection has been recognized in chickens over ten decades ago, the mechanistic basis for cell transformation became clearer until 1970. The discoveries of oncogenes/proto-oncogenes and tumor suppressors, as well as insights into cell growth factors, cell cycle regulation, checkpoints, and their operative protein factors, further promoted the understanding of infectious agents-associated cancers. Today in 2017, although close association between viruses and cancer has been established only in seven human viruses (HBV, HCV, HPV, HTLV, EBV, KSHV, and MCV), other infectious microbes including HIV, bacteria (*H. pylori*), parasites (blood flukes/liver flukes), and prions as presented in many of the chapters in the book illustrate a potential association with a variety of human cancers. The interplay between microbes and various microenvironment factors, including stress, inflammation, and deregulation of immune responses, is currently a hot topic in the field of microbe-related cancer.

A few important scientific concepts detailed in the various chapters include the animal tumor models, the coinfection of different microbes, the interplay between microbe and microenvironmental stress, the multistepped process for cell transformation caused by infectious microbe, and the common and various mechanisms used by different types of infectious microbes. From such mechanistic and translational research, we hope that safe and more effective therapeutic drugs or vaccines against specific cancers will ensue in the future.

This book emerged from a desire to provide an up-to-date progress of human cancers and their infectious causes. The editors have made great efforts to bring together teams of expert authors from all aspects of infectious microbes associated with cancer in this book. As tumor virologists, they have personally witnessed many pivotal advances in infectious causes of human cancers. It is therefore appropriate for them and all authors to review current and past exciting discoveries on human infectious oncology, to share with a larger audience, particularly for colleagues and students. To make the list of infectious agents complete, this book includes not only viruses but also bacteria, parasites, and prions. However, the emphasis is on viruses, as viruses have more complex interactions with hosts, and new breakthroughs in oncovirology are to be expected.

I believe this book will help the readers to better understand the pathogenesis of human cancers, particularly those that are associated with infectious microbes, and assist in the development of new and more effective strategies toward prevention and treatment of cancer in the future.

Key Laboratory of Medical Molecular Virology
(Ministries of Education and Health)
School of Basic Medical Science
Fudan University
Shanghai, People's Republic of China

Yumei Wen

Contents

Contributors

Lei Bai Key Laboratory of Molecular Virology and Immunology, Institute Pasteur of Shanghai, Chinese Academy of Sciences, Shanghai, People's Republic of China

Cankun Cai Key Laboratory of Medical Molecular Virology (Ministries of Education and Health), School of Basic Medical Sciences, Fudan University, Shanghai, People's Republic of China

Qiliang Cai Key Laboratory of Medical Molecular Virology (Ministries of Education and Health), School of Basic Medical Science, Fudan University, Shanghai, People's Republic of China

Chi-Ping Chan School of Biomedical Sciences, The University of Hong Kong, Pokfulam, Hong Kong

Xunjia Cheng Department of Medical Microbiology and Parasitology, Institute of Biomedical Sciences, School of Basic Medical Sciences, Fudan University, Shanghai, People's Republic of China

Zhijun Cheng Department of the First Abdominal Surgery, Jiangxi Tumor Hospital, Nanchang, People's Republic of China

Jiazhen Dong Key Laboratory of Molecular Virology and Immunology, Institute Pasteur of Shanghai, Chinese Academy of Sciences, Shanghai, People's Republic of China

Sihan Dong Key Laboratory of Molecular Virology & Immunology, Institut Pasteur of Shanghai, Chinese Academy of Sciences, Shanghai, People's Republic of China

Meng Feng Department of Medical Microbiology and Parasitology, Institute of Biomedical Sciences, School of Basic Medical Sciences, Fudan University, Shanghai, People's Republic of China

J. Craig Forrest Department of Microbiology and Immunology and Center for Microbial Pathogenesis and Host Inflammatory Responses, University of Arkansas for Medical Sciences, Little Rock, AR, USA

Zhenxing Gao State Key Laboratory of Virology, Institute of Virology, Chinese Academy of Sciences, Wuhan, Hubei Province, People's Republic of China

Feng Gu Key Laboratory of Medical Molecular Virology (Ministries of Education and Health), School of Basic Medical Sciences, Fudan University, Shanghai, People's Republic of China

Yongjia Ji Department of Infectious Diseases, Shanghai Public Health Clinical Center, Fudan University, Shanghai, People's Republic of China

Dong-Yan Jin School of Biomedical Sciences, The University of Hong Kong, Pokfulam, Hong Kong

Kin-Hang Kok Department of Microbiology, The University of Hong Kong, Pokfulam, Hong Kong

Ke Lan State Key Laboratory of Virology, College of Life Science, Wuhan University, Wuhan, People's Republic of China

Alexandria E. Lewis Department of Otorhinolaryngology-Head and Neck Surgery, and Microbiology, Tumor Virology Program, Abramson Comprehensive Cancer Center, Perelman School of Medicine at the University of Pennsylvania, Philadelphia, PA, USA

Chaoyang Li State Key Laboratory of Virology, Wuhan Institute of Virology, Chinese Academy of Sciences, Wuhan, Hubei, People's Republic of China

Hubei Collaborative Innovation Center for Industrial Fermentation, Wuhan, Hubei, People's Republic of China

Shasha Li State Key Laboratory of Virology, College of Life Sciences, Wuhan University, Wuhan, People's Republic of China

Yanyun Li Hospital and Institute of Obstetrics and Gynecology, Shanghai Medical College, Fudan University, Shanghai, People's Republic of China

Xiaozhen Liang Key Laboratory of Molecular Virology & Immunology (Chinese Academy of Sciences), Institut Pasteur of Shanghai, Shanghai, People's Republic of China

Lin Jia School of Biomedical Sciences, Li Ka Shing Faculty of Medicine, The University of Hong Kong, Pokfulam, Hong Kong

Chun Lu Key Laboratory of Pathogen Biology (Jiangsu Province), Nanjing Medical University, Nanjing, People's Republic of China

Hongzhou Lu Department of Infectious Diseases, Shanghai Public Health Clinical Center, Fudan University, Shanghai, People's Republic of China

Department of Infectious Diseases, Huashan Hospital Affiliated to Fudan University, Shanghai, People's Republic of China

Department of Internal Medicine, Shanghai Medical College, Fudan University, Shanghai, People's Republic of China

Margo MacDonald Department of Microbiology, Perelman School of Medicine, University of Pennsylvania, Philadelphia, PA, USA

Yonggang Pei Department of Otorhinolaryngology-Head and Neck Surgery, and Microbiology, Tumor Virology Program, Abramson Comprehensive Cancer Center, Perelman School of Medicine at the University of Pennsylvania, Philadelphia, PA, USA

Jie Qin Key Laboratory of Pathogen Biology (Jiangsu Province), Nanjing Medical University, Nanjing, People's Republic of China

Erle S. Robertson Department of Otorhinolaryngology-Head and Neck Surgery, and Microbiology, Tumor Virology Program, Abramson Comprehensive Cancer Center, Perelman School of Medicine at the University of Pennsylvania, Philadelphia, PA, USA

Run Shi Wuhan Institute of Virology, Chinese Academy of Sciences, University of Chinese Academy of Sciences, Wuhan, Hubei Province, People's Republic of China

Rui Sun Key Laboratory of Molecular Virology and Immunology (Chinese Academy of Sciences), Institut Pasteur of Shanghai, Shanghai, People's Republic of China

Chi Man Tsang School of Biomedical Sciences, Li Ka Shing Faculty of Medicine, The University of Hong Kong, Pokfulam, Hong Kong

Sai Wah Tsao School of Biomedical Sciences, Li Ka Shing Faculty of Medicine, The University of Hong Kong, Pokfulam, Hong Kong

Yuyan Wang Key Laboratory of Medical Molecular Virology (Ministries of Education and Health), School of Basic Medical Sciences, Fudan University, Shanghai, People's Republic of China

Fang Wei Sheng Yushou Center of Cell Biology and Immunology, School of Life Sciences and Biotechnology, Shanghai Jiao Tong University, Shanghai, People's Republic of China

Guiru Wu Wuhan Institute of Virology, Chinese Academy of Sciences, University of Chinese Academy of Sciences, Wuhan, Hubei Province, People's Republic of China

Youhua Xie Key Laboratory of Medical Molecular Virology (Ministries of Education and Health), Shanghai Medical College, Fudan University, Shanghai, People's Republic of China

Congjian Xu Hospital and Institute of Obstetrics and Gynecology, Shanghai Medical College, Fudan University, Shanghai, People's Republic of China

Xiaowen Yang Department of the First Abdominal Surgery, Jiangxi Tumor Hospital, Nanchang, People's Republic of China

Zhigang Yi Key Laboratory of Medical Molecular Virology (Ministries of Education and Health), School of Basic Medical Sciences, Fudan University, Shanghai, People's Republic of China

Jianxin You Department of Microbiology, Perelman School of Medicine, University of Pennsylvania, Philadelphia, PA, USA

Zhenghong Yuan Key Laboratory of Medical Molecular Virology (Ministries of Education and Health), School of Basic Medical Science, Fudan University, Shanghai, People's Republic of China

Jun Zhang School of Biomedical Sciences, Li Ka Shing Faculty of Medicine, The University of Hong Kong, Pokfulam, Hong Kong

Lihua Zhang Department of Pathology, Zhongda Hospital, Southeast University, Nanjing, People's Republic of China

Caixia Zhu Key Laboratory of Medical Molecular Virology (Ministries of Education and Health), School of Basic Medical Sciences, Fudan University, Shanghai, People's Republic of China

Qing Zhu Key Laboratory of Medical Molecular Virology (Ministries of Education and Health), School of Basic Medical Sciences, Fudan University, Shanghai, People's Republic of China

Chapter 1
Overview of Infectious Causes of Human Cancers

Qiliang Cai and Zhenghong Yuan

Abstract Emerging evidence has demonstrated that almost each person is infected at least one potentially cancer-causing organism; however, only a small proportion of infected individual develops cancer. In this review, to elucidate the potential role of infectious organisms in the development and progression of human cancers, we summarize the previous history and current understandings of infection-associated cancers and highlight the common molecular mechanisms of cancers caused by infectious agents and their potential cofactors, which may bring us to effectively prevent and reduce the infection-associated cancers in the future.

Keywords Infection-associated cancer • Molecular mechanisms • Epidemiological distribution

1.1 Introduction

It has been estimated that over 99% people worldwide are infected with at least one potential cancer-causing organism during whole lifetime and about six million people die each year of cancer [1]. Globally, at least 16% of all cancers are associated with chronic infections, while in the developing countries the proportion could be greater than 25% [1], which is underestimated due to absence of cases of infections acting as cocarcinogens.

Although it has been debated for over two centuries whether cancer is an infectious disease or not, the interval between the first recognition of the virus (tobacco mosaic by Ivanovsky in 1892) and the first proposal that animal virus plays a critical role in some cancer formation (foot-and-mouth virus by Loeffler and Frosch in 1898) was short [2]. After yellow fever as the first human virus identified by Reed

Q. Cai (✉) • Z. Yuan
Key Laboratory of Medical Molecular Virology (Ministries of Education and Health), School of Basic Medical Science, Fudan University, Shanghai 200032, People's Republic of China
e-mail: qiliang@fudan.edu.cn

Q. Cai et al. (eds.), *Infectious Agents Associated Cancers: Epidemiology and Molecular Biology*, Advances in Experimental Medicine and Biology 1018,
DOI 10.1007/978-981-10-5765-6_1

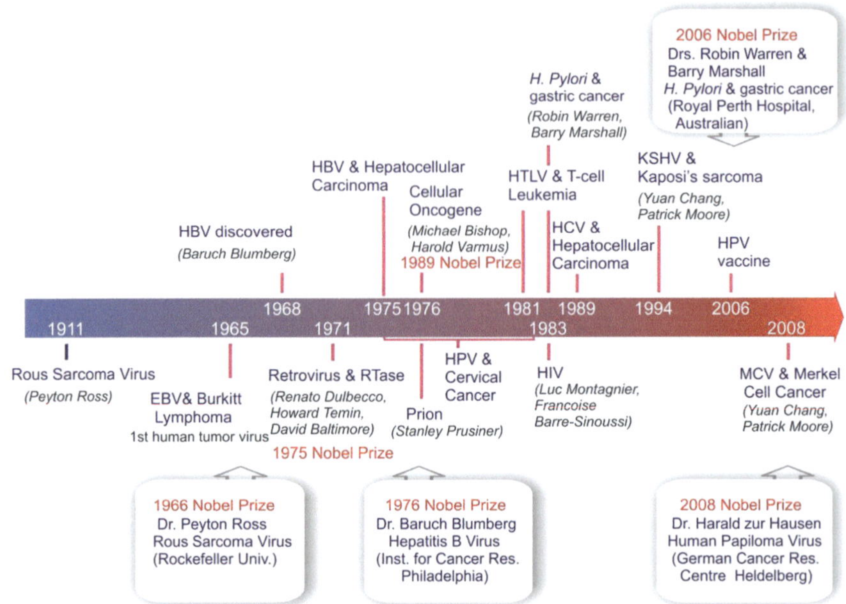

Fig. 1.1 Timeline of milestones of discoveries in infectious causes of human cancer

in 1901, viruses have been proposed as common causes of cancer 2 years later, which led to establishment of how important and widespread viruses are as human carcinogens and the appearance of concepts of the oncogene and tumor suppressor (like p53) [2]. Several key milestones occur in history of infection-associated cancer as shown in Fig. 1.1. The first tumor virus was discovered by Peyton Rous from Rockefeller Institute in 1911, which he provided the first experimental proof of the malignant avian tumor is dependent on a filterable virus [3, 4]. However, despite Rous went on to confirm that other avian tumors were also transmissible in a similar fashion, there were many arguments against that filtrations could be inadequate to remove all cell-fragments, or no relevance for human cancers. Until the 1930s, two major events further stimulate the notion of infection as a significant cause of cancer—one is that Richard Shope from Rockefeller Institute also reported that papillomaviruses can induce tumors in rabbits [5] and another is that Nobel Prize in 1926 was awarded to Johannes Fibiger, a Danish medical researcher who has demonstrated that a nematode worm causes stomach cancer in laboratory rats, albeit it has been proved later that diet (vitamin A-deficient rat) not nematodes was the crucial factor causing the cancer [6, 7]. Due to Fibiger's "mistaken" Nobel award, Rous eventually received a Nobel Prize for his viral cancer-related work 40 years later (just 4 years before he died) [8]. In addition, it is also worthy to mention John Bittner's discoveries of nonchromosomal influence in the incidence of murine mammary tumor [9], which leads to the identifications of the first retroviruses called MMTV and reverse transcriptase later.

In human, Burkitt's lymphoma (BL) was first described as a sarcoma in African children in 1958 by Denis Burkitt who is a surgeon of Uganda [10]. Three years later, virologist Anthony Epstein occasionally attended Burkitt's lecture and was intrigued by the possibility of a viral cause and started to prove his speculation based on frozen tumor samples which are kindly provided by Dr. Burkitt [11]. Another 3 years later (in 1964), Epstein, along with colleagues Bert Achong and Yvonne Barr, identified the first human tumor virus named as Epstein-Barr virus (EBV) from Burkitt lymphoma cell line by using electron microscopy [12]. However, the viral genome of EBV B95.8 strain was fully sequenced until 20 years later [13].

In contrast to EBV, *Helicobacter pylori* was the first bacterium bug found to associate with gastritis and peptic ulceration by Robin Warren and Barry Marshall in 1984 [14, 15], which led to the award of Nobel Prize for Physiology or Medicine in 2006. Initially, Warren failed to culture the organism by standard 48-h culture protocol and succeed by a chance for 5 days of culture due to leftovers from Easter holiday [16]. Although stomach ulcers were previously ascribed to diet (too many alcohol and spicy food), increasing evidence demonstrated that chronic inflammation is linked with cancer, and *H. pylori* was considered as a carcinogen that directly contributes to malignant transformation of stomach ulcers.

During the development history of infectious cause of cancers, we have to mention two key events. One is the discoveries of retroviruses and reverse transcriptase by Baltimore, Dulbecco, and Temin in 1975; another is the understanding of the oncogene and the tumor suppressor gene, which were initially introduced as of virus origin by Huebner and Todaro in 1969 [17] and were later termed cellular oncogenes or proto-oncogenes (their normal functions are to promote cell growth and division, while malignant cell occurs due to they are expressed aberrantly) by Bishop and Vermus in 1976 [18], which subsequently led them to receive a Nobel Prize in 1989.

Based on the discovery of cellular oncogenes, Harald zur Hausen began series of studies to demonstrate the relationship between HPV and cervical cancer in the 1970s [19], which eventually led to a Nobel Prize in 2008. In addition, it is worthy to mention that two important tumor viruses, namely, Kaposi's sarcoma-associated virus (KHSV) and Merkel cell polyomavirus (MCV), were discovered by Drs. Yuan Chang and Patrick Moore (a couple who from the University of Pittsburgh) in 1994 and 2008, which are the etiology causes of Kaposi's sarcoma and Merkel carcinoma, respectively [20–22]. This indicates that more and more new tumor viruses will be discovered as our researches are ongoing in the future.

1.2 Basic Molecular Mechanisms of Cancer Caused by Infection

It is well known now that carcinogenesis caused by infection may be direct or indirect. For example, insertion of viral genes into the host cell's genome will trigger cell malignant transformation, while induction of chronic inflammation (i.e., cirrhosis, chronic gastritis) will create a local microenvironment with a greatly increased risk of cell transformation, or when the infective organism suppresses the host immune response, it will trigger tumorigenesis. In general, direct cause increases the risk of individual cells to malignant transformation, while indirect cause usually acts at tissue microenvironment level to increase the risk of emergence of a malignant clone.

1.2.1 Direct Cause by Infection

To date, it is well known that chromosome instability is a virtual feature of cancer cells. To understand the molecular biology of infectious causes of cancer, we need to know normal cell biology and then to interpret how the malignant cell subverts the normal processes. Some key concepts have been demonstrated in cell malignant transformation, which include *apoptosis* (one type of cell death, which is distinct from necrosis that induces inflammatory response) and *the cell cycle* (cells divide in an ordered sequence of G1, S, G2, and M phase under the control of genes including cyclins and cyclin-dependent kinases, while some cells are not dividing or preparing to divide, called G0 or rest phase). During the process of cell malignant transformation, a normal cell usually occurs as a complex of genetic changes, which regulate cell growth, division, and death, as well as escape from localization controls of basement membrane integrity (metastases). The major types of these essential genes usually included oncogenes (which drive pathological cell division), tumor suppressor genes (which normally inhibit growth and division), and DNA repair genes (which lose ability to maintain genomic integrity).

Since oncogene was discovered in about four decades ago, it has been well demonstrated that oncogenes are frequently active and typically act in a dominant fashion to drive forward the cell cycle and cell division. The normal functions of these oncogenes are usually acting as growth factors (messages communicate between cells in blood), cell surface receptors (receive and pass chemical messages from one to other cells), transcription factors (regulate genes on or off), or signal transmission proteins (carry the signal from the cell surface receptors to the nucleus). In contrast to oncogene functions as an accelerator of cell division, tumor suppressor gene is a braker [23]. In general, oncogene and tumor suppressor gene operate cooperatively during the cell cycle. The main function of tumor suppressor genes is to activate DNA repair process once any abnormal DNA occurs. If repair is unsuccessful, the cell will initiate apoptosis and sacrifice itself. Therefore, it is not surprising that the arrest of the cell division process will fail and lead to generation of faulty daughter cells with high-potential malignant property, once tumor suppressor genes are absent or defective. Many cellular tumor suppressors have been discovered so far. One of the most important tumor suppressors is p53. It is well known that p53 is the key gene (also called the "guardian of the genome") to initiate cellular repair pathways in response to DNA damage and to ensure apoptosis of any cell with irreparably defective genome. Other prominent examples include BRCA1, BRCA2, and retinoblastoma (Rb). Given that the DNA repair genes are of great importance to the cell and highly conserved across species in eukaryote evolution, the dysfunction of DNA repair process by infection of organisms through various different pathways will lead to chromosome abnormality. To directly increase the cell dividing and risk of a cell acquiring mutation, one of common strategies used by pathogen infection is to deregulate the promoters of oncogene and tumor suppressor genes and another is to encode oncoproteins which may directly deregulate cell cycle, alter apoptotic or other cell signal pathways. These are basic molecular mechanisms why infection of organism could trigger cell transformation.

1.2.2 Indirect Cause by Tissue Microenvironment

Another important trigger of cancer malignancy caused by infection is induction of chronic inflammation within tumor tissue microenvironment. Inflammation is a protective response of body tissues to different harmful stimuli by activating immune cell, blood vessels, and molecular mediators. It can be classified as two types: acute or chronic. Acute inflammatory response will increase movement of plasma and leukocyte (especially granulocytes) from the blood into the injured tissues to eliminate the initial cause of cell injury and repair, while prolonged inflammation, also known as chronic inflammation, will result into a progressive shift of cell types such as mononuclear cells at the injury tissues and stimulate destruction and healing of the tissue during inflammatory process. The role of chronic inflammation in cell malignancy is to create a background in which oncogenesis is more likely to occur, while inflammation alone is not sufficient to induce malignant diseases. It is known that persistent organism infection may cause chronic inflammation by producing cytotoxic molecules into tissue microenvironment, which will alter cellular immune response and eventually lead to inflammation. For instances, the persistent infection of hepatitis viruses can cause cirrhosis and potentially hepatocellular carcinoma. Another example is that the chronic inflammation induced by *H. pylori* infection and chronic gastritis.

During the inflammatory process of tumor development, one of key events is angiogenesis (also named vasculogenesis). It is a normal physiological process in which new blood vessels form from preexisting vessels in tissue growth and development as well as wound healing. However, as tumors grow, to overcome the requirement for oxygen and nutrient supply, the cancer cells were also found by Judah Folkman in 1971 to release different cytokines to induce angiogenesis around tumor. It has been demonstrated that one of the most famous cytokines associated with angiogenesis is called VEGF. It has been well demonstrated that many organism infection could greatly enhance expression of VEGF through the HIF1α pathway and highly associate with tumor angiogenesis [24]. Given the role of angiogenesis in driving cancer development, it has long been proposed as a potential target for anticancer therapy. In 1994, although Judah Folkman and his team in the first time reported promising results of endostatin as an antiangiogenic compound in inhibition of new vessel development both in vitro and in murine models, there is no effective anti-angiogenesis agent applied in clinical cancer therapy.

1.3 Epidemiological Distribution of Infection-Associated Cancers

The chief contributors to the burden of infection-associated cancer discovered so far are viruses including EBV (Epstein-Barr virus, review in Chaps. 5 and 6) [25], hepatitis virus (HBV and HCV, review in Chaps. 2 and 8, respectively) [26], human papillomaviruses (HPV 16 and 18, review in Chap. 3) [27], HLTV-1 (human T lymphotropic virus-1, review in Chap. 9), KSHV (Kaposi's sarcoma-associated herpesvirus, review in Chap. 7), and MCV (Merkel cell polyomavirus, review in Chap. 4). These are estimated to account for over 90 % of infection-associated cancers. In addition to viruses, nonviral infections including bacterium such as *Helicobacter pylori* (gastric cancer and lymphoma, review in Chap. 11) [28] and parasite including *Schistosoma* species and liver flukes (review in Chap. 12) as well as prion (review in Chap. 13) are also known to associate with cancer. Although some pathogens like parasites are rare in the developed world, knowledge of their associated cancer is necessary as it may be encountered in any clinical environment. In many cases, parasite-associated cancers can be found due to very brief exposure, tourists who are vulnerable, or population movement after the long latency between infection and cancer diagnosis. Thus, those who have experience in the developing world may be diagnosed with a parasite-link cancer after they spent many years living in the developed world.

In addition to those infectious organisms definitely identified as causes of specific cancers, there are many others which are suspected of carcinogenic potential and various pathways by which infection may lead to cancer. For example, the patient who is infected with human immunodeficiency virus (HIV) not only cause acquired immunodeficiency syndrome (AIDS) but also increase the risk of EBV- or KSHV-induced lymphoma due to coinfection. Chapters 10 and 15 will review the basic concepts of HIV-associated cancer biology and microbiology, to explore the current understanding of HIV infections which may induce or drive malignant transformation. Given the critical role of tumor microenvironment and animal model in studying infectious causes of cancer, we not only address the recent progress on interplay between microenvironmental abnormalities and infectious agents in tumorigenesis in Chap. 16 but also include Chap. 14 to address the recent understanding of murine gammaherpesvirus 68 (MHV68) as an animal model in studying EBV- and KSHV-associated disease in vivo.

Virtually, most cases of infection-associated cancers occur after an extended latency – sometimes decades long from initial infection to eventual diagnosis of cancer. Commonly, only a small percentage of infected individuals will develop cancer. For instance, in the case of *Helicobacter pylori*, India has one of the highest infection rates in the world, but the incidence of gastric cancer is very low [29]. Therefore, the risk of gastric cancer in *Helicobacter pylori* carriers appears to be determined by a combination of several factors including host genetics, bacterial genetics, and habits of diet, smoking, etc. [30] In conclusion, there is geographical variation in incidence rates of each infection-associated cancer due to several cofactors. The reason why only small population and certain population develop cancer after infection could be due to the consequence of different interactions among several factors as follows: (1) different incidence of relevant infections, (2) timing of infection, (3) biological variability of the infectious agents, (4) genetic variation in host susceptibility to infection, and (5) incidence of external cofactors, e.g., diet and smoking.

1.4 Future Perspective

Given the incredible amount of infectious organisms in the world, it is almost impossible to estimate the amount of infectious organism within an average person's body in his/her whole life. Despite the rate of host cell malignant transformation which is low, the incidence of many of the infections is very high. Due to infection-associated cancers which usually have a very long latency between infection and development of malignancy, reduction of the burden of infection-associated cancer will require a combination of primary prevention (blocking transmission route of infection, boosting host immune resistance against infection by vaccination), and secondary prevention (preventing progression from chronic infection to malignant transformation), based on different infection-associated cancers. For example, in hepatitis B, a compelling evidence has been found that infection during early infancy carries a high risk of eventual liver cancer, while infection in adult confers a much lower risk. Therefore, the priority strategy for prevention of HBV-associated liver cancer is to block transmission from mother to child. Another case is high-risk HPV-associated cervical cancer and, in almost all cases of infection, is acquired early after a woman becomes sexually active. The best strategy for prevention is to ensure vaccination before young women first experience penetrative sex. In contrast, the tropical infections of schistosomiasis and fluke appear to be potentially carcinogenic at any age of population, and the effective interruption of the associated cancer requires a program and may take decades. Thus, to effectively prevent the infection-associated cancers, it requires all medical scientists and healthcare professionals continue to work together and explore the nature of each infection-associated cancers.

References

1. de Martel C, Ferlay J, Franceschi S, Vignat J, Bray F, Forman D et al (2012) Global burden of cancers attributable to infections in 2008: a review and synthetic analysis. Lancet Oncol 13:607–615
2. Javier RT, Butel JS (2008) The history of tumor virology. Cancer Res 68:7693–7706
3. Rous P (1911) A sarcoma of the fowl transmissible by an agent separable from the tumor cells. J Exp Med 13:397–411
4. Rous P (1910) A transmissible avian neoplasm. (Sarcoma of the common fowl.) J Exp Med 12:696–705
5. Shope RE, Hurst EW (1933) Infectious papillomatosis of rabbits : with a note on the histopathology. J Exp Med 58:607–624
6. Hitchcock CR, Bell ET (1952) Studies on the nematode parasite, Gongylonema neoplasticum (spiroptera neoplasticum), and avitaminosis A in the forestomach of rats: comparison with Fibiger's results. J Natl Cancer Inst 12:1345–1387
7. Raju TN (1998) The nobel chronicles. 1926: Johannes Andreas Grib Fibiger (1867–1928). Lancet 352:1635
8. Vogt PK (1996) Peyton Rous: homage and appraisal. FASEB J Off Publ Fed Am Soc Exp Biol 10:1559–1562
9. Jackson RB, Little CC (1933) The existence of non-chromosomal influence in the incidence of mammary tumors in mice. Science 78:465–466
10. Burkitt D (1958) A sarcoma involving the jaws in African children. Br J Surg 46:218–223
11. Coakley D (2006) Denis Burkitt and his contribution to haematology/oncology. Br J Haematol 135:17–25
12. Epstein MA, Achong BG, Barr YM (1964) Virus particles in cultured lymphoblasts from Burkitt's lymphoma. Lancet 1:702–703
13. Baer R, Bankier AT, Biggin MD, Deininger PL, Farrell PJ, Gibson TJ et al (1984) DNA sequence and expression of the B95-8 Epstein-Barr virus genome. Nature 310:207–211
14. Marshall BJ, Warren JR (1984) Unidentified curved bacilli in the stomach of patients with gastritis and peptic ulceration. Lancet 1:1311–1315
15. Warren JR, Marshall B (1983) Unidentified curved bacilli on gastric epithelium in active chronic gastritis. Lancet 1:1273–1275
16. Warren JR (2006) Helicobacter: the ease and difficulty of a new discovery (Nobel lecture). ChemMedChem 1:672–685
17. Huebner RJ, Todaro GJ (1969) Oncogenes of RNA tumor viruses as determinants of cancer. Proc Natl Acad Sci U S A 64:1087–1094
18. Stehelin D, Varmus HE, Bishop JM, Vogt PK (1976) DNA related to the transforming gene(s) of avian sarcoma viruses is present in normal avian DNA. Nature 260:170–173
19. Zur Hausen H, Meinhof W, Scheiber W, Bornkamm GW (1974) Attempts to detect virus-secific DNA in human tumors. I. Nucleic acid hybridizations with complementary RNA of human wart virus. Int J Cancer 13:650–656
20. Chang Y, Cesarman E, Pessin MS, Lee F, Culpepper J, Knowles DM et al (1994) Identification of herpesvirus-like DNA sequences in AIDS-associated Kaposi's sarcoma. Science 266:1865–1869
21. Shuda M, Feng H, Kwun HJ, Rosen ST, Gjoerup O, Moore PS et al (2008) T antigen mutations are a human tumor-specific signature for Merkel cell polyomavirus. Proc Natl Acad Sci U S A 105:16272–16277
22. Feng H, Shuda M, Chang Y, Moore PS (2008) Clonal integration of a polyomavirus in human Merkel cell carcinoma. Science 319:1096–1100
23. Knudson AG (2001) Two genetic hits (more or less) to cancer. Nat Rev Cancer 1:157–162
24. Zhu C, Zhu Q, Wang C, Zhang L, Wei F, Cai Q (2016) Hostile takeover: manipulation of HIF-1 signaling in pathogen-associated cancers (review). Int J Oncol 49:1269–1276

25. Andersson J (2000) An overview of Epstein-Barr virus: from discovery to future directions for treatment and prevention. Herpes: J IHMF 7:76–82
26. McGlynn KA, London WT (2005) Epidemiology and natural history of hepatocellular carcinoma. Best Pract Res Clin Gastroenterol 19:3–23
27. Zur Hausen H (2002) Papillomaviruses and cancer: from basic studies to clinical application. Nat Rev Cancer 2:342–350
28. Ahmed N, Sechi LA (2005) Helicobacter pylori and gastroduodenal pathology: new threats of the old friend. Ann Clin Microbiol Antimicrob 4:1
29. Prabhu SR, Amrapurkar AD, Amrapurkar DN (1995) Role of Helicobacter pylori in gastric carcinoma. Natl Med J India 8:58–60
30. Megraud F, Lehours P (2004) Helicobacter pylori and gastric cancer prevention is possible. Cancer Detect Prev 28:392–398

Chapter 2
Hepatitis B Virus-Associated Hepatocellular Carcinoma

Youhua Xie

Abstract Liver cancer is the fifth most common cancer worldwide in men and the ninth in women. It is also the second most common cause of cancer mortality. Hepatocellular carcinoma (HCC) is the most common type of liver cancer. About 350 million people globally are chronically infected with HBV. Chronic hepatitis B virus (HBV) infection accounts for at least 50% cases of HCC worldwide. Other non-HBV factors may increase HCC risk among persons with chronic HBV infection. Both indirect and direct mechanisms are involved in HCC oncogenesis by HBV. HCC-promoting HBV factors include long-lasting infection, high levels of HBV replication, HBV genotype, HBV integration, specific HBV mutants, and HBV-encoded oncoproteins (e.g., HBx and truncated preS2/S proteins). Recurrent liver inflammation caused by host immune responses during chronic HBV infection can lead to liver fibrosis and cirrhosis and accelerate hepatocyte turnover rate and promote accumulation of mutations. Major breakthroughs have been achieved in the prevention of HBV-associated HCC with HBV vaccines and antiviral therapies.

Keywords Chronic infection • Cirrhosis • Genotype • Hepatitis B virus • Hepatocellular carcinoma • HBeAg • HBsAg • HBx • Integration • Mutation • PreS/S

2.1 Introduction

According to a survey conducted in 2012, liver cancer is the fifth most common cancer worldwide in men (7.5% of the total new cancer cases in 2012) and the ninth in women (3.4%) [1]. It is also the second most common cause of cancer

Y. Xie (✉)
Key Laboratory of Medical Molecular Virology (Ministries of Education and Health), Shanghai Medical College, Fudan University, Shanghai, People's Republic of China
e-mail: yhxie@fudan.edu.cn

Q. Cai et al. (eds.), *Infectious Agents Associated Cancers: Epidemiology and Molecular Biology*, Advances in Experimental Medicine and Biology 1018,
DOI 10.1007/978-981-10-5765-6_2

mortality [1]. Hepatocellular carcinoma (HCC) is the most common type of liver cancer. The majority of HCC is associated with chronic infection of hepatitis B virus (HBV) or hepatitis C virus (HCV). This concise review focuses on HBV-associated HCC.

About 350 million people globally are chronically infected with HBV [2]. Chronic HBV infection accounts for at least 50% cases of HCC worldwide [3] and is the dominant risk factor for HCC in areas with endemic HBV infection such as Eastern and Southeastern Asia and sub-Saharan Africa [4].

Other non-HBV factors may increase HCC risk among persons with chronic HBV infection, including older age [5], male sex [6], cirrhosis [7], diabetes mellitus [8], exposure to environmental carcinogens (aflatoxin B1 (AFB1), heavy alcohol and tobacco consumption) [9, 10], HIV coinfection [11], and possibly HDV superinfection [12].

HBV infection is transmitted mainly vertically in endemic HBV areas, in contrast to horizontally in HBV low prevalent areas. More than 90% of vertical HBV transmission cases lead to chronic infection, whereas only 5–10% of horizontal HBV transmission cases do so. Accordingly, the average age of HBV chronic carriers who develop HCC is younger in endemic HBV areas. Men are more susceptible to HBV-associated HCC than women, probably as a result of stimulation of HBV replication by androgens and a protective role of estrogens against HBV replication [13–15]. In most cases, HBV-associated HCC develops progressively from chronic liver disease, with cirrhosis in the majority of patients (70–90%) [5]. However, cirrhosis is not a prerequisite for the development of HBV-associated HCC [7]. HBV carriers without cirrhosis, especially those who have long-lasting infection, may also develop HCC. AFB1 is the foremost environmental risk factor of HCC in some Eastern Asian areas with endemic HBV infection. AFB1 causes a specific p53 mutation and predisposes mutant hepatocytes to DNA damage [9]. AFB1 was reported to exert a synergistic carcinogenic effect with chronic HBV infection, resulting in a 60-fold increased HCC risk [16].

2.2 HBV Oncogenic Factors for HCC Development

Both indirect and direct mechanisms are involved in HCC oncogenesis by HBV. HCC-promoting HBV factors include long-lasting infection, high levels of HBV replication, HBV genotype, HBV integration, specific HBV mutants, and HBV-encoded oncoproteins. In addition, recurrent liver inflammation caused by host immune responses during chronic HBV infection can lead to liver fibrosis and cirrhosis and accelerate hepatocyte turnover rate and promote accumulation of mutations.

2.2.1 Long-Lasting Infection and High Levels of Viral Replication

Long-lasting chronic HBV infection is associated with HCC development. As aforementioned, there is a much higher rate of chronic HBV infection in endemic HBV areas due to vertical viral transmission. The lengthened HBV infection period is thought to provide more opportunities for various viral and nonviral risk factors to promote HCC oncogenesis.

Hepatitis B e antigen (HBeAg) seropositivity and higher levels of serum HBV load are associated with high risk of HCC. A long-term follow-up study among 11,893 male HBV carriers in Taiwan who were without HCC at study entry showed that the relative risk of HCC was 9.6 among men who were positive for hepatitis B surface antigen (HBsAg) alone and 60.2 among those who were positive for both HBsAg and HBeAg, as compared with men who were negative for both [17]. HBeAg seropositivity was also found associated with higher risk of early recurrence and poorer survival in patients after curative tumor resection [18]. With the routine application of HBV DNA quantification, HBeAg as a surrogate of HBV replication indicator is less utilized. The REVEAL-HBV study reported that the incidence of cirrhosis and HCC is positively and quantitatively correlated to the serum HBV DNA load in a cohort of 3653 participants with chronic HBV infection [19, 20]. Similar results were observed in a follow-up study among a prospective cohort of 1006 patients with chronic HBV infection from Hong Kong [21].

2.2.2 HBV Genotype

There are at least eight HBV genotypes (A–H), which display distinct geographical distributions [22]. Both genotypes B and C are prevalent in Eastern Asian areas. Infection with genotype C was reported to more likely result in severe liver disease, cirrhosis, and HCC than infection with genotype B [21, 23, 24]. However, a study from Taiwan reported that genotype B was associated with HCC in children with chronic HBV infection [25]. In Europe where genotypes A and D are dominant, infection with genotype D is associated with more severe liver disease or HCC than infection with genotype A [26].

2.2.3 HBV Integration

HBV replicates through reverse transcription using its pregenomic RNA as template. Progeny viral DNA in nascent capsids can be trafficked to nucleus to supplement nuclear cccDNA pool, which constitutes a reservoir of templates for HBV gene expression and replication. Unlike retroviruses, chromosomal DNA integration is not required for HBV replication. Nevertheless, DNA integration into the genomes of host hepatocytes likely contributes to oncogenesis by HBV.

HBV DNA integration in host chromosomes has been found in the majority (85–90%) of HBV-associated HCC and probably occurs early during HBV infection [27, 28]. The genomic sites of HBV DNA integration appear random [27]. However, it is thought that HBV DNA integration into some specific genomic sites may allow the integrant-containing cells to obtain a growth advantage so that they may expand clonally. The integrated HBV DNA may induce chromosomal instability or alter the expression of host genes through *cis*-acting mechanisms. In addition, the integrated viral DNA may allow the continuous expression of viral oncoproteins such as HBx and truncated preS2/S proteins.

Recurrent HBV DNA integration occurs near actively transcribed gene-coding chromosomal regions, as well as within or near fragile genomic sites or repetitive regions, such as the Alu sequences and long interspersed nuclear elements (LINEs) [29–31]. Sequence analysis has revealed integration sites that are in the proximity of many genes involved in cell survival, proliferation, metabolism, and cell cycle regulation [29–31]. Among these genes, insertion of HBV DNA near the *hTERT* gene, encoding the catalytic subunit of telomerase, has been frequently found in HCC [29, 32]. The integration of HBV DNA into fragile genomic sites or repetitive regions may induce genomic instability or alter the expression of noncoding RNAs [33]. A HBV-human fusion transcript (HBx-LINE1) was reported to function as a long noncoding RNA (lncRNA) to influence the epithelial-mesenchymal transition and correlate with reduced patient survival and tumor formation in mice [34].

2.2.4 HBV Mutations

The reverse transcriptase of HBV lacks of proofreading activity. As a result, mutations are accumulated during chronic HBV infection and selected under the pressure of host immunity and antiviral drugs during treatment. Due to the compact and overlapping properties of HBV genome, many mutations generate defective viruses. HBV mutations that have been identified to be associated with HCC are enriched in the basal core promoter (BCP)/preC region and the preS region.

Among the many mutations in the BCP/preC region, the most common one that is significantly associated with HCC development in genotypes B and C is the T1762 and A1764 double mutation (BCP double mutation) [35, 36]. The G1896A mutation in the preC region is a common HBV mutation that creates a premature stop codon that abolishes HBeAg translation. No association exists between the G1896A mutation and HCC development [37, 38]. Several other mutations in the BCP/preC region (C1653T, T1753V) may also be associated with HCC development [38]. It is unclear how these mutations contribute to HCC development. Since the BCP/preC region contains essential HBV regulatory elements, these mutations may alter HBV gene expression and replication. In addition, because the HBx open reading frame overlaps the BCP/preC region, some mutations may affect HBx expression or activity.

HBV mutants with point mutations, deletions, or insertions in the preS region have been frequently found in HCC [39, 40] and are associated with an increased risk of HCC [38, 39]. The preS mutations may alter the expression and secretion of HBV envelope proteins, resulting in intracellular accumulation of HBV envelope proteins, which can cause endoplasmic reticulum (ER) stress, leading to cell transformation [41, 42].

2.2.5 *HBx Protein*

The viral regulatory protein HBx contributes critically to HBV replication [43] and is thought to be closely related to HBV oncogenicity. It probably does not bind directly to DNA but rather acts on many cellular and viral promoters through protein-protein interactions. In the cytoplasm, HBx modulates multiple signaling pathways. These nuclear and cytoplasmic interactions result in the activation or repression of a large number of signaling pathways that play important roles in chromatin dynamics, DNA damage response, cell proliferation, viability, metabolism and migration, angiogenesis, and immune response. However, precautions should be taken concerning HBx's multiple activities. Due to the low-level expression of HBx during HBV infection and a lack of sensitive detection tools, many findings have been derived from in vitro HBx overexpression experiments and need to be verified in models that more closely mimic HBV infection and HBV-associated HCC.

HBx causes chromosomal instability by binding with different cellular proteins (Crm1, HBXIP, DDB1, p53, hBubR1) to dysregulate centriole replication, mitotic checkpoint, mitotic spindle formation, and chromosome segregation [44–47]. HBx promotes cell proliferation, viability, and migration through modulating multiple signaling pathways. HBx binds with p53 to impair p53-mediated apoptosis and checkpoint functions [48, 49]. HBx may upregulate *TERT* expression [50], but conflict results have been shown in HBx transgenic mice [51]. HBx induces CREB-dependent transcriptional activation through interacting with the CBP/p300 acetyltransferases and preventing CREB inactivation by PP1 phosphatase, resulting in expression of CREB-responsive genes involved in hepatocyte metabolism and proliferation [52, 53]. HBx can recruit DNMT3a DNA methyltransferase to suppress *all-trans* retinoic acid (ATRA)-mediated induction of p16 and p21 in HepG2 and Hep3B cells via promoter hypermethylation, resulting in inactivation of retinoblastoma protein [54]. HBx may promote cell migration and HCC cell invasive and metastatic capacity by increasing the expression of matrix metalloproteinase 3 and 9 [55, 56] and epigenetically suppressing E-cadherin expression [57]. HBx can also block tumor necrosis factor-α-mediated apoptosis [58]. On the other hand, HBx can increase cellular reactive oxygen species (ROS) levels that lead to apoptosis by promoting mitochondria membrane depolarization or Ca2+ accumulation in mitochondria [59, 60]. HBx may also promote stemness of HCC cells [61].

HBx has been shown to promote HCC angiogenesis. HBx was reported to upregulate the stability and transcriptional activity of hypoxia-inducible factor-1α (HIF1α) and the expression of vascular endothelial growth factor (VEGF) and angiopoietin 2 (ANG2), which leads to enhanced angiogenesis [62, 63].

2.2.6 *PreS/S Proteins*

The PreS/S open reading frame of HBV uses alternative start codons for translation and encodes three envelope proteins (large, middle, and small) that share the 226-amino-acid sequence of the small envelope polypeptide. The contribution of wild-type or mutant PreS/S proteins to HCC development is not fully understood. Wild-type large envelope protein accumulated in the ER of hepatocytes of transgenic mice could induce ER stress and consequently cause inflammation, hyperplasia, and aneuploidy [64]. PreS2/S mutant proteins frequently found in HBV-associated HCC also accumulate in ER and may trigger a similar process [42], resulting in the upregulation of cyclin A that in turn promotes cell proliferation and chromosome instability [65, 66]. In addition, PreS2/S mutant proteins have been shown to transcriptionally activate the TERT expression [67].

2.3 Prevention

HBV-associated HCC can be prevented by vaccination against HBV infection. Vaccination of newborns against HBV has been incorporated into universal hepatitis B immunization programs of many countries and regions, which has greatly reduced the incidence of HCC in children [68]. Hepatitis B immune globulin (HBIG), in addition to hepatitis B vaccine, administered within 12–24 h after birth, has been shown to achieve 90–100% protective efficacy against perinatal transmission from mothers who are positive for HBsAg and HBeAg [69]. Recent studies showed that tenofovir treatment of HBeAg-positive mothers can successfully prevent vertical HBV transmission [70, 71].

Antiviral therapy can significantly suppress HBV replication in chronic HBV patients. Studies with patients treated with lamivudine or adefovir have shown to help prevent HCC in patients with chronic hepatitis [72, 73]. Nevertheless, nucleos(t)ide analogue therapy does not completely eliminate the risk of HCC [73]. The current first-line anti-HBV drugs, namely, entecavir and tenofovir, have been shown to improve the prevention of HCC in responders with cirrhosis [74].

2.4 Conclusions

HCC will continue to be one of the major cancers worldwide as chronic HBV infection remains a public health threat. A great deal of knowledge has been gained on the epidemiologic features and pathogenesis of HBV-associated HCC in the past three decades. However, the oncogenic mechanisms of HBV and HBV-related risk factors are not fully understood, in large part owing to a lack of animal models that recuperate clinical HBV-associated HCC. Nevertheless, major breakthroughs have been achieved in the prevention of HBV-associated HCC with HBV vaccines and antiviral therapies. With the advances in HBV virology and pathology, there will be novel prophylactic and therapeutic means for HBV-associated HCC.

Acknowledgments The author would like to apologize for citing selected papers due to the space limit. This work was supported by the National Key Project for Infectious Diseases of China (2012ZX10002-006, 2012ZX10002012-003), National Basic Research Program of China (2012CB519002), and Natural Science Foundation of China (81472226).

No potential conflict of interest was disclosed.

References

1. Globocan (2012) Estimated cancer incidence, mortality and prevalence worldwide in 2012. International agency for research on cancer, World Health Organization, http://globocan.iarc.fr
2. Ganem D, Prince AM (2004) Hepatitis B virus infection – natural history and clinical consequences. N Engl J Med 350:1118–1129
3. Parkin DM (2006) The global health burden of infection-associated cancers in the year 2002. Int J Cancer 118:3030–3044
4. Mittal S, El-Serag HB (2013) Epidemiology of hepatocellular carcinoma: consider the population. J Clin Gastroenterol 47(Suppl):S2–S6
5. Sherman M (2010) Hepatocellular carcinoma: epidemiology, surveillance, and diagnosis. Semin Liver Dis 30:3–16
6. Yuan JM, Ross RK, Stanczyk FZ, Govindarajan S, Gao YT, Henderson BE, Yu MC (1995) A cohort study of serum testosterone and hepatocellular carcinoma in Shanghai, China. Int J Cancer 63:491–493
7. Yang JD, Kim WR, Coelho R, Mettler TA, Benson JT, Sanderson SO, Therneau TM, Kim B, Roberts LR (2011) Cirrhosis is present in most patients with hepatitis B and hepatocellular carcinoma. Clin Gastroenterol Hepatol 9:64–70
8. El-Serag HB, Tran T, Everhart JE (2004) Diabetes increases the risk of chronic liver disease and hepatocellular carcinoma. Gastroenterology 126:460–468
9. Bressac B, Kew M, Wands J, Ozturk M (1991) Selective G to T mutations of p53 gene in hepatocellular carcinoma from southern Africa. Nature 350:429–431
10. Marrero JA, Fontana RJ, Fu S, Conjeevaram HS, Su GL, Lok AS (2005) Alcohol, tobacco and obesity are synergistic risk factors for hepatocellular carcinoma. J Hepatol 42:218–224
11. Brau N, Fox RK, Xiao P, Marks K, Naqvi Z, Taylor LE, Trikha A, Sherman M, Sulkowski MS, Dieterich DT, Rigsby MO, Wright TL, Hernandez MD, Jain MK, Khatri GK, Sterling RK, Bonacini M, Martyn CA, Aytaman A, Llovet JM, Brown ST, Bini EJ, North American Liver Cancer in HIVSG (2007) Presentation and outcome of hepatocellular carcinoma in HIV-infected patients: a U.S.-Canadian multicenter study. J Hepatol 47:527–537

12. Farci P, Niro GA (2012) Clinical features of hepatitis D. Semin Liver Dis 32:228–236
13. Wang SH, Yeh SH, Lin WH, Wang HY, Chen DS, Chen PJ (2009) Identification of androgen response elements in the enhancer I of hepatitis B virus: a mechanism for sex disparity in chronic hepatitis B. Hepatology 50:1392–1402
14. Naugler WE, Sakurai T, Kim S, Maeda S, Kim K, Elsharkawy AM, Karin M (2007) Gender disparity in liver cancer due to sex differences in MyD88-dependent IL-6 production. Science 317:121–124
15. Wang SH, Yeh SH, Lin WH, Yeh KH, Yuan Q, Xia NS, Chen DS, Chen PJ (2012) Estrogen receptor alpha represses transcription of HBV genes via interaction with hepatocyte nuclear factor 4alpha. Gastroenterology 142:989–998. e984
16. Qian GS, Ross RK, Yu MC, Yuan JM, Gao YT, Henderson BE, Wogan GN, Groopman JD (1994) A follow-up study of urinary markers of aflatoxin exposure and liver cancer risk in Shanghai, People's Republic of China. Cancer Epidemiol Biomark Prev 3:3–10
17. Yang HI, Lu SN, Liaw YF, You SL, Sun CA, Wang LY, Hsiao CK, Chen PJ, Chen DS, Chen CJ, Taiwan Community-Based Cancer Screening Project G (2002) Hepatitis B e antigen and the risk of hepatocellular carcinoma. N Engl J Med 347:168–174
18. Sun HC, Zhang W, Qin LX, Zhang BH, Ye QH, Wang L, Ren N, Zhuang PY, Zhu XD, Fan J, Tang ZY (2007) Positive serum hepatitis B e antigen is associated with higher risk of early recurrence and poorer survival in patients after curative resection of hepatitis B-related hepatocellular carcinoma. J Hepatol 47:684–690
19. Chen CJ, Yang HI, Su J, Jen CL, You SL, Lu SN, Huang GT, Iloeje UH, Group R-HS (2006) Risk of hepatocellular carcinoma across a biological gradient of serum hepatitis B virus DNA level. JAMA 295:65–73
20. Chen CJ, Yang HI, Iloeje UH, Group R-HS (2009) Hepatitis B virus DNA levels and outcomes in chronic hepatitis B. Hepatology 49:S72–S84
21. Chan HL, Tse CH, Mo F, Koh J, Wong VW, Wong GL, Lam Chan S, Yeo W, Sung JJ, Mok TS (2008) High viral load and hepatitis B virus subgenotype ce are associated with increased risk of hepatocellular carcinoma. J Clin Oncol 26:177–182
22. Lin CL, Kao JH (2011) The clinical implications of hepatitis B virus genotype: recent advances. J Gastroenterol Hepatol 26(Suppl 1):123–130
23. Kao JH, Chen PJ, Lai MY, Chen DS (2000) Hepatitis B genotypes correlate with clinical outcomes in patients with chronic hepatitis B. Gastroenterology 118:554–559
24. Chan HL, Hui AY, Wong ML, Tse AM, Hung LC, Wong VW, Sung JJ (2004) Genotype C hepatitis B virus infection is associated with an increased risk of hepatocellular carcinoma. Gut 53:1494–1498
25. Ni YH, Chang MH, Wang KJ, Hsu HY, Chen HL, Kao JH, Yeh SH, Jeng YM, Tsai KS, Chen DS (2004) Clinical relevance of hepatitis B virus genotype in children with chronic infection and hepatocellular carcinoma. Gastroenterology 127:1733–1738
26. Sanchez-Tapias JM, Costa J, Mas A, Bruguera M, Rodes J (2002) Influence of hepatitis B virus genotype on the long-term outcome of chronic hepatitis B in western patients. Gastroenterology 123:1848–1856
27. Brechot C, Gozuacik D, Murakami Y, Paterlini-Brechot P (2000) Molecular bases for the development of hepatitis B virus (HBV)-related hepatocellular carcinoma (HCC). Semin Cancer Biol 10:211–231
28. Minami M, Daimon Y, Mori K, Takashima H, Nakajima T, Itoh Y, Okanoue T (2005) Hepatitis B virus-related insertional mutagenesis in chronic hepatitis B patients as an early drastic genetic change leading to hepatocarcinogenesis. Oncogene 24:4340–4348
29. Murakami Y, Saigo K, Takashima H, Minami M, Okanoue T, Brechot C, Paterlini-Brechot P (2005) Large scaled analysis of hepatitis B virus (HBV) DNA integration in HBV related hepatocellular carcinomas. Gut 54:1162–1168
30. Sung WK, Zheng H, Li S, Chen R, Liu X, Li Y, Lee NP, Lee WH, Ariyaratne PN, Tennakoon C, Mulawadi FH, Wong KF, Liu AM, Poon RT, Fan ST, Chan KL, Gong Z, Hu Y, Lin Z, Wang G, Zhang Q, Barber TD, Chou WC, Aggarwal A, Hao K, Zhou W, Zhang C, Hardwick J, Buser

C, Xu J, Kan Z, Dai H, Mao M, Reinhard C, Wang J, Luk JM (2012) Genome-wide survey of recurrent HBV integration in hepatocellular carcinoma. Nat Genet 44:765–769
31. Ding D, Lou X, Hua D, Yu W, Li L, Wang J, Gao F, Zhao N, Ren G, Li L, Lin B (2012) Recurrent targeted genes of hepatitis B virus in the liver cancer genomes identified by a next-generation sequencing-based approach. PLoS Genet 8:e1003065
32. Ferber MJ, Montoya DP, Yu C, Aderca I, McGee A, Thorland EC, Nagorney DM, Gostout BS, Burgart LJ, Boix L, Bruix J, McMahon BJ, Cheung TH, Chung TK, Wong YF, Smith DI, Roberts LR (2003) Integrations of the hepatitis B virus (HBV) and human papillomavirus (HPV) into the human telomerase reverse transcriptase (hTERT) gene in liver and cervical cancers. Oncogene 22:3813–3820
33. Liang HW, Wang N, Wang Y, Wang F, Fu Z, Yan X, Zhu H, Diao W, Ding Y, Chen X, Zhang CY, Zen K (2016) Hepatitis B virus-human chimeric transcript HBx-LINE1 promotes hepatic injury via sequestering cellular microRNA-122. J Hepatol 64:278–291
34. Lau CC, Sun T, Ching AK, He M, Li JW, Wong AM, Co NN, Chan AW, Li PS, Lung RW, Tong JH, Lai PB, Chan HL, To KF, Chan TF, Wong N (2014) Viral-human chimeric transcript predisposes risk to liver cancer development and progression. Cancer Cell 25:335–349
35. Baptista M, Kramvis A, Kew MC (1999) High prevalence of 1762(T) 1764(A) mutations in the basic core promoter of hepatitis B virus isolated from black Africans with hepatocellular carcinoma compared with asymptomatic carriers. Hepatology 29:946–953
36. Kuang SY, Jackson PE, Wang JB, Lu PX, Munoz A, Qian GS, Kensler TW, Groopman JD (2004) Specific mutations of hepatitis B virus in plasma predict liver cancer development. Proc Natl Acad Sci U S A 101:3575–3580
37. Yang HI, Yeh SH, Chen PJ, Iloeje UH, Jen CL, Su J, Wang LY, Lu SN, You SL, Chen DS, Liaw YF, Chen CJ, Group R-HS (2008) Associations between hepatitis B virus genotype and mutants and the risk of hepatocellular carcinoma. J Natl Cancer Inst 100:1134–1143
38. Liu S, Zhang H, Gu C, Yin J, He Y, Xie J, Cao G (2009) Associations between hepatitis B virus mutations and the risk of hepatocellular carcinoma: a meta-analysis. J Natl Cancer Inst 101:1066–1082
39. Pollicino T, Cacciola I, Saffioti F, Raimondo G (2014) Hepatitis B virus PreS/S gene variants: pathobiology and clinical implications. J Hepatol 61:408–417
40. Chen BF, Liu CJ, Jow GM, Chen PJ, Kao JH, Chen DS (2006) High prevalence and mapping of pre-S deletion in hepatitis B virus carriers with progressive liver diseases. Gastroenterology 130:1153–1168
41. Hsieh YH, Su IJ, Wang HC, Chang WW, Lei HY, Lai MD, Chang WT, Huang W (2004) Pre-S mutant surface antigens in chronic hepatitis B virus infection induce oxidative stress and DNA damage. Carcinogenesis 25:2023–2032
42. Wang HC, Huang W, Lai MD, Su IJ (2006) Hepatitis B virus pre-S mutants, endoplasmic reticulum stress and hepatocarcinogenesis. Cancer Sci 97:683–688
43. Lucifora J, Arzberger S, Durantel D, Belloni L, Strubin M, Levrero M, Zoulim F, Hantz O, Protzer U (2011) Hepatitis B virus X protein is essential to initiate and maintain virus replication after infection. J Hepatol 55:996–1003
44. Forgues M, Difilippantonio MJ, Linke SP, Ried T, Nagashima K, Feden J, Valerie K, Fukasawa K, Wang XW (2003) Involvement of Crm1 in hepatitis B virus X protein-induced aberrant centriole replication and abnormal mitotic spindles. Mol Cell Biol 23:5282–5292
45. Wen Y, Golubkov VS, Strongin AY, Jiang W, Reed JC (2008) Interaction of hepatitis B viral oncoprotein with cellular target HBXIP dysregulates centrosome dynamics and mitotic spindle formation. J Biol Chem 283:2793–2803
46. Martin-Lluesma S, Schaeffer C, Robert EI, van Breugel PC, Leupin O, Hantz O, Strubin M (2008) Hepatitis B virus X protein affects S phase progression leading to chromosome segregation defects by binding to damaged DNA binding protein 1. Hepatology 48:1467–1476
47. Kim S, Park SY, Yong H, Famulski JK, Chae S, Lee JH, Kang CM, Saya H, Chan GK, Cho H (2008) HBV X protein targets hBubR1, which induces dysregulation of the mitotic checkpoint. Oncogene 27:3457–3464

48. Knoll S, Furst K, Thomas S, Villanueva Baselga S, Stoll A, Schaefer S, Putzer BM (2011) Dissection of cell context-dependent interactions between HBx and p53 family members in regulation of apoptosis: a role for HBV-induced HCC. Cell Cycle 10:3554–3565
49. Ahn JY, Jung EY, Kwun HJ, Lee CW, Sung YC, Jang KL (2002) Dual effects of hepatitis B virus X protein on the regulation of cell-cycle control depending on the status of cellular p53. J Gen Virol 83:2765–2772
50. Liu H, Shi W, Luan F, Xu S, Yang F, Sun W, Liu J, Ma C (2010) Hepatitis B virus X protein upregulates transcriptional activation of human telomerase reverse transcriptase. Virus Genes 40:174–182
51. Kojima H, Kaita KD, Xu Z, Ou JH, Gong Y, Zhang M, Minuk GY (2003) The absence of up-regulation of telomerase activity during regeneration after partial hepatectomy in hepatitis B virus X gene transgenic mice. J Hepatol 39:262–268
52. Cougot D, Wu Y, Cairo S, Caramel J, Renard CA, Levy L, Buendia MA, Neuveut C (2007) The hepatitis B virus X protein functionally interacts with CREB-binding protein/p300 in the regulation of CREB-mediated transcription. J Biol Chem 282:4277–4287
53. Cougot D, Allemand E, Riviere L, Benhenda S, Duroure K, Levillayer F, Muchardt C, Buendia MA, Neuveut C (2012) Inhibition of PP1 phosphatase activity by HBx: a mechanism for the activation of hepatitis B virus transcription. Sci Signal 5:ra1
54. Park SH, Jung JK, Lim JS, Tiwari I, Jang KL (2011) Hepatitis B virus X protein overcomes all-trans retinoic acid-induced cellular senescence by downregulating levels of p16 and p21 via DNA methylation. J Gen Virol 92:1309–1317
55. Chung TW, Lee YC, Kim CH (2004) Hepatitis B viral HBx induces matrix metalloproteinase-9 gene expression through activation of ERK and PI-3K/AKT pathways: involvement of invasive potential. FASEB J 18:1123–1125
56. Yu FL, Liu HJ, Lee JW, Liao MH, Shih WL (2005) Hepatitis B virus X protein promotes cell migration by inducing matrix metalloproteinase-3. J Hepatol 42:520–527
57. Arzumanyan A, Friedman T, Kotei E, Ng IO, Lian Z, Feitelson MA (2012) Epigenetic repression of E-cadherin expression by hepatitis B virus x antigen in liver cancer. Oncogene 31:563–572
58. Pan J, Lian Z, Wallett S, Feitelson MA (2007) The hepatitis B x antigen effector, URG7, blocks tumour necrosis factor alpha-mediated apoptosis by activation of phosphoinositol 3-kinase and beta-catenin. J Gen Virol 88:3275–3285
59. Rahmani Z, Huh KW, Lasher R, Siddiqui A (2000) Hepatitis B virus X protein colocalizes to mitochondria with a human voltage-dependent anion channel, HVDAC3, and alters its transmembrane potential. J Virol 74:2840–2846
60. Yang B, Bouchard MJ (2012) The hepatitis B virus X protein elevates cytosolic calcium signals by modulating mitochondrial calcium uptake. J Virol 86:313–327
61. Arzumanyan A, Friedman T, Ng IO, Clayton MM, Lian Z, Feitelson MA (2011) Does the hepatitis B antigen HBx promote the appearance of liver cancer stem cells? Cancer Res 71:3701–3708
62. Yoo YG, Oh SH, Park ES, Cho H, Lee N, Park H, Kim DK, Yu DY, Seong JK, Lee MO (2003) Hepatitis B virus X protein enhances transcriptional activity of hypoxia-inducible factor-1alpha through activation of mitogen-activated protein kinase pathway. J Biol Chem 278:39076–39084
63. Sanz-Cameno P, Martin-Vilchez S, Lara-Pezzi E, Borque MJ, Salmeron J, Munoz de Rueda P, Solis JA, Lopez-Cabrera M, Moreno-Otero R (2006) Hepatitis B virus promotes angiopoietin-2 expression in liver tissue: role of HBV x protein. Am J Pathol 169:1215–1222
64. Chisari FV, Klopchin K, Moriyama T, Pasquinelli C, Dunsford HA, Sell S, Pinkert CA, Brinster RL, Palmiter RD (1989) Molecular pathogenesis of hepatocellular carcinoma in hepatitis B virus transgenic mice. Cell 59:1145–1156
65. Wang HC, Chang WT, Chang WW, Wu HC, Huang W, Lei HY, Lai MD, Fausto N, Su IJ (2005) Hepatitis B virus pre-S2 mutant upregulates cyclin A expression and induces nodular proliferation of hepatocytes. Hepatology 41:761–770

66. Wang LH, Huang W, Lai MD, Su IJ (2012) Aberrant cyclin A expression and centrosome overduplication induced by hepatitis B virus pre-S2 mutants and its implication in hepatocarcinogenesis. Carcinogenesis 33:466–472
67. Luan F, Liu H, Gao L, Liu J, Sun Z, Ju Y, Hou N, Guo C, Liang X, Zhang L, Sun W, Ma C (2009) Hepatitis B virus protein preS2 potentially promotes HCC development via its transcriptional activation of hTERT. Gut 58:1528–1537
68. Chang MH, Chen CJ, Lai MS, Hsu HM, Wu TC, Kong MS, Liang DC, Shau WY, Chen DS (1997) Universal hepatitis B vaccination in Taiwan and the incidence of hepatocellular carcinoma in children. Taiwan Child Hepatoma Study Group New England J Med 336:1855–1859
69. Greenberg DP (1993) Pediatric experience with recombinant hepatitis B vaccines and relevant safety and immunogenicity studies. Pediatr Infect Dis J 12:438–445
70. Greenup AJ, Tan PK, Nguyen V, Glass A, Davison S, Chatterjee U, Holdaway S, Samarasinghe D, Jackson K, Locarnini SA, Levy MT (2014) Efficacy and safety of tenofovir disoproxil fumarate in pregnancy to prevent perinatal transmission of hepatitis B virus. J Hepatol 61:502–507
71. Pan CQ, Duan Z, Dai E, Zhang S, Han G, Wang Y, Zhang H, Zou H, Zhu B, Zhao W, Jiang H, China Study Group for the Mother-to-Child Transmission of Hepatitis B (2016) Tenofovir to prevent hepatitis B transmission in mothers with high viral load. N Engl J Med 374:2324–2334
72. Liaw YF, Sung JJ, Chow WC, Farrell G, Lee CZ, Yuen H, Tanwandee T, Tao QM, Shue K, Keene ON, Dixon JS, Gray DF, Sabbat J, Cirrhosis Asian Lamivudine Multicentre Study G (2004) Lamivudine for patients with chronic hepatitis B and advanced liver disease. N Engl J Med 351:1521–1531
73. Papatheodoridis GV, Lampertico P, Manolakopoulos S, Lok A (2010) Incidence of hepatocellular carcinoma in chronic hepatitis B patients receiving nucleos(t)ide therapy: a systematic review. J Hepatol 53:348–356
74. Papatheodoridis GV, Chan HL, Hansen BE, Janssen HL, Lampertico P (2015) Risk of hepatocellular carcinoma in chronic hepatitis B: assessment and modification with current antiviral therapy. J Hepatol 62:956–967

Chapter 3
Human Papillomavirus-Related Cancers

Yanyun Li and Congjian Xu

Abstract Human papillomavirus (HPV) infection is associated with several cancers such as cancer in the cervix, vagina, and vulva and oropharyngeal, anal, penile, and cutaneous carcinomas, which is regarded as a great public health concern. The association between HPV is the strongest with cervical cancer because almost all such malignancies contain viral DNA, notably HPV types 16 and 18. The present chapter summarizes recent progresses of the HPV-associated cancers regarding epidemiology, molecular biology, HPV testing, vaccination, and treatment of HPV-related cancers.

Keywords Human papillomavirus • Cancer • Epidemiology • Molecular biology • Vaccine

3.1 Introduction

Human papillomaviruses (HPVs) are non-enveloped, double-stranded circular DNA viruses and are associated with a wide spectrum of benign and malignant neoplasia. Until now, more than 120 HPV subtypes have been identified and characterized. HPV has an exclusively intraepithelial infectious cycle and infects both the cutaneous and mucosal squamous epithelia. According to oncogenic potentials epidemiologically, some of them are classified into "high-risk (HR)" group and "low-risk (LR)" group.

Until now, HPV subtypes 16, 18, 31, 33, 35, 39, 45, 51, 52, 56, 58, 59, 68, 69, 73, and 82 are considered high risk. They are carcinogenic viruses, causing not only most cervical cancers in women but also a substantial proportion of other anogenital cancers, head and neck cancers, and notably carcinoma of the oropharynx, tonsils, and tongue in men and women [1–3]. Among all the high-risk genotypes of HPV, HPV16 and HPV18 are considered to have the highest ability to cause cancer and

Y. Li • C. Xu (✉)
Hospital and Institute of Obstetrics and Gynecology, Shanghai Medical College, Fudan University, Shanghai 200032, People's Republic of China
e-mail: xucj@hotmail.com

Q. Cai et al. (eds.), *Infectious Agents Associated Cancers: Epidemiology and Molecular Biology*, Advances in Experimental Medicine and Biology 1018,
DOI 10.1007/978-981-10-5765-6_3

are responsible for approximately ~70% of invasive cervical cancers worldwide [4]. In contrast with the HR types, the LR HPV types, notably types 6 and 11, cause almost all clinically visible benign lesions, that is, genital warts (flat condyloma and condyloma acuminata) and laryngeal papillomas.

3.2 Epidemiology

In general, HPV is associated with more than 90% of anal and cervical cancers, about 70% of vaginal and vulvar cancers, 70% of oropharyngeal cancers, more than 60% of penile cancers, and more than 10% of oral cavity cancers [5]. In addition, specific HPV types also have been linked to certain cutaneous cancers [1]. On a global scale, HPV accounts for approximately 600,000 cases of cervical cancer, oropharyngeal cancers, anal cancers, vulvovaginal and penile cancers, as well as a genital wart and recurrent papillomatosis of the lungs worldwide [5].

HPV could be transmitted through direct skin-to-skin, skin-to-mucosa, and mucosa-to-mucosa contact with an HPV-infected individual. The transmission is correlated with sexual intercourse, both vaginal and anal. In addition, transmission can also occur following non-penetrative sexual activity such as oral–genital sex [6]. It is less common that HPV can occur via vertical transmission from mother to infant during the intrapartum period [2], and occasionally transmission even through fomites has been known to occur [7].

Risk factors for the development of HPV infection include sexual activity, multiple sex partners, lack of condom use, use of oral contraceptives, age at first sexual intercourse, non-monogamous sexual relationships, and history of prior infection [8]. In addition, low socioeconomic status and smoking habit of individuals have reported to increase the risk of acquiring the infection.

The estimated global prevalence rate of genital HPV infection is 12% [9]. However, in most cases, the infection of HPV is subclinical, and 60–70% of new infections are cleared within 1 year and 91% are cleared within 2 years [10]. Only a small proportion of HPV infections progress to persistent infection, often involving HR HPV types. Among sexually active individuals, the lifetime risk of HPV infection is ~75%. The HPV risk and prevalence vary by age and gender. Genital HPV infection is very common in the younger age group (<25 years old). Sexually active adolescents and young adult females are at a higher risk compared to women and men greater than 25 years of age [10, 11]. However, the more clinical cases were studied in women than that in men, although the natural rate of the HPV infection is similar in both populations.

3.3 Molecular Biology

It is known that HPVs replicate and assemble exclusively in the nucleus. Virus infects the keratinocytes in the basal layers of a stratified squamous epithelium. The expression of HPV genes from polycistronic pre-mRNA depends on cell differentiation and is tightly regulated at the transcriptional and posttranscriptional levels. The genome of HPV consists of double-stranded cDNA and encodes DNA sequences for six nonstructural viral regulatory proteins (E1, E2, E4, E5, E6, and E7) from the early region of the viral genome in undifferentiated or intermediately differentiated keratinocytes and two structural viral capsid proteins (L1 and L2) from the late region of the genome in keratinocytes undergoing terminal differentiation [12].

The E1 and E2 proteins are the early viral proteins required for viral DNA replication and the regulation of early transcription. The E4 protein continues to be expressed in the terminally differentiated keratinocytes, and E5 helps in viral assembly and growth stimulation, whereas late proteins L1 and L2 form minor and major capsid proteins [13]. E5, E6, and E7 are viral oncogenes, and their expression induces cell immortalization and transformation. In particular, E6 and E7 are two viral oncoproteins, where among other functions, E6 destabilizes p53 and prevents apoptosis and E7 promotes cell proliferation by degrading the retinoblastoma protein pRb, inducing epithelial cell malignant transformation [14].

3.4 Pathogenesis

Human papillomaviruses are small, non-enveloped double-stranded DNA viruses that infect the mucosal or cutaneous epithelium. The HPV infects squamous epithelial cells, which have the capacity to proliferate and get access to basal cell during trauma or abrasion. In the basal cells, HPV infection induces the expression of viral genes that helps in the viral replication. The virus may persist in the basal layer in latent form or may continue to replicate as the basal layers differentiate and rise through the epithelium at which point histological and cytological changes may occur. The early proteins E1 and E2 are required for the initiation of replication. The protein E2, being the transcriptional repressor of E6 and E7, controls the expression of E6 and E7. The mode of replication is the rolling circle mechanism during which the virus gets integrated into the human genome. The integration disturbs the E2 gene, thereby resulting in a higher expression of E6 and E7 oncoproteins and leading to cell transformation. After the viral replication, the L1 and L2 gene products form the virus capsid and the mature virus is produced. Finally, the virus is released with the help of E4 protein [15–17].

Despite the high prevalence of HPV infection, many cases resolve spontaneously due to a gradual development of an immune response against HPV DNA. Some HPV infections can be subclinical and consequently the person is asymptomatic. It is possible for the virus to remain in a non-detectable, dormant state and then reactivate years later [18].

3.5 Molecular Diagnostics

To date, highly sensitive and reproducible laboratory techniques to detect oncogenic HPV and cervical cancer have been developed and are being used or considered in place of cervical cytology for primary screening. The US Food and Drug Administration has approved five of the many tests available for routine laboratory service including [19]:

- Hybrid Capture 2 detects 13 oncogenic types of HPV (16, 18, 31, 33, 35, 39, 45, 51, 52, 56, 58, 59, and 68).
- Cervista HPV HR detects 14 HPV types (16, 18, 31, 33, 35, 39, 45, 51, 52, 56, 58, 59, 66, and 68).
- Cervista HPV16/18 detects only HPV16 and 18.
- Aptima (transcription-mediated amplification test) detects RNA from 14 HPV types (16, 18, 31, 33, 35, 39, 45, 51, 52, 56, 58, 59, 66, and 68).
- Cobas 4800 (real-time polymerase chain reaction [PCR]-based test) detects 14 HPV types (16, 18, 31, 33, 35, 39, 45, 51, 52, 56, 58, 59, 66, and 68).

Until now, most of HPV diagnostics are mainly based on viral DNA detection for which signal and target amplification technique are required. For example, PCR techniques can be divided into type-specific and consensus PCRs. While the Hybrid Capture 2 (HC2) test became the gold standard in routine HPV testing, due to its high clinical sensitivity and its relatively high specificity. The HC2 method hybridizes 13 (near) full-length stabilized synthetic RNA probes of high-risk HPV types to denature target DNA followed by detection with antibodies and chemiluminescence [20].

To avoid costly validation studies for new HPV tests, internationally accepted standards for evaluation have been defined. Meanwhile, several new HPV detection assays have been commercialized. Three tests have received Food and Drug Administration approval (Cervista™, signal amplification; Cobas™ HPV test, real-time PCR; APTIMA™ HPV RNA test). Cervista can detect 14 high-risk HPV types and is based on signal amplification technique using two simultaneous isothermal reactions [20]. This offers an advantage over hybrid capture because it determines the presence of sufficient DNA for reliable results [21]. Cobas™ HPV test uses DNA technology or hybrid capture to increase the DNA signal to detectable levels. Unlike other Food and Drug Administration (FDA)-approved DNA-based test, APTIMA™ detects mRNA overexpressed from E6 and E7 viral oncogenes that are entangled to carcinogenesis [22]. Among them, the Cobas HPV test has been most broadly validated for use in triage and as an adjunct to cytology.

HPV RNA testing is another promising option with potentially higher specificity. The HPV test is already in use for primary screening in several countries, although in the United States, primary HPV testing has been recommended only in combination with cytology in primary screening or for triage of cytologic abnormalities. Accumulating data support that HPV testing without cytology might be sufficiently sensitive for primary screening [23]. However, developing countries face several challenges in widespread adoption of these tests for screening purposes.

3.5.1 HPV and Cervical Cancer

Cervical cancer, caused by HPV, is the third leading malignancy among women in the world, after breast cancer and colorectal cancer, with an estimated 527,624 new cases and 265,653 deaths in 2012. Incidence and mortality rates have been declining in most areas of the world in the past 30 years, at a worldwide rate of about 1.6% per year [24].

Almost every cervical cancer is positive for some HR HPV [25]. In a worldwide survey, HPV16 was the most prevalent type in cervical cancer (61%), followed by HPV18 (10%), HPV45 (6%), HPV31 (4%), HPV33 (4%), HPV52 (3%), HPV35 (2%), and HPV58 (2%) [26]. About 90% of cervical cancers are squamous cell carcinoma (SCC), whereas 10% are adenocarcinoma [27]. Both types of cancer are mainly caused by HPV type 16 (62% and 50%, respectively), but adenocarcinoma is significantly associated more with HPV types 18 (32%) and 45 (12%). Another study that evaluated HPV infection in 10,575 histologically confirmed cases of invasive cancer from 38 countries in Asia, Europe, Latin America and the Caribbean, North America, Oceania, and sub-Saharan Africa over a 60-year period found that 85% ($n = 8977$) of the cases were positive for HPV DNA [27]. HPV types 16, 18, and 45 were the three most common types in each histologic form of cervical cancer (squamous cell, adenocarcinoma, and adenosquamous carcinoma), accounting for 61%, 10%, and 6%, respectively.

Good evidence suggests that HPV infection precedes the development of cervical cancer by decades and that persistent infection with HPV is necessary for the development and progression of precancerous lesions of the cervix, either to higher grades of precancerous disease or to cancer. Cervical cancer progresses slowly from a preinvasive state to invasive cervical cancer, a process that can take 10–30 years. Although HR-HPV infection may result in cervical low-grade squamous intraepithelial lesion (LSIL) and HPV infections are very common, particularly among young women, most HR HPV infections resolve in instances of spontaneous regression without appearance of any clinical manifestations. Only a small proportion (10–30%) of HR HPV infection that persists for a long time, with a high viral load, eventually progress to high-grade squamous intraepithelial lesion HSIL and/or invasive cervical cancer [28–33]. A small proportion (~1%) of LSIL and ~12% of HSIL will progress into invasive cancer, if left untreated [34]. Progression of precursor lesions to invasive cancer usually takes place over a period of more than a decade, allowing time for the identification and treatment. Cervical cancer precursor lesions progress more quickly in women with HPV16 and/or 18 infections than in women with other HR HPV types [35]. HPV16 and HPV18 viral loads have been reported to be a stronger predictor for the persistence of lesions than the load of other HR HPV types [36]. Studies have shown that HPV16 and HPV18 load increased with the increased lesion grade [37]. In this regard, viral load per unit amount of genomic DNA is a potential HPV-related biomarker, which could predict those at risk of cervical cancer development [38]. Factors that may influence progression include coinfection with other sexually transmitted infections such as Chlamydia trachomatis, herpes simplex virus or HIV, tobacco smoking, high parity, and immune suppression [25].

The prognostic value of HPV genotypes has also been studied in patients with cervical cancer treated with radiotherapy. In one study including 327 patients with cervical cancer treated with radiotherapy alone or concurrent chemoradiation, of the 22 genotypes detected in 98.8% patients, the most common genotypes were HPV 16, 58, 18, and 33. A significant improvement was reported in the chemoradiotherapy arm of patients with HPV 18 and HPV 58 positive tumors [39]. Another study implies that viral DNA status including intactness of E2 gene was evaluated as a marker for optimization of radiation treatment [40].

3.5.2 HPV and Oropharyngeal Cancer

Oral infection with HPV is recognized as an independent cause of oropharyngeal cancer, although the occurrence of HPV-associated head and neck cancer is lower than those of the genital tract. Studies have shown that 63% of oropharyngeal cancers each year are associated with HPV infection [41] and 95% of HPV-associated oropharyngeal cancers are HPV16 related [42]. HPV-associated oropharyngeal cancers typically develop near the base of the tongue and in the tonsils.

A higher incidence of HPV-associated oropharyngeal cancers has been related to an increased number of sexual partners and younger adults. A fourfold higher incidence has been observed in men (48,900 cases) as compared to women (12,600 cases). HPV-positive oropharyngeal cancers are associated with oral sex, age younger than 60 years, infrequent p53 gene mutation, and a more favorable clinical outcome, whereas HPV-negative cancers are associated with smoking, excessive alcohol use, age older than 60 years, frequent p53 gene mutation, and poor prognosis [43].

Studies have shown that as compared to those with environmentally related cancers, patients with HPV oropharyngeal cancer present with a better performance status, are healthier, and have a higher likelihood of a complete response to treatment [44, 45]. Thus, HPV-associated cancers have a favorable outcome with radiotherapy, and treatment may be optimized depending on the HPV status so as to achieve the best possible treatment outcome and circumvent treatment-related toxicity and morbidity. Hence, less intensive treatment regimens could be used to achieve a similar treatment efficacy along with decreased toxicity and an improved quality of life. Several ongoing clinical trials are currently under investigation to evaluate the possibility of de-escalation of radiotherapy doses in the treatment of HPV-associated oropharyngeal cancers [46].

3.5.3 HPV and Anal, Penile, and Vulvovaginal Cancers

Similar to the cervix, the anus has a transformation zone that is highly susceptible to HPV infection. The association of HPV is well documented in the development of anal cancers. Of anal cancer cases, 97% are HPV positive, mostly with HPV16 (75%) followed by HPV18 (3%) [47]. The incidence of this relatively uncommon cancer has been reported increasing over the last few decades. The risk for anal carcinoma is increased among men having sex with men (MSM) and in the immunocompromised population (individuals with AIDS and organ graft recipients) [26]. Though there is emerging evidence that anal intraepithelial neoplasia (AIN) is a precursor of anal cancer unlike cervical cancer, the evidence is mainly from small studies with a follow-up duration of only 5–10 years. Larger studies are required to evaluate the progression of AIN to anal cancers and to study its impact on treatment outcome [48].

HPV contribution in penile cancer is 45%, mostly attributed to the two most common HR HPV types, HPV16 (60%) and 18 (13%), but also to the two most common LR HPV types, 6 and 11 (together 8%) [49].

Most vulvar and vaginal cancers are squamous cervical cancer in older women, and mechanisms similar to those of cervical cancer development have been documented. A systematic review has reported a progression rate of 3.3% from vulval intraepithelial neoplasia to SCC of the vulva [50]. The HPV contribution in vulvar cancer is 40%, mostly HPV16 (32%) and then HPV18 (4%) [49]. In vaginal cancer, the HPV contribution is higher, being 70% and mostly HPV16 (54%) followed by HPV18 (8%) [49].

3.5.4 HPV and Skin Cancers

In addition to the HR α-HPV types, several β-HPV types, notably HPV types 5 and 8, are associated with skin cancer and are thus considered possibly carcinogenic. The clinical relevance of β-HPV infection has clearly been demonstrated in patients suffering from epidermodysplasia verruciformis (EV). EV is a rare genetically heterogeneous disease, either autosomal recessive or X linked, but also associated with a high risk for nonmelanoma skin cancer [51]. EV is a unique model where genetic susceptibility to HPVs is demonstrated [52]. In the normal population, beta-PV is suspected to have an etiologic role in skin carcinogenesis as well, but this is still controversially discussed. Their oncogenic potency has been investigated in mouse models and in vitro. In 2009, the International Agency for Research on Cancer

(IARC) classified the genus beta HPV types 5 and 8 as "possible carcinogenic" biological agents in EV disease.

3.6 HPV-Related Cancer Prevention

So far, HPV vaccines have been used mainly in the prevention of cervical cancer. Prophylactic HPV vaccines were primarily designed and produced to prevent infection with the most common HR HPVs, types 16 and 18, which cause about 70% of cervical cancer cases. The quadrivalent vaccine (4vHPV) Gardasil, marketed by Merck, protects against initial infection with HPV types 6, 11, 16, and 18. Another bivalent vaccine is Cervarix by GlaxoSmithKline that is protective against HPV16 and 18 [53–55]. Both vaccines are highly effective in preventing cervical dysplasia. Countries that have achieved high coverage with the 4vHPV vaccine have seen dramatic reductions in genital warts and infection with HPV16 and 18 [56, 57]. The Advisory Committee on Immunization Practices (ACIP) recommends routine HPV vaccination starting at age 11 or 12 years, though the series can be started as early as 9 years of age. Vaccination is also recommended for female ages 13 through 26 years and for males ages 13 through 21 years who have not completed the three-dose series. Men up to age 26 should also be vaccinated if they have sex with men or are immunocompromised [2].

The two HPV vaccines, i.e., Gardasil and Cervarix, approved for the use in cervical cancer prevention by the FDA have been also been approved for other indications. Gardasil is now approved for the prevention of genital warts and the HPV-associated precancerous lesions in the anogenital region besides prevention of vulvar, vaginal, and anal cancers. Cervarix is approved for the prevention of precancerous cervical lesions caused by HPV infection besides cervical cancer prevention. These vaccines have not been approved as yet in the prevention of penile or oropharyngeal cancer. The vaccine is however not effective if infections or lesions in the cervix have already been reportedly caused by HPV [53].

A new 9-valent HPV vaccine (9vHPV) was approved by the FDA in December 2014 for females ages 9–26 and males ages 9–15. In addition to the four HPV subtypes (6, 11, 16, 18) found in the quadrivalent vaccine, it includes five additional oncogenic HPV subtypes (31, 33, 45, 52, 58), which cause an additional 15% of cervical cancer. In March 2015, the ACIP updated their guidelines to allow substitution of the 9vHPV vaccine for the quadrivalent vaccine [58].

Cervical screening should be sought by women, even if she has received the vaccine. Furthermore, the recommendations for screening continue to remain the same even for females who have received the HPV vaccine. Other preventive and cost-effective strategies continue to play an important role, especially in the context of developing countries [59].

Great progress has also been made to develop and improve therapeutic HPV vaccines to treat existing HPV infections and diseases. These are being targeted to E6 and E7 oncoproteins that are expressed throughout the life cycle of thevirus [60, 61].

3.7 Conclusion

HPV infection has been shown to play an important role in various benign and malignant diseases. HPV-related cancers constitute a distinct entity compared to non-HPV-related lesions and are usually associated with a favorable prognosis. HPV type 16 is the most common type found in all anogenital, oral, and oropharyngeal malignancies. For cervical cancer, well-established cytology and HPV test screening allow early detection and successful treatment of precancerous cervical lesions [62]. For other HPV-associated disorders, the early detection of precancerous lesion is either difficult or almost impossible. Effective preventive measures could help reduce the burden of these cancers, which have shown an increase in the incidence over the last few decades. Many questions regarding HPV testing, vaccination, and treatment of HPV-related cancers continue to remain unanswered. Large multicenter trials are required in order to study the biological behavior and treatment strategies in the management of HPV-related cancers.

References

1. Humans IWGotEoCRt (2008) IARC monographs on the evaluation of carcinogenic risks to humans. Volume 97. 1,3-butadiene, ethylene oxide and vinyl halides (vinyl fluoride, vinyl chloride and vinyl bromide). IARC Monogr Eval Carcinog Risks Hum 97:3–471
2. Markowitz LE, Dunne EF, Saraiya M, Chesson HW, Curtis CR, Gee J, Bocchini JA Jr, Unger ER, Centers for Disease C, Prevention (2014) Human papillomavirus vaccination: recommendations of the Advisory Committee on Immunization Practices (ACIP). MMWR Recomm Rep: Morb Mortal Wkly Rep Recomm Rep 63:1–30
3. Bansal A, Singh MP, Rai B (2016) Human papillomavirus-associated cancers: a growing global problem. Int J Appl Basic Med Res 6:84–89
4. de Sanjose S, Diaz M, Castellsague X, Clifford G, Bruni L, Munoz N, Bosch FX (2007) Worldwide prevalence and genotype distribution of cervical human papillomavirus DNA in women with normal cytology: a meta-analysis. Lancet Infect Dis 7:453–459
5. Arbyn M, de Sanjose S, Saraiya M, Sideri M, Palefsky J, Lacey C, Gillison M, Bruni L, Ronco G, Wentzensen N, Brotherton J, Qiao YL, Denny L, Bornstein J, Abramowitz L, Giuliano A, Tommasino M, Monsonego J (2012) EUROGIN 2011 roadmap on prevention and treatment of HPV-related disease. Int J Cancer 131:1969–1982
6. Valentino K, Poronsky CB (2016) Human papillomavirus infection and vaccination. J Pediatr Nurs 31:e155–e166
7. Mayeaux EJ Jr (2008) Reducing the economic burden of HPV-related diseases. J Am Osteopath Assoc 108:S2–S7
8. Chelimo C, Wouldes TA, Cameron LD, Elwood JM (2013) Risk factors for and prevention of human papillomaviruses (HPV), genital warts and cervical cancer. J Infect 66:207–217
9. Bruni L, Diaz M, Castellsague X, Ferrer E, Bosch FX, de Sanjose S (2010) Cervical human papillomavirus prevalence in 5 continents: meta-analysis of 1 million women with normal cytological findings. J Infect Dis 202:1789–1799
10. Dunne EF, Unger ER, Sternberg M, McQuillan G, Swan DC, Patel SS, Markowitz LE (2007) Prevalence of HPV infection among females in the United States. JAMA 297:813–819
11. Matkins PP (2013) Sexually transmitted infections in adolescents. N C Med J 74:48–52

12. Zheng ZM, Baker CC (2006) Papillomavirus genome structure, expression, and post-transcriptional regulation. Front Biosci: J Virtual Libr 11:2286–2302
13. Doorbar J, Egawa N, Griffin H, Kranjec C, Murakami I (2015) Human papillomavirus molecular biology and disease association. Rev Med Virol 25(Suppl 1):2–23
14. Zheng ZM (2010) Viral oncogenes, noncoding RNAs, and RNA splicing in human tumor viruses. Int J Biol Sci 6:730–755
15. Wiest T, Schwarz E, Enders C, Flechtenmacher C, Bosch FX (2002) Involvement of intact HPV16 E6/E7 gene expression in head and neck cancers with unaltered p53 status and perturbed pRb cell cycle control. Oncogene 21:1510–1517
16. Evander M, Frazer IH, Payne E, Qi YM, Hengst K, McMillan NA (1997) Identification of the alpha6 integrin as a candidate receptor for papillomaviruses. J Virol 71:2449–2456
17. Joyce JG, Tung JS, Przysiecki CT, Cook JC, Lehman ED, Sands JA, Jansen KU, Keller PM (1999) The L1 major capsid protein of human papillomavirus type 11 recombinant virus-like particles interacts with heparin and cell-surface glycosaminoglycans on human keratinocytes. J Biol Chem 274:5810–5822
18. Juckett G, Hartman-Adams H (2010) Human papillomavirus: clinical manifestations and prevention. Am Fam Physician 82:1209–1213
19. Ikenberg H (2014) Laboratory diagnosis of human papillomavirus infection. Curr Probl Dermatol 45:166–174
20. Wiwanitkit V (2013) Cervista HPV HR test kit in cervical cancer screening. J Low Genit Tract Dis 17:99
21. Johnson LR, Starkey CR, Palmer J, Taylor J, Stout S, Holt S, Hendren R, Bock B, Waibel E, Tyree G, Miller GC (2008) A comparison of two methods to determine the presence of high-risk HPV cervical infections. Am J Clin Pathol 130:401–408
22. Rebolj M, Preisler S, Ejegod DM, Bonde J, Rygaard C, Lynge E (2013) Prevalence of Human Papillomavirus infection in unselected SurePath samples using the APTIMA HPV mRNA assay. J Mol Diagn: JMD 15:670–677
23. Wright TC, Stoler MH, Behrens CM, Sharma A, Zhang G, Wright TL (2015) Primary cervical cancer screening with human papillomavirus: end of study results from the ATHENA study using HPV as the first-line screening test. Gynecol Oncol 136:189–197
24. Forouzanfar MH, Foreman KJ, Delossantos AM, Lozano R, Lopez AD, Murray CJ, Naghavi M (2011) Breast and cervical cancer in 187 countries between 1980 and 2010: a systematic analysis. Lancet (London, England) 378:1461–1484
25. Walboomers JM, Jacobs MV, Manos MM, Bosch FX, Kummer JA, Shah KV, Snijders PJ, Peto J, Meijer CJ, Munoz N (1999) Human papillomavirus is a necessary cause of invasive cervical cancer worldwide. J Pathol 189:12–19
26. Schiffman M, Kjaer SK (2003) Chapter 2: natural history of anogenital human papillomavirus infection and neoplasia. J Natl Cancer Inst Monogr 31:14–19
27. de Sanjose S, Quint WG, Alemany L, Geraets DT, Klaustermeier JE, Lloveras B, Tous S, Felix A, Bravo LE, Shin HR, Vallejos CS, de Ruiz PA, Lima MA, Guimera N, Clavero O, Alejo M, Llombart-Bosch A, Cheng-Yang C, Tatti SA, Kasamatsu E, Iljazovic E, Odida M, Prado R, Seoud M, Grce M, Usubutun A, Jain A, Suarez GA, Lombardi LE, Banjo A, Menendez C, Domingo EJ, Velasco J, Nessa A, Chichareon SC, Qiao YL, Lerma E, Garland SM, Sasagawa T, Ferrera A, Hammouda D, Mariani L, Pelayo A, Steiner I, Oliva E, Meijer CJ, Al-Jassar WF, Cruz E, Wright TC, Puras A, Llave CL, Tzardi M, Agorastos T, Garcia-Barriola V, Clavel C, Ordi J, Andujar M, Castellsague X, Sanchez GI, Nowakowski AM, Bornstein J, Munoz N, Bosch FX (2010) Human papillomavirus genotype attribution in invasive cervical cancer: a retrospective cross-sectional worldwide study. Lancet Oncol 11:1048–1056
28. Giuliano AR, Harris R, Sedjo RL, Baldwin S, Roe D, Papenfuss MR, Abrahamsen M, Inserra P, Olvera S, Hatch K (2002) Incidence, prevalence, and clearance of type-specific human papillomavirus infections: The Young Women's Health Study. J Infect Dis 186:462–469
29. Castle PE, Dockter J, Giachetti C, Garcia FA, McCormick MK, Mitchell AL, Holladay EB, Kolk DP (2007) A cross-sectional study of a prototype carcinogenic human papillomavirus E6/

E7 messenger RNA assay for detection of cervical precancer and cancer. Clin Cancer Res: Off J Am Assoc Cancer Res 13:2599–2605
30. Khan MJ, Castle PE, Lorincz AT, Wacholder S, Sherman M, Scott DR, Rush BB, Glass AG, Schiffman M (2005) The elevated 10-year risk of cervical precancer and cancer in women with human papillomavirus (HPV) type 16 or 18 and the possible utility of type-specific HPV testing in clinical practice. J Natl Cancer Inst 97:1072–1079
31. Dalstein V, Riethmuller D, Pretet JL, Le Bail CK, Sautiere JL, Carbillet JP, Kantelip B, Schaal JP, Mougin C (2003) Persistence and load of high-risk HPV are predictors for development of high-grade cervical lesions: a longitudinal French cohort study. Int J Cancer 106:396–403
32. Josefsson AM, Magnusson PK, Ylitalo N, Sorensen P, Qwarforth-Tubbin P, Andersen PK, Melbye M, Adami HO, Gyllensten UB (2000) Viral load of human papilloma virus 16 as a determinant for development of cervical carcinoma in situ: a nested case-control study. Lancet (London, England) 355:2189–2193
33. Lillo FB, Lodini S, Ferrari D, Stayton C, Taccagni G, Galli L, Lazzarin A, Uberti-Foppa C (2005) Determination of human papillomavirus (HPV) load and type in high-grade cervical lesions surgically resected from HIV-infected women during follow-up of HPV infection. Clin Infect Dis : Off Publ Infect Dis Soc Am 40:451–457
34. Ostor AG (1993) Natural history of cervical intraepithelial neoplasia: a critical review. Int J Gynecol Pathol: Off J Int Soc Gynecol Pathol 12:186–192
35. Human papillomaviruses. IARC Monogr Eval Carcinog Risks Hum 90:1–636
36. Monnier-Benoit S, Dalstein V, Riethmuller D, Lalaoui N, Mougin C, Pretet JL (2006) Dynamics of HPV16 DNA load reflect the natural history of cervical HPV-associated lesions. J Clin Virol: Off Publ Pan Am Soc Clin Virol 35:270–277
37. Peitsaro P, Johansson B, Syrjanen S (2002) Integrated human papillomavirus type 16 is frequently found in cervical cancer precursors as demonstrated by a novel quantitative real-time PCR technique. J Clin Microbiol 40:886–891
38. Saunier M, Monnier-Benoit S, Mauny F, Dalstein V, Briolat J, Riethmuller D, Kantelip B, Schwarz E, Mougin C, Pretet JL (2008) Analysis of human papillomavirus type 16 (HPV16) DNA load and physical state for identification of HPV16-infected women with high-grade lesions or cervical carcinoma. J Clin Microbiol 46:3678–3685
39. Wang S, Wei H, Wang N, Zhang S, Zhang Y, Ruan Q, Jiang W, Xiao Q, Luan X, Qian X, Zhang L, Gao X, Sun X (2012) The prevalence and role of human papillomavirus genotypes in primary cervical screening in the northeast of China. BMC Cancer 12:160
40. Kahla S, Kochbati L, Maalej M, Oueslati R (2014) Situation of HPV16 E2 gene status during radiotherapy treatment of cervical carcinoma. Asian Pac J Cancer Prev 15:2869–2873
41. Gillison ML (2004) Human papillomavirus-associated head and neck cancer is a distinct epidemiologic, clinical, and molecular entity. Semin Oncol 31:744–754
42. Katki HA, Kinney WK, Fetterman B, Lorey T, Poitras NE, Cheung L, Demuth F, Schiffman M, Wacholder S, Castle PE (2011) Cervical cancer risk for women undergoing concurrent testing for human papillomavirus and cervical cytology: a population-based study in routine clinical practice. Lancet Oncol 12:663–672
43. Leemans CR, Braakhuis BJ, Brakenhoff RH (2011) The molecular biology of head and neck cancer. Nat Rev Cancer 11:9–22
44. Hong AM, Dobbins TA, Lee CS, Jones D, Harnett GB, Armstrong BK, Clark JR, Milross CG, Kim J, O'Brien CJ, Rose BR (2010) Human papillomavirus predicts outcome in oropharyngeal cancer in patients treated primarily with surgery or radiation therapy. Br J Cancer 103:1510–1517
45. Sathish N, Wang X, Yuan Y (2014) Human Papillomavirus (HPV)-associated oral cancers and treatment strategies. J Dent Res 93:29s–36s
46. Mirghani H, Amen F, Blanchard P, Moreau F, Guigay J, Hartl DM, Lacau SGJ (2015) Treatment de-escalation in HPV-positive oropharyngeal carcinoma: ongoing trials, critical issues and perspectives. Int J Cancer 136:1494–1503
47. Abramowitz L, Jacquard AC, Jaroud F, Haesebaert J, Siproudhis L, Pradat P, Aynaud O, Leocmach Y, Soubeyrand B, Dachez R, Riethmuller D, Mougin C, Pretet JL, Denis F (2011)

Human papillomavirus genotype distribution in anal cancer in France: the EDiTH V study. Int J Cancer 129:433–439
48. Stanley MA, Winder DM, Sterling JC, Goon PK (2012) HPV infection, anal intra-epithelial neoplasia (AIN) and anal cancer: current issues. BMC Cancer 12:398
49. De Vuyst H, Clifford GM, Nascimento MC, Madeleine MM, Franceschi S (2009) Prevalence and type distribution of human papillomavirus in carcinoma and intraepithelial neoplasia of the vulva, vagina and anus: a meta-analysis. Int J Cancer 124:1626–1636
50. van Seters M, van Beurden M, de Craen AJ (2005) Is the assumed natural history of vulvar intraepithelial neoplasia III based on enough evidence? A systematic review of 3322 published patients. Gynecol Oncol 97:645–651
51. Orth G (2010) Genetics and susceptibility to human papillomaviruses: epidermodysplasia verruciformis, a disease model. Bull Acad Natl Med 194:923–940. discussion 941
52. Lazarczyk M, Cassonnet P, Pons C, Jacob Y, Favre M (2009) The EVER proteins as a natural barrier against papillomaviruses: a new insight into the pathogenesis of human papillomavirus infections. Microbiol Mol Biol Rev: MMBR 73:348–370
53. Armstrong EP (2010) Prophylaxis of cervical cancer and related cervical disease: a review of the cost-effectiveness of vaccination against oncogenic HPV types. J Manage Care Pharm: JMCP 16:217–230
54. Quadrivalent vaccine against human papillomavirus to prevent high-grade cervical lesions. New Engl J Med 356:1915–1927
55. Paavonen J, Naud P, Salmeron J, Wheeler CM, Chow SN, Apter D, Kitchener H, Castellsague X, Teixeira JC, Skinner SR, Hedrick J, Jaisamrarn U, Limson G, Garland S, Szarewski A, Romanowski B, Aoki FY, Schwarz TF, Poppe WA, Bosch FX, Jenkins D, Hardt K, Zahaf T, Descamps D, Struyf F, Lehtinen M, Dubin G (2009) Efficacy of human papillomavirus (HPV)-16/18 AS04-adjuvanted vaccine against cervical infection and precancer caused by oncogenic HPV types (PATRICIA): final analysis of a double-blind, randomised study in young women. Lancet (London, England) 374:301–314
56. Hariri S, Markowitz LE, Dunne EF, Unger ER (2013) Population impact of HPV vaccines: summary of early evidence. J Adolesc Health: Off Publ Soc Adolesc Med 53:679–682
57. Ali H, Donovan B, Wand H, Read TR, Regan DG, Grulich AE, Fairley CK, Guy RJ (2013) Genital warts in young Australians five years into national human papillomavirus vaccination programme: national surveillance data. BMJ (Clin Res Ed) 346:f2032
58. Petrosky E, Bocchini JA Jr, Hariri S, Chesson H, Curtis CR, Saraiya M, Unger ER, Markowitz LE (2015) Use of 9-valent human papillomavirus (HPV) vaccine: updated HPV vaccination recommendations of the advisory committee on immunization practices. MMWR Morb Mortal Wkly Rep 64:300–304
59. Nigam A, Saxena P, Acharya AS, Mishra A, Batra S (2014) HPV Vaccination in India: Critical Appraisal. ISRN Obstet Gynecol 2014:394595
60. Tahamtan A, Ghaemi A, Gorji A, Kalhor HR, Sajadian A, Tabarraei A, Moradi A, Atyabi F, Kelishadi M (2014) Antitumor effect of therapeutic HPV DNA vaccines with chitosan-based nanodelivery systems. J Biomed Sci 21:69
61. McNamara M, Batur P, Walsh JM, Johnson KM (2016) HPV update: vaccination, screening, and associated disease. J Gen Intern Med 31:1360–1366
62. Grce M (2009) Primary and secondary prevention of cervical cancer. Expert Rev Mol Diagn 9:851–857

Chapter 4
Merkel Cell Polyomavirus: A New DNA Virus Associated with Human Cancer

Margo MacDonald and Jianxin You

Abstract Merkel cell polyomavirus (MCPyV or MCV) is a novel human polyomavirus that has been discovered in Merkel cell carcinoma (MCC), a highly aggressive skin cancer. MCPyV infection is widespread in the general population. MCPyV-associated MCC is one of the most aggressive skin cancers, killing more patients than other well-known cancers such as cutaneous T-cell lymphoma and chronic myelogenous leukemia (CML). Currently, however, there is no effective drug for curing this cancer. The incidence of MCC has tripled over the past two decades. With the widespread infection of MCPyV and the increase in MCC diagnoses, it is critical to better understand the biology of MCPyV and its oncogenic potential. In this chapter, we summarize recent discoveries regarding MCPyV molecular virology, host cellular tropism, mechanisms of MCPyV oncoprotein-mediated oncogenesis, and current therapeutic strategies for MCPyV-associated MCC. We also present epidemiological evidence for MCPyV infection in HIV patients and links between MCPyV and non-MCC human cancers.

Keywords Merkel cell polyomavirus • Merkel cell carcinoma • Oncogenesis • HIV patients • Host cellular tropism

4.1 Introduction

Merkel cell polyomavirus (MCPyV), a member of the *Polyomaviridae* family, was first isolated from Merkel cell carcinoma (MCC) in 2008 using digital transcriptome subtraction [1, 2]. Merkel cell carcinoma (MCC) typically presents as a neuroendocrine carcinoma of the skin. Historically, MCC has been thought to arise from Merkel cells, a unique cell type of the skin bearing both epithelial and

M. MacDonald • J. You (✉)
Department of Microbiology, Perelman School of Medicine, University of Pennsylvania, Philadelphia, PA 19104, USA
e-mail: jianyou@mail.med.upenn.edu

Q. Cai et al. (eds.), *Infectious Agents Associated Cancers: Epidemiology and Molecular Biology*, Advances in Experimental Medicine and Biology 1018,
DOI 10.1007/978-981-10-5765-6_4

neurosecretory characteristics [3]. The incidence rates of MCC have tripled in the last decades, and it is incredibly lethal, with a disease-associated mortality of 46% [4]. The major risk factors for MCC include advanced age, immunosuppression, and prolonged exposure to sunlight and ultraviolet (UV) radiation [5, 6]. Since the initial discovery of MCPyV in MCC tumors, a compelling line of evidence has established it as the causative agent of MCC. As many as 80% of all MCC tumors have clonally integrated MCPyV genomes [1, 7]. Integration of MCPyV genome into the host genome appears to occur before clonal expansion of the tumor, while persistent expression of the viral tumor antigens is required to maintain MCC tumor growth [1, 8]. This evidence strongly supports an important oncogenic role of MCPyV in MCC tumors. However, MCPyV infection of the skin is ubiquitous and largely asymptomatic in the general population [9–11]. Therefore, there is a growing interest in understanding the basic molecular virology of MCPyV and its role in oncogenesis. In this chapter, we present our current knowledge on the first polyomavirus linked to human cancer.

4.2 MCPyV Genome and Encoded Proteins

MCPyV, like other members of the polyomavirus family, is a small, non-enveloped, icosahedral, double-stranded circular DNA virus [12]. The 5.3 kb viral genome is composed of a multiply spliced early "tumor antigen" region, a late region, and a noncoding regulatory region (NCRR) which divides the two coding regions (Fig. 4.1). The NCRR contains the viral origin of replication (Ori) flanked by the promoters that drive early and late gene expression.

The early region of MCPyV encodes large tumor antigen (LT), small tumor antigen (sT), 57-kilodalton tumor antigen (57kT), and the overprinting gene

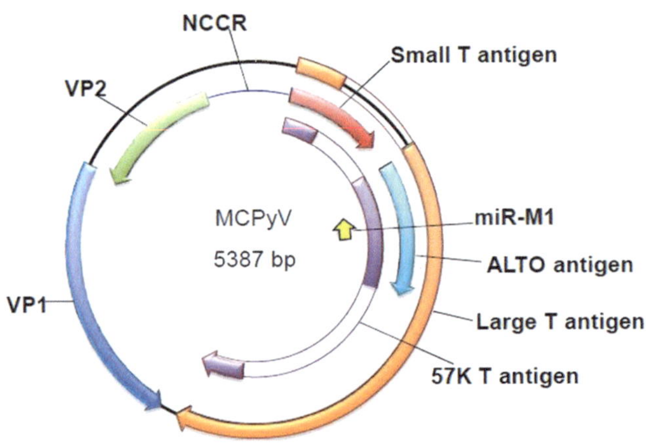

Fig. 4.1 MCPyV genome. This schematic diagram shows the noncoding regulatory region (*NCRR*), early genes, late genes, and a microRNA (*miR-M1*) encoded by the MCPyV genome

alternate LT ORF (ALTO) (Fig. 4.1). The MCPyV LT antigen is a multifunctional protein that plays important roles in host cell-cycle regulation as well as viral genome replication (reviewed in [13]). The N-terminal region of LT contains a conserved region 1 (CR1), a DnaJ domain (for binding heat-shock proteins), and an LxCxE motif that interacts with retinoblastoma protein (RB) to stimulate host cell proliferation (Fig. 4.2) [14]. The C-terminal region of LT contains an Ori binding domain (OBD) necessary for LT binding to the viral Ori and a helicase domain that stimulates replication of the viral genome (Fig. 4.2) [15, 16]. The sT protein shares the LT N-terminal region, including the CR1 and DnaJ domains, but has a unique C-terminus carrying a protein phosphatase 2A (PP2A) binding site [17]. Unlike other polyomavirus sTs, MCPyV sT appears to play a central role in inducing oncogenesis [18]. MCPyV sT has been shown to stimulate cellular proliferation by inducing hyper-phosphorylation of the eukaryotic translation initiation factor 4E-binding protein 1 (4E-BP1) independent of PP2A binding [18]. It also binds the ubiquitin ligase SCF^{Fbw7} and disrupts proteasomal degradation of both LT and certain cellular cell-cycle regulators [19]. The unique C-terminal domain of MCPyV sT also contains highly conserved iron-sulfur clusters that are important for stimulating LT-mediated viral replication [20]. In contrast to LT and sT, the functions and physiological significance of both 57kT and ALTO remain to be elucidated [12, 21, 22].

The late region of MCPyV encodes VP1 and VP2 (Fig. 4.1), which function as major and minor subunits of the viral capsid, respectively. VP1 and VP2 form capsids around the MCPyV genome. While the major capsid protein VP1 is necessary and sufficient for producing pseudovirions, the minor protein VP2 may confer specificity in host cell targeting [11, 23–26].

Like many polyomaviruses, MCPyV encodes a microRNA, termed miR-M1 (Fig. 4.1), which has been shown to downregulate expression of LT [27, 28]. This regulation of LT was shown to be important for long-term MCPyV episome maintenance in cell culture and potentially for establishing persistent infection in vivo [27, 28].

4.3 Mechanisms of MCPyV Oncoprotein-Mediated Oncogenesis

Like papillomavirus-induced cancers, MCPyV-associated MCC tumors typically carry the viral genome integrated into the host genome [1, 29, 30]. MCPyV-associated MCC tumors demonstrate a clonal integration pattern of the viral genome, suggesting that the integration event occurs prior to the initiation of oncogenesis and expansion of tumor cells. These tumors typically express both of the major viral tumor antigens, LT and sT [8, 31]. However, the MCC tumors carrying the integrated viral genome do not support a productive viral life cycle. Both LT and sT play unique and important roles in driving MCC oncogenesis.

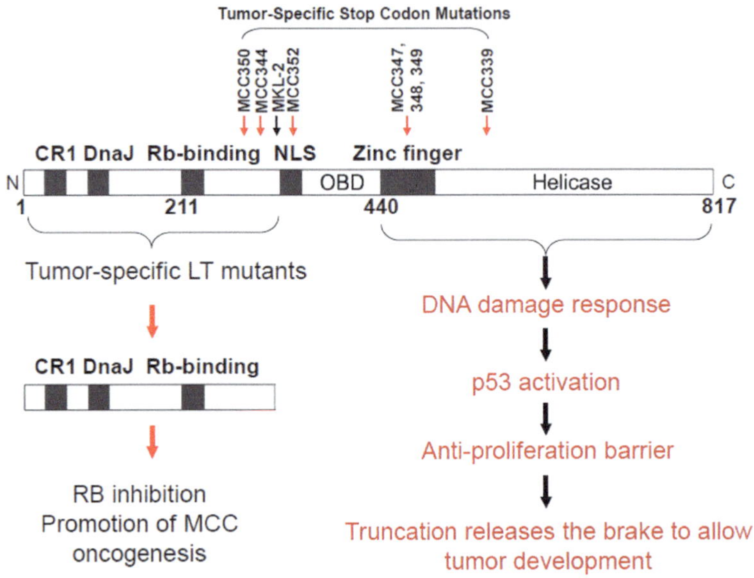

Fig. 4.2 Selective deletion of the C-terminus of MCPyV LT is a critical event during the oncogenic progression of MCPyV-associated cancers. Shown are the domain structures of MCPyV LT antigen and mutations found in MCPyV-associated MCCs that introduce premature stop codons to delete the LT C-terminus. The LT C-terminus can induce DNA damage response to activate the p53 tumor suppressor. This growth inhibitory property may function as an anticancer brake to inhibit cell proliferation and transformation. Deletion of the replication domains in the C-terminus of LT releases this antitumor brake to allow oncogenic progression. The truncated tumor-specific LT mutants retain RB tumor-suppressor inhibiting activities to drive cellular transformation

A common feature of MCPyV genomes integrated into the MCC genome is the selection for mutations in the LT coding sequence that introduce premature stop codons which delete the LT C-terminal Ori binding and helicase domains (Fig. 4.2) [14]. Therefore, the MCPyV LT protein is typically expressed in a truncated form in MCC tumors [14]. In contrast, these tumor-specific mutations do not disrupt the expression of sT. The N-terminal portion of LT still expressed in these tumors is referred to as LTT (tumor-derived LT). LTT retains the CR1, DnaJ, and RB-binding motifs, allowing these tumor-specific LT mutants to robustly disrupt the host cell cycle (Fig. 4.2) [32].

The selection for premature stop mutations in MCPyV LT is remarkably common in MCPyV-associated tumors, suggesting that deletion of the replication domains in the C-terminus of LT is required for tumorigenesis. One selective pressure that may exist is the elimination of viral DNA replication activity after the genome has been integrated into the host DNA [14]. It is conceivable that continuous LT-mediated replication from the integrated viral Ori could result in replication fork collisions and double-strand breaks in the host DNA; disrupting LT's OBD and helicase domains would alleviate this genotoxic stress. In addition, other functional activities of the C-terminal domain may need to be negatively selected for during tumorigenesis. For example, expression of just the OBD and helicase domains of MCPyV LT induces a host cellular DNA damage response (DDR). This activation stimulates p53 activity and can arrest the host cell cycle [33]. This growth inhibitory property of the MCPyV LT C-terminus may function as a barrier to oncogenic progression (Fig. 4.2) [33, 34]. Since replicative stress, DNA damage responses, and cell-cycle arrest all pose challenges to oncogenesis, together they provide a strong rationale for the selection of truncated LT proteins which retain RB tumor-suppressor inhibiting activities while avoiding potentially antagonistic activity conferred by the C-terminal domain of LT.

Compared to MCPyV LT, sT plays a more dominant role in MCPyV-induced carcinogenesis [18]. Contrary to other polyomaviruses, expression of MCPyV LT alone is not sufficient to transform cells [18]. MCPyV sT, however, has been suggested to transform immortalized rat fibroblasts in cell culture even when expressed alone [18]. MCPyV sT also demonstrates robust transforming activity in vivo [35]. sT's oncogenic activity is mostly mediated through induction of the hyper-phosphorylated and inactivated state of 4E-BP1, leading to dysregulation of cap-dependent translation that accelerates cell proliferation and malignant transformation [18]. In addition, sT inhibits the E3 ubiquitin ligase SCF^{Fbw7} to prevent proteasomal degradation of MCPyV LT and key cellular proliferative proteins like c-Myc and cyclin E [19]. Unlike LT, sT is commonly expressed in MCPyV-associated tumors, and almost no mutations have been found in the sT-coding regions integrated into the genome of MCC tumors [13], again highlighting the important role this protein plays in MCPyV-associated cancers.

MCPyV-positive MCC cells are dependent on MCPyV LT/sT oncoproteins. Persistent expression of these oncogenes from the integrated viral genome is required to sustain growth of MCPyV-associated tumors, in both in vitro and xenograft models [8, 31]. Knockdown of LT/sT antigens induces growth arrest and cell death in all MCPyV-positive MCC cell lines tested [8, 31] and leads to tumor regression in xenotransplantation models [32].

4.4 Genetic Basis of MCPyV-Associated MCC

Recent studies have begun to delineate the differences in the causes of MCPyV-positive and MCPyV-negative MCCs. Genetic studies have shown that UV radiation is the primary cause of MCPyV-negative MCCs, which constitute about 20% of all MCC cases [36–38]. Compared to MCPyV-positive MCCs, MCPyV-negative tumors demonstrate much higher mutational burdens, which are associated with a prominent UV-induced DNA damage signature [36–38]. Both MCPyV-positive and MCPyV-negative MCC tumors are commonly found on sun-exposed regions of the body, such as the head, neck, and limbs [37]. However, the lower number of genetic mutations found in the genomes of MCPyV-positive MCCs compared to MCPyV-negative tumors, along with the lack of a definitive UV mutation signature in MCPyV-positive MCCs, indicates that UV plays a primary etiologic role in

MCPyV-negative MCC tumorigenesis [36]. In MCPyV-positive MCCs, UV may simply promote tumor growth through immunosuppressive effects on the tumor microenvironment or through inducing the mutations needed for MCPyV integration and generation of the truncated viral LT antigen [36].

Compared to MCPyV-positive MCCs, MCPyV-negative tumors also contain a much higher number of cancer-promoting mutations [36–38]. Some of the common mutations frequently observed in MCPyV-negative MCCs include mutations in *RB1*, *TP53*, and *PIK3CA*, along with mutations in host DDR and chromatin modulation pathways [36–38]. Inactivating mutations in the NOTCH signaling pathway were also detected, supporting a tumor-suppressor role for this pathway in MCC [38]. The lower levels of cancer-promoting mutations observed in MCPyV-positive MCCs confirm that MCPyV oncogenes are the primary oncogenic drivers for these tumors [36–38]. Activating mutations of *HRAS* were among the very few frequently observed in MCPyV-positive tumors, suggesting that these genetic mutations may cooperate with the viral oncogenes to drive MCC tumorigenic progression [36, 38]. In several studies, inactivating mutations in *RB1* were observed in MCPyV-negative tumors, but not in MCPyV-positive tumors [36–38]. This is consistent with the fact that the truncated MCPyV LT antigen interacts with and inactivates RB1, suggesting that RB1 disruption is required for all MCC tumorigenesis [37].

Potentially due to their higher mutational burden, MCPyV-negative MCCs typically display a more aggressive subtype, with patients having an increased risk of disease progression and death [39]. MCPyV-negative tumors are also more likely to recur after treatment than MCPyV-positive tumors [39]. There are a variety of possible reasons for the more aggressive behavior observed in MCPyV-negative subtype of MCCs, including the fact that fewer oncoproteins are expressed as targets for T-cell-infiltrating lymphocytes (TILs), their advanced stage at presentation, and a higher number of mutations in oncogenic pathways [39].

4.5 MCPyV Host Cellular Tropism and the Origin of MCC

4.5.1 MCPyV Entry into the Host Cells

An important part of the MCPyV life cycle that is particularly useful for the development of antivirals and vaccines is viral entry into the host cell. MCPyV dsDNA genome is encapsidated in an icosahedral shell of viral capsid consisting of the structural proteins VP1 and VP2 at a ratio of 5:2 [1, 25]. For most polyomaviruses, the major capsid protein VP1 determines antigenicity and receptor specificity. It initiates viral entry into host cells and has a significant impact on attachment, tissue tropism, and viral pathogenicity [40]. In line with findings from other polyomaviruses, MCPyV's entry into host cells is mediated by binding of the major capsid protein VP1 pentamer to cellular receptors [41]. The minor capsid protein VP2 is essential for infectious MCPyV entry in some cell types, but others could be

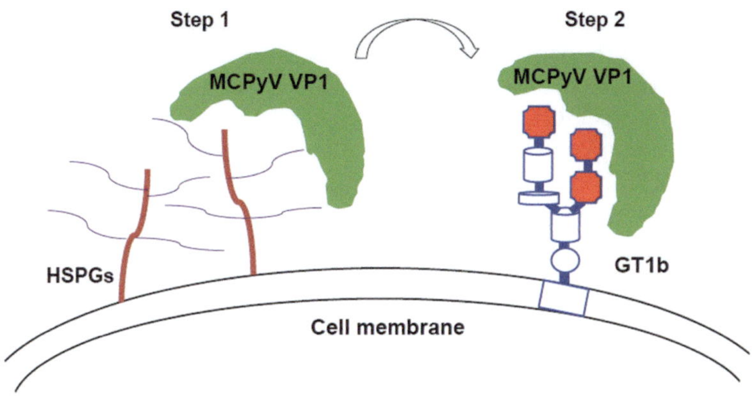

Fig. 4.3 MCPyV entry into host cells. MCPyV viral entry is a two-step process mediated primarily by the major capsid protein VP1. The primary binding partner for MCPyV VP1 is the glycosaminoglycan (GAG) heparan sulfate. This GAG is found on the host cell membrane in the form of heparan sulfate proteoglycans (*HSPGs*). While this interaction is all that is necessary for VP1 binding to occur, a secondary interaction with the sialylated ganglioside GT1b is required for entry. MCPyV interacts with the sialic acids in both arms of GT1b, which are shown in red. MCPyV minor capsid protein VP2 (not shown) has been suggested to play a role in post-attachment entry

transduced with MCPyV pseudovirions deficient in VP2, indicating that in cell types where VP2 is necessary for MCPyV entry, there is some barrier that it helps to overcome [25, 42]. Virion assembly, packaging of DNA, and attachment to target cells were not significantly affected by knockdown of VP2, indicating that the role of VP2 is in post-attachment entry [25]. While most polyomaviruses also contain another minor capsid protein VP3, this minor capsid protein is not detectable in either MCPyV-infected cells or native MCPyV virions [25]. Phylogenetic analysis suggests that MCPyV belongs to a member of a divergent clade of polyomaviruses lacking the conserved VP3 N-terminus [25].

The functional receptors of most human polyomaviruses are sialylated glycans [43]. Sialic acids are found mostly in glycoproteins and gangliosides [43]. The ganglioside GT1b, which has sialic acids on both arms, was the first proposed to be the receptor for MCPyV VP1 (Fig. 4.3) [44]. VP1 interacts with the sialic acids on both branches of the GT1b carbohydrate chain (Fig. 4.3) [44]. In a later study, where MCPyV reporter vectors and native MCPyV virions were used to transduce human cells, it was discovered that the initial attachment receptors of MCPyV VP1 are sulfated glycosaminoglycans (GAGs), specifically heparan sulfate (HS) proteoglycans [26]. While VP1 can bind to GAGs such as HS and chondroitin sulfate (CS), only the N-sulfated and 6-O-sulfated forms of HS mediate infectious entry, and both CS and other forms of HS were dispensable for this process [26]. In addition, Neu et al. used X-ray structures to show that a shallow binding site on the apical surface of the VP1 capsomer recognizes the linear sialylated disaccharide Neu5Ac-α2,3-Gal, which is present in GT1b [41]. Previous studies were not able to show GD1a binding with VP1, yet this study showed VP1 interactions with GT1b, GD1a, 3SLN, and DSL oligosaccharides, all of which contain the Neu5Ac motif [41, 44]. This indicates that GT1b, and possibly these other sialylated glycans, are the post-attachment co-receptors for VP1 that enable the secondary entry step after primary attachment through GAGs. More importantly, mutagenesis studies revealed that the VP1 sialic acid binding site plays a role in post-attachment infection, not initial attachment [41]. Together, current knowledge supports a two-step attachment and entry process for MCPyV, with sulfated GAGs being the initial attachment receptors for VP1. The primary binding, mediated mostly by HS, is followed by secondary interactions with a sialylated glycan post-attachment co-receptor (Fig. 4.3) [26, 40, 42, 43, 45]. These glycans are not required for initial attachment of MCPyV virions, but they are necessary for viral entry into the cell (Fig. 4.3) [26, 41, 43].

MCPyV enters its target cells in a slow and asynchronous motion [45]. After entry, MCPyV must travel through the cytoplasm to the nucleus in order to use the host cell replication machinery. Host cell surface glycoproteins and glycolipids play a role in both the entry stage of viral infection and channeling the virions to specific intracellular membrane-bound compartments and ultimately to the nucleus [43]. However, the molecular events that occur after MCPyV penetrates the cell membrane and allow delivery of the encapsidated viral DNA to the host nucleus have not been elucidated. This is largely due to a lack of cell culture model for MCPyV infection [45].

4.5.2 MCPyV Host Cellular Tropism

While the MCPyV binding factors, such as sialic acid and heparan sulfate, which mediate attachment and entry, have been actively discovered [26, 41], much remains to be elucidated with respect to MCPyV natural infection and MCPyV host cellular tropism. It is unclear how MCPyV targets specific cell types given both sialic acid and heparan sulfate are ubiquitous. Studies of basic MCPyV virology have been hampered by the facts that MCPyV replicates poorly in the majority of cell lines tested thus far and its natural host cell up until very recently had not been described. The lack of a robust cell culture system for MCPyV infection has limited our understanding of this important tumor virus.

Multiple lines of evidence point toward the skin being the major site of MCPyV replication in humans. First, various deep sequencing studies have provided evidence of persistent and asymptomatic infection of MCPyV in adult skin [46, 47]. Additionally, the cell types which support MCPyV replication have been either epithelial or fibroblast in origin [15, 23, 48]. Finally, MCC is a tumor of the dermis, and the presumed cells of origin for MCC, Merkel cells, are a resident of the epidermis. Following this line of reasoning, our group examined the ability of various skin cell types to support MCPyV infection and discovered that human dermal fibroblasts (HDFs) are a natural host cell of MCPyV [49]. We demonstrated that both epidermal growth factor (EGF) and fibroblast growth factor (FGF) were required to promote efficient MCPyV infection of dermal fibroblasts; these factors may stimulate expression of cellular factors necessary for infection, producing an environment conducive to MCPyV infection and replication [49]. Interestingly, these growth factors are stimulated upon wounding [50], suggesting that wounding processes may facilitate MCPyV infection in the human skin. We also found that induction of matrix metalloproteinases (MMPs) mediated by the WNT/β-catenin signaling pathway is critical for MCPyV infection of HDFs. WNT signaling is crucial for the formation of hair follicles [51]. Interestingly, we showed that MCPyV could efficiently infect the HDFs surrounding hair follicles [49]. This finding is in line with the observation that MCPyV is frequently detected in eyebrow hair bulbs [52]. Remarkably, several MCC risk factors, including UV exposure and aging, are known to upregulate MMPs [53–58], once again linking risk of MCC incidence with MCPyV infection.

4.5.3 Origin of MCC

The relationship between the cells that MCPyV infects and those that it transforms to cause MCC remains a central question for MCPyV research. The establishment of dermal fibroblasts as a natural host cell for MCPyV may help resolve a long-standing puzzle in the MCC field regarding the cells of origin for MCC [59]. Historically, MCC has been thought to arise from Merkel cells due to its expression of cytokeratin 20, a unique marker of Merkel cells. However, this assumption has been challenged by a number of recent studies. First, Merkel cells are postmitotic and do not have robust proliferative potential, making them less likely to support MCPyV infection, replication, and associated tumorigenesis [60, 61]. Additionally, Merkel cells are of epidermal origin, while MCC tumors are thought to derive from the dermis [62–64]. Because MCC tumors also express markers common to pro-/pre-B cells, such as paired box gene 5 (PAX5) and terminal deoxynucleotidyl transferase (TdT), it has been suggested that MCC tumors may derive from the B-cell lineage [65]. The finding that dermal fibroblasts support MCPyV infection provides new alternative hypotheses [49]. For example, MCPyV infection of dermal fibroblasts can, over time, induce their transformation and upregulate genes commonly expressed in other cell types, including B cells and Merkel cells. This hypothesis is in line with the finding that MCC tumors are dermal in origin. Alternatively, Merkel cells residing at the boundary of the epidermis and dermis may be infected as a nonproductive bystander of dermal fibroblast infection. Along this line, the fact that Merkel cells do not support the full MCPyV life cycle may predispose this infection toward genome integration, which eventually lead to oncogenesis. Both of these models – dermal fibroblast origin or infection of bystander Merkel cells – remain to be tested in vivo using animal and skin explant models.

4.6 Current Therapeutic Strategies for MCPyV-Associated MCC

4.6.1 Surgery, Radiation Therapy, and Chemotherapy

Early-stage, localized MCCs are usually treated with surgical excision [66]. Adjuvant radiation therapy applied after the initial surgery has been shown to improve local and regional recurrence rates and therefore has also been recommended for primary tumors [66]. However, MCC frequently undergoes metastasis, increasing the probability that tumors may arise in areas that are harder to reach and to fully eradicate with radiotherapy [67]. Thus, chemotherapy has been used to treat advanced stage MCC. Although MCC tumors are responsive to chemotherapy in the short term, the duration of the response is usually transient, and many tumors often develop resistance to chemotherapy [66, 68, 69]. Additionally, chemotherapy has little overall survival benefit for MCC tumors due to its immunosuppressive effect, which counteracts the cellular immune reaction to MCC tumors. Currently, there are very few viable options for patients with advanced MCCs [69].

4.6.2 Immune Checkpoint Inhibitors and Immunotherapy

MCC patients with robust immune responses and higher level of intratumoral TILs generally showed better prognoses and increased rates of regression [70–73]. Intratumoral CD8+ and CD3+ lymphocytes, which predict better survival, are typically more commonly found in MCPyV-positive MCCs [74]. This tight correlation between prognosis and immune function suggests that immunotherapies may have great potential for treating metastatic MCCs. Methods of increasing interferon production, such as stimulation by the targeted delivery of the IL-12 gene using vaccine and electroporation, are currently being investigated [75]. A promising immunotherapy strategy for MCC treatment targets the programmed cell death receptor 1/ programmed cell death ligand 1 (PD-1/PD-L1) checkpoint. PD-L1 is often overexpressed in MCC tumors, especially in MCPyV-positive cases [37]. MCPyV-specific T cells also express elevated levels of PD-1 [76]. Interaction of PD-L1 with the PD-1 receptor on the surface of MCPyV-specific T cells activates an immune checkpoint pathway, which inhibits the antitumor immune response [77, 78]. Therefore, anti-PD-1 monoclonal antibody treatment has become an attractive treatment option for MCC [79]. A response rate of 56% was observed in patients treated with an anti-PD-1 antibody called pembrolizumab [77]. However, these responses were not lasting, ranging in duration from 2.2 to 9.7 months [77]. Studies with this drug and another anti-PD-1 antibody called avelumab both showed more success following fewer first-line treatments, suggesting that they should be used as a first-line therapy for advanced MCC rather than the last in a long line of treatments [75]. While these and some other studies have shown improvements in patients with various immunotherapies, the success has for the most part been minimal, further highlighting the need for new treatments for this cancer.

4.6.3 MCPyV DNA Vaccine

One potential treatment option to explore for MCC is a MCPyV DNA vaccine. As described above, it has been repeatedly shown that CD8+ T-cell tumor infiltration in MCC is associated with better prognosis, decreased likelihood of metastasis, and prolonged survival [70, 74]. Therefore, DNA vaccines capable of generating potent antigen-specific CD8+ T-cell immunity are a promising option for MCC treatment [80]. DNA vaccines are an attractive therapeutic option due to their safety, simplicity, stability, and the possibility to be administered multiple times [80].

MCPyV LT is truncated by stop codons in nearly all MCPyV-positive MCC tumors, losing its C-terminal domain responsible for viral replication, while retaining the N-terminal RB-interacting domain for driving cancer development [8]. Because the MCPyV LT amino terminus plays an important role in oncogenesis and is expressed in all MCPyV-positive tumors, it is an ideal vaccine target. In addition, as a foreign antigen, MCPyV LT avoids the issue of immune tolerance that could impede the development of antitumor immunity. Stop codons introduced by MCPyV integration into MCC tumor genome typically truncate LT at amino acid 258 (aa258) or beyond, so the Hung laboratory created a DNA vaccine to specifically target the MCPyV LT aa1-258 region [80]. When tested in mice injected with the B16/LT murine melanoma cell line stably expressing LT, this vaccine demonstrated both protective and therapeutic effects against LT-expressing tumors in vivo [80]. Compared to mice vaccinated with control empty vector, MCPyV LT-vaccinated mice injected with B16/LT cells exhibited smaller tumors and better survival, and the tumor-bearing mice given the MCPyV LT vaccine as a treatment showed smaller tumor volume and longer survival [80]. These antitumor effects appear to be mediated by CD4+ T-cell induction, while no significant CD8+ T-cell induction was observed [80].

Due to CD8+ T cells' association with better prognosis and disease clearance [70, 74], the Hung group went on to construct a DNA vaccine tailored to eliciting LT-specific CD8+ T-cell responses [81]. This vaccine encodes LT fused to a damage-associated molecular pattern protein, calreticulin (CRT), which has been shown to promote induction of CD8+ T cells when fused to other antigens [80, 81]. The new vaccine, named CRT/LT, was also tested on the B16/LT mice and showed prolonged survival after tumor challenge compared to mice vaccinated with the original MCPyV LT vaccine [81]. Compared to MCPyV LT vaccine or control empty vector, this vaccine also resulted in the best survival when given to tumor-bearing mice [81]. Confirming that this better performance was due to CD8+ T-cell induction, CD8+-depleted mice were not protected by the CRT/LT vaccine [81]. The results indicate that CD8+ T cells were the main mediator of the antitumor effects of the CRT/LT vaccine [81].

A MCPyV DNA vaccine was also created to target the sT antigen, which is a key driver of MCC oncogenesis [82]. MCPyV sT shares the same N-terminus with LT but contains a different C-terminus with a PP2A binding site that is important for virus-induced transformation in other PyVs [17, 82]. When tested against a murine tumor model that expresses MCPyV sT antigen, this vaccine demonstrated successful protection and treatment, leading to increased survival and decreased tumor volume in vivo [83]. As in the case of CRT/LT vaccine, CD8+ T-cell induction was essential for the sT vaccine antitumor effect, which was diminished upon CD8+ T-cell depletion [83]. These preliminary results are promising, but testing in an actual MCC cell line and a MCC animal model would be needed to confirm the efficacy of these MCPyV-targeted vaccines for the control of MCC.

4.7 MCPyV Infection in HIV Patients

Immunosuppression is one of the most important risk factors for the development of MCPyV-associated MCC skin cancer, with immunocompromised individuals making up about 10% of the MCC patient population [84]. This relationship is likely linked to the causative role played by MCPyV in MCC tumorigenesis. A significant portion of these patients experiences immunosuppression as a result of HIV/AIDS infection. HIV-infected individuals have a 13.4-fold increased risk of developing MCC compared to the general population [85]. While this association between HIV infection and MCC has been observed for some time, various recent studies have started to validate the link between HIV and MCPyV infection. The elevated MCPyV prevalence in HIV patients was confirmed by a study looking at MCPyV status in HIV-positive men [86]. This study showed that 59.0% of HIV-positive men had MCPyV DNA in their forehead swabs, compared to only 49.4% of HIV-negative men [86]. However, the level of viral DNA loads in HIV-positive and HIV-negative men did not differ significantly [86]. Another study confirmed that there is no difference in MCPyV viral load between HIV-positive and HIV-negative populations of women [87]. Nonetheless, within the HIV-positive subset of patients, men with poorly controlled HIV infection had higher viral loads compared to those with well-controlled infection [86].

The majority of healthy adults (45–85%, increasing with age) are positive for MCPyV immunoglobulin G (IgG). Using VLP-based enzyme-linked immunosorbent assay (ELISA) to measure MCPyV IgG titers, it was recently shown that levels of MCPyV IgG were higher in HIV/AIDS patients than in either non-AIDS/HIV patients or uninfected controls [88]. Again, MCPyV viral loads did not differ significantly between the tested populations, and there was not much of a difference between uninfected controls and HIV patients without severe immunosuppression [88].

MCPyV detection in the skin is frequent, but the virus is rarely detected in the blood [89]. One study found that only 5.5% of the general population had MCPyV-positive blood serum, while MCPyV DNA was found in the sera of 39.1% of untreated HIV-positive patients [89]. For those in each population whose sera were MCPyV positive, the copy number did not differ significantly between the HIV-positive and HIV-negative groups [89].

MCPyV DNA is usually found in the skin, and MCC typically arises in sun-exposed areas of the body [84, 90]. However, in HIV-positive individuals, MCC often arises in sites not exposed to the sun [84, 90, 91]. One HIV-positive patient even had an oral MCC tumor that tested positive for MCPyV DNA [90]. In some other HIV-positive patients, MCPyV DNA has been detected not only on the skin but also in oral and anogenital mucosa as well as in plucked eyebrow hairs [91].

MCC in HIV-positive individuals is also unusual in the sense that it typically has a much earlier onset in HIV/AIDS patients, with a mean age of diagnosis of 49 years – 20 years younger than the average for immunocompetent patients [84, 91]. In addition, MCCs in AIDS patients are characterized by aggressive clinical course with higher-grade lesions, more advanced tumor stage, and shortened survival [84]. These differences suggest that viral oncogenesis is more rapid and aggressive in patients with HIV-induced immunosuppression [91]. One reason could be that MCPyV infectivity may be exacerbated by these patients' impaired immune response [84]. In addition, the elevated MCPyV DNA loads associated with HIV-induced immunosuppression could explain the increased likelihood of MCC development observed in HIV-infected individuals [91]. Also, the increased viral infection in HIV-positive individuals could make integration of MCPyV into the host cell genome more likely and therefore increase the risk of tumorigenesis [91].

In summary, significantly increased risk of developing MCPyV-associated MCC has been observed among immunocompromised individuals, including HIV-infected patients [84]. This data suggests that screening for early detection of MCC in HIV-positive patients and MCPyV antiviral therapy could both be beneficial to the survival of these patients [91].

4.8 Epidemiological Evidence for MCPyV in Non-MCC Cancers

While MCPyV has an established correlation with MCC, with 80% of this cancer being MCPyV positive, its potential association with a variety of other cancers has been a common topic of exploration recently. There is some evidence suggesting that, in addition to MCC, MCPyV may be associated with extrapulmonary small cell carcinoma (ESCC), cervical cancer, other types of skin cancer, lung cancer, and even some types of leukemia.

One of these cancers, ESCC, was investigated because it shows histological similarities to both small cell lung cancer (SCLC) and MCC, although ESCC is negative for the CK20 marker [92]. ESCC tumors were tested for MCPyV DNA through the use of qPCR, and 19% of the tumors were MCPyV positive [92]. While this prevalence is not high, it is significant enough to suggest that MCPyV may be the driver of a small number of ESCC cases. On the other hand, this same study found no MCPyV DNA in any SCLC samples tested [92].

Among the lung cancers investigated, non-small cell lung cancer (NSCLC) has shown the most evidence supporting an association with MCPyV. In one study, 18% of NSCLC patients had MCPyV DNA present in their tumors, and viral infection was significantly correlated with poorer cancer prognosis within subgroups [93]. Another study found MCPyV DNA in various types of NSCLC in Japanese patients, including squamous cell carcinomas, adenocarcinomas, and others, with some tumors expressing LT RNA transcripts [94]. Prevalence was low in this study as well, but the virus's presence suggests that MCPyV is at least partially associated with NSCLC pathogenesis in some patients. A study of MCPyV in NSCLC in Chilean patients also found a small but not insignificant portion of patients with MCPyV-positive tumors, with 4.7% of patients' cancer testing positive for the virus [95].

While these recent studies suggest that lung cancer may be associated with MCPyV, others show contradictory results. One study looked for MCPyV and other human PyV antibodies in lung cancer samples from patients in China but found no association between MCPyV or other human PyV antibodies and incident lung cancer [96]. Another study examined PyV infection and the risk of lung cancer in never smokers but also found no association. MCPyV seropositivity was detected in 59.3% of lung cancer samples and 61.6% of controls, indicating that there is no difference in MCPyV infection rates in populations with and without lung cancer [97].

In addition to lung cancer, there is contradictory evidence regarding whether MCPyV is associated with various nonmelanoma skin cancers. One paper suggested that 36% of immunocompetent cutaneous squamous cell carcinoma (SCC) patients that participated in their study had one or more samples test positive for MCPyV, while MCPyV positivity in SCCs overall was found to be only 15% [98]. However, most other studies found no correlation between MCPyV and common nonmelanoma skin cancers such as SCC and basal cell carcinoma (BCC). For example, in a study examining SCC and BCC samples from Japanese patients for MCPyV DNA, only 13% of SCCs were found to be MCPyV positive, and none of the BCC samples tested were [99]. One case study tested the SCC tumors of a patient who had both SCC and MCC but only found MCPyV LT in the MCC tumor [100]. Another study detected MCPyV DNA in both BCC and SCC samples but observed a low immunohistochemical detection rate of MCPyV and a lack of MCC-specific MCPyV mutations in the samples [101]. Similarly, Reisinger et al. tested BCC and SCC samples for MCPyV LT and found that none of the samples contained this protein [102]. These results indicate that frequent MCPyV detection in these cancers could simply be due to the ubiquitous spread of MCPyV in the general population, not a result of a causative relationship between MCPyV and these cancers.

However, MCPyV DNA was found in the SCC lesions of a patient with both MCC and SCC, along with HPV coinfection in both lesions, indicating a potential for co-oncogenesis between the two viruses [103]. It was suggested that a low viral copy number in SCC cases has led to difficulties with immunohistochemical detection and may be the reason why many studies do not detect MCPyV in these cancers [103]. To help elucidate whether MCPyV is truly associated with SCC and BCC, larger epidemiological studies are likely necessary.

Additional studies have also investigated the role of MCPyV in non-cutaneous squamous cell carcinomas. A study of esophageal SCCs in Northern Iran detected MCPyV DNA at a low viral copy number in both cancerous and noncancerous esophageal samples but did not find a statistically significant difference in detection rates [104]. Imajoh et al. investigated MCPyV in cervical SCCs and cervical adenocarcinomas (ACs) in Japanese women. They detected MCPyV DNA in 19% of cervical SCCs and 25% of cervical ACs [105]. MCPyV LT was detected in virus-positive tumors [105].

There is also evidence that MCPyV may be associated with some blood cancers. For example, 50% of follicular mycosis fungoides, a lymphoma of the skin, contained MCPyV DNA [106]. However, MCPyV LT was not detected in these samples [106]. In addition, the complete DNA sequence of MCPyV was found in a patient with acute myeloid leukemia [107]. Since there is an established epidemiological link between chronic lymphocytic leukemia (CLL) and MCC, one group investigated the potential role of MCPyV in CLL oncogenesis [108]. They discovered that 13% of T cells in CLL patients tested positive for MCPyV, while none of the patients' B cells did, suggesting that MCPyV may have tropism for T cells [108]. Another study detected MCPyV in 27.1% of CLL patients and even observed LT expression and deletions in some of these patients, suggesting that MCPyV may play a role in a subset of CLL cases [109].

Despite its association with immunosuppression, no correlation between Kaposi's sarcoma and MCPyV has been demonstrated [110, 111]. On the other hand, some recent studies have shown possible correlations between MCPyV and various rare cancers. One group investigated porocarcinoma, a rare malignant neoplasm that arises from the intraepidermal ductal portion of the eccrine sweat glands [112]. MCPyV was found in 68% of primary porocarcinomas, compared to 30% of healthy controls, suggesting that MCPyV may play a role in oncogenesis of this cancer [112]. Another study investigated epidermodysplasia verruciformis-associated (EV-associated) skin neoplasms and detected MCPyV in the in situ carcinomas of all congenital EV patients tested, revealing a strong association between the disease and MCPyV [113].

In summary, most of the recent studies suggest a possible link between MCPyV and various non-MCC cancers. However, there is conflicting information regarding whether MCPyV truly is involved in the pathogenesis of other cancers beside MCC. Therefore, larger epidemiological studies and more definitive data are necessary to further elucidate MCPyV's role in tumorigenesis outside of MCC.

4.9 Conclusions and Future Perspectives

Accumulating evidence demonstrates a role for MCPyV in the development of MCC, making MCPyV the first polyomavirus to be clearly associated with human cancer [1, 12]. MCPyV infection is prevalent in the general population. During the course of its persistent infection, integration of viral genomes into the host genome induces LT truncation mutations that antagonize the tumor-suppressor function. Proliferation of cells carrying integrated viral genomes expressing LT truncation mutations and the sT viral oncogene is selected for during viral oncogenesis. Immune downregulation by viral proteins likely allows MCPyV to establish a persistent infection; it may also play a key role in allowing virally induced early-stage MCC tumors to persist and expand. MCPyV oncogenes are not only persistently expressed as foreign viral antigens in MCC tumors but also required for the growth of the tumors cells. These key features make them ideal targets for developing novel immunotherapy to treat MCC tumors. Elucidation of the mechanism by which MCPyV escapes host immune surveillance and modulates the host immune system to drive cellular transformation will offer important leads for developing viral-targeted therapeutic strategies to treat MCPyV-associated cancers. The recent discovery of HDFs as a target of MCPyV infection in human skin provides an exciting opportunity to study the infectious life cycle of this important oncogenic human polyomavirus [49]. Identification of the target cells of MCPyV natural infection will also facilitate establishment of better animal models to fully elucidate the MCPyV infectious life cycle and MCPyV-induced tumorigenesis in vivo.

Acknowledgments The authors would like to thank the members of our laboratory for helpful discussion and Dr. Wei Liu for helping with figure preparation. This work was supported in part by the National Institutes of Health (NIH) Grants R01CA148768, R01CA142723, and R01CA187718. Margo MacDonald is a Vagelos scholar in the Molecular Life Sciences and is supported by a Vagelos MLS summer stipend.

References

1. Feng H, Shuda M, Chang Y, Moore PS (2008) Clonal integration of a polyomavirus in human Merkel cell carcinoma. Science 319:1096–1100
2. Feng H, Taylor JL, Benos PV, Newton R, Waddell K, Lucas SB, Chang Y, Moore PS (2007) Human transcriptome subtraction by using short sequence tags to search for tumor viruses in conjunctival carcinoma. J Virol 81:11332–11340
3. Wang TS, Byrne PJ, Jacobs LK, Taube JM (2011) Merkel cell carcinoma: update and review. Semin Cutan Med Surg 30:48–56
4. Lemos BD, Storer BE, Iyer JG, Phillips JL, Bichakjian CK, Fang LC, Johnson TM, Liegeois-Kwon NJ, Otley CC, Paulson KG, Ross MI, Yu SS, Zeitouni NC, Byrd DR, Sondak VK, Gershenwald JE, Sober AJ, Nghiem P (2010) Pathologic nodal evaluation improves prognostic accuracy in Merkel cell carcinoma: analysis of 5823 cases as the basis of the first consensus staging system. J Am Acad Dermatol 63:751–761

5. Chang Y, Moore PS (2012) Merkel cell carcinoma: a virus-induced human cancer. Annu Rev Pathol 7:123–144
6. Heath M, Jaimes N, Lemos B, Mostaghimi A, Wang LC, Peñas PF, Nghiem P (2008) Clinical characteristics of Merkel cell carcinoma at diagnosis in 195 patients: the AEIOU features. J Am Acad Dermatol 58:375–381
7. Sastre-Garau X, Peter M, Avril MF, Laude H, Couturier J, Rozenberg F, Almeida A, Boitier F, Carlotti A, Couturaud B, Dupin N (2009) Merkel cell carcinoma of the skin: pathological and molecular evidence for a causative role of MCV in oncogenesis. J Pathol 218:48–56
8. Houben R, Shuda M, Weinkam R, Schrama D, Feng H, Chang Y, Moore PS, Becker JC (2010) Merkel cell polyomavirus-infected Merkel cell carcinoma cells require expression of viral T antigens. J Virol 84:7064–7072
9. Kean JM, Rao S, Wang M, Garcea RL (2009) Seroepidemiology of human polyomaviruses. PLoS Pathog 5:e1000363
10. Pastrana DV, Tolstov YL, Becker JC, Moore PS, Chang Y, Buck CB (2009) Quantitation of human seroresponsiveness to Merkel cell polyomavirus. PLoS Pathog 5:e1000578
11. Tolstov YL, Pastrana DV, Feng H, Becker JC, Jenkins FJ, Moschos S, Chang Y, Buck CB, Moore PS (2009) Human Merkel cell polyomavirus infection II. MCV is a common human infection that can be detected by conformational capsid epitope immunoassays. Int J Cancer 125:1250–1256
12. Gjoerup O, Chang Y (2010) Chapter 1 – update on human polyomaviruses and cancer. In: George FVW, George K (eds) Adv Cancer Res, vol. Volume 106. Academic Press, Cambridge, MA, pp 1–51
13. Liu W, MacDonald M, You J (2016) Merkel cell polyomavirus infection and Merkel cell carcinoma. Curr Opin Virol 20:20–27
14. Shuda M, Feng H, Kwun HJ, Rosen ST, Gjoerup O, Moore PS, Chang Y (2008) T antigen mutations are a human tumor-specific signature for Merkel cell polyomavirus. Proc Natl Acad Sci U S A 105:16272–16277
15. Kwun HJ, Guastafierro A, Shuda M, Meinke G, Bohm A, Moore PS, Chang Y (2009) The minimum replication origin of Merkel cell polyomavirus has a unique large T-antigen loading architecture and requires small T-antigen expression for optimal replication. J Virol 83:12118–12128
16. Diaz J, Wang X, Tsang SH, Jiao J, You J (2014) Phosphorylation of large T antigen regulates merkel cell polyomavirus replication. Cancers (Basel) 6:1464–1486
17. Pallas DC, Shahrik LK, Martin BL, Jaspers S, Miller TB, Brautigan DL, Roberts TM (1990) Polyoma small and middle T antigens and SV40 small t antigen form stable complexes with protein phosphatase 2A. Cell 60:167–176
18. Shuda M, Kwun HJ, Feng H, Chang Y, Moore PS (2011) Human Merkel cell polyomavirus small T antigen is an oncoprotein targeting the 4E-BP1 translation regulator. J Clin Invest 121:3623–3634
19. Kwun HJ, Shuda M, Feng H, Camacho CJ, Moore PS, Chang Y (2013) Merkel cell polyomavirus small T antigen controls viral replication and oncoprotein expression by targeting the cellular ubiquitin ligase SCFFbw7. Cell Host Microbe 14:125–135
20. Tsang SH, Wang R, Nakamaru-Ogiso E, Knight SA, Buck CB, You J (2016) The oncogenic small tumor antigen of Merkel cell polyomavirus is an iron-sulfur cluster protein that enhances viral DNA replication. J Virol 90:1544–1556
21. Carter JJ, Daugherty MD, Qi X, Bheda-Malge A, Wipf GC, Robinson K, Roman A, Malik HS, Galloway DA (2013) Identification of an overprinting gene in Merkel cell polyomavirus provides evolutionary insight into the birth of viral genes. Proc Natl Acad Sci U S A 110:12744–12749
22. Spurgeon ME, Lambert PF (2013) Merkel cell polyomavirus: a newly discovered human virus with oncogenic potential. Virology 435:118–130
23. Feng H, Kwun HJ, Liu X, Gjoerup O, Stolz DB, Chang Y, Moore PS (2011) Cellular and viral factors regulating Merkel cell polyomavirus replication. PLoS One 6:e22468

24. Neumann F, Borchert S, Schmidt C, Reimer R, Hohenberg H, Fischer N, Grundhoff A (2011) Replication, gene expression and particle production by a consensus Merkel cell polyomavirus (MCPyV) genome. PLoS One 6:e29112
25. Schowalter RM, Buck CB (2013) The Merkel cell polyomavirus minor capsid protein. PLoS Pathog 9:e1003558
26. Schowalter RM, Pastrana DV, Buck CB (2011) Glycosaminoglycans and sialylated glycans sequentially facilitate Merkel cell polyomavirus infectious entry. PLoS Pathog 7:e1002161
27. Seo GJ, Chen CJ, Sullivan CS (2009) Merkel cell polyomavirus encodes a microRNA with the ability to autoregulate viral gene expression. Virology 383:183–187
28. Theiss JM, Gunther T, Alawi M, Neumann F, Tessmer U, Fischer N, Grundhoff A (2015) A comprehensive analysis of replicating Merkel cell polyomavirus genomes delineates the viral transcription program and suggests a role for mcv-miR-M1 in episomal persistence. PLoS Pathog 11:e1004974
29. Doorbar J (2006) Molecular biology of human papillomavirus infection and cervical cancer. Clin Sci (Lond) 110:525–541
30. Doorbar J, Quint W, Banks L, Bravo IG, Stoler M, Broker TR, Stanley MA (2012) The biology and life-cycle of human papillomaviruses. Vaccine 30(Suppl 5):F55–F70
31. Shuda M, Chang Y, Moore PS (2014) Merkel cell polyomavirus-positive Merkel cell carcinoma requires viral small T-antigen for cell proliferation. J Invest Dermatol 134:1479–1481
32. Houben R, Adam C, Baeurle A, Hesbacher S, Grimm J, Angermeyer S, Henzel K, Hauser S, Elling R, Brocker EB, Gaubatz S, Becker JC, Schrama D (2012) An intact retinoblastoma protein-binding site in Merkel cell polyomavirus large T antigen is required for promoting growth of Merkel cell carcinoma cells. Int J Cancer 130:847–856
33. Li J, Wang X, Diaz J, Tsang SH, Buck CB, You J (2013) Merkel cell polyomavirus large T antigen disrupts host genomic integrity and inhibits cellular proliferation. J Virol 87:9173–9188. PMCID: PMC3754048
34. Cheng J, Rozenblatt-Rosen O, Paulson KG, Nghiem P, DeCaprio JA (2013) Merkel cell polyomavirus large T antigen has growth-promoting and inhibitory activities. J Virol 87:6118–6126
35. Verhaegen ME, Mangelberger D, Harms PW, Vozheiko TD, Weick JW, Wilbert DM, Saunders TL, Ermilov AN, Bichakjian CK, Johnson TM, Imperiale MJ, Dlugosz AA (2014) Merkel cell polyomavirus small T antigen is oncogenic in transgenic mice. J Invest Dermatol 88:3144–3160
36. Goh G, Walradt T, Markarov V, Blom A, Riaz N, Doumani R, Stafstrom K, Moshiri A, Yelistratova L, Levinsohn J, Chan TA, Nghiem P, Lifton RP, Choi J (2016) Mutational landscape of MCPyV-positive and MCPyV-negative Merkel cell carcinomas with implications for immunotherapy. Oncotarget 7:3403–3415
37. Wong SQ, Waldeck K, Vergara IA, Schroder J, Madore J, Wilmott JS, Colebatch AJ, De Paoli-Iseppi R, Li J, Lupat R, Semple T, Arnau GM, Fellowes A, Leonard JH, Hruby G, Mann GJ, Thompson JF, Cullinane C, Johnston M, Shackleton M, Sandhu S, Bowtell DD, Johnstone RW, Fox SB, McArthur GA, Papenfuss AT, Scolyer RA, Gill AJ, Hicks RJ, Tothill RW (2015) UV-associated mutations underlie the etiology of MCV-negative Merkel cell carcinomas. Cancer Res 75:5228–5234
38. Harms PW, Vats P, Verhaegen ME, Robinson DR, Wu YM, Dhanasekaran SM, Palanisamy N, Siddiqui J, Cao X, Su F, Wang R, Xiao H, Kunju LP, Mehra R, Tomlins SA, Fullen DR, Bichakjian CK, Johnson TM, Dlugosz AA, Chinnaiyan AM (2015) The distinctive mutational spectra of polyomavirus-negative Merkel cell carcinoma. Cancer Res 75:3720–3727
39. Moshiri AS, Doumani R, Yelistratova L, Blom A, Lachance K, Shinohara MM, Delaney M, Chang O, McArdle S, Thomas H, Asgari MM, Huang ML, Schwartz SM, Nghiem P (2016) Polyomavirus-negative Merkel cell carcinoma: a more aggressive subtype based on analysis of 282 cases using multimodal tumor virus detection. J Invest Dermatol 136:2128–2130
40. Barth H, Solis M, Kack-Kack W, Soulier E, Velay A, Fafi-Kremer S (2016) In vitro and in vivo models for the study of human polyomavirus infection. Virus 8:292

41. Neu U, Hengel H, Blaum BS, Schowalter RM, Macejak D, Gilbert M, Wakarchuk WW, Imamura A, Ando H, Kiso M, Arnberg N, Garcea RL, Peters T, Buck CB, Stehle T (2012) Structures of Merkel cell polyomavirus VP1 complexes define a sialic acid binding site required for infection. PLoS Pathog 8:e1002738
42. Schafer G, Blumenthal MJ, Katz AA (2015) Interaction of human tumor viruses with host cell surface receptors and cell entry. Virus 7:2592–2617
43. Bhattacharjee S, Chattaraj S (2017) Entry, infection, replication, and egress of human polyomaviruses: an update. Can J Microbiol 63:193–211
44. Erickson KD, Garcea RL, Tsai B (2009) Ganglioside GT1b is a putative host cell receptor for the Merkel cell polyomavirus. J Virol 83:10275–10279
45. Schowalter RM, Reinhold WC, Buck CB (2012) Entry tropism of BK and Merkel cell polyomaviruses in cell culture. PLoS One 7:e42181
46. Foulongne V, Sauvage V, Hebert C, Dereure O, Cheval J, Gouilh MA, Pariente K, Segondy M, Burguière A, Manuguerra J-C, Caro V, Eloit M (2012) Human skin microbiota: high diversity of DNA viruses identified on the human skin by high throughput sequencing. PLoS One 7:e38499
47. Schowalter RM, Pastrana DV, Pumphrey KA, Moyer AL, Buck CB (2010) Merkel cell polyomavirus and two previously unknown polyomaviruses are chronically shed from human skin. Cell Host Microbe 7:509–515
48. Wang X, Li J, Schowalter RM, Jiao J, Buck CB, You J (2012) Bromodomain protein Brd4 plays a key role in Merkel cell polyomavirus DNA replication. PLoS Pathog 8:e1003021. PMCID:PMC3493480
49. Liu W, Yang R, Payne AS, Schowalter RM, Spurgeon ME, Lambert PF, Xu X, Buck CB, You J (2016) Identifying the target cells and mechanisms of Merkel cell polyomavirus infection. Cell Host Microbe 19:775–787
50. Werner S, Grose R (2003) Regulation of wound healing by growth factors and cytokines. Physiol Rev 83:835–870
51. Chen D, Jarrell A, Guo C, Lang R, Atit R (2012) Dermal beta-catenin activity in response to epidermal Wnt ligands is required for fibroblast proliferation and hair follicle initiation. Development 139:1522–1533
52. Peretti A, Borgogna C, Rossi D, De Paoli L, Bawadekar M, Zavattaro E, Boldorini R, De Andrea M, Gaidano G, Gariglio M (2014) Analysis of human beta-papillomavirus and Merkel cell polyomavirus infection in skin lesions and eyebrow hair bulbs from a cohort of patients with chronic lymphocytic leukaemia. Br J Dermatol 171:1525–1528
53. Cho S, Shin MH, Kim YK, Seo JE, Lee YM, Park CH, Chung JH (2009) Effects of infrared radiation and heat on human skin aging in vivo. J Investig Dermatol Symp Proc 14:15–19
54. Fisher GJ, Datta SC, Talwar HS, Wang ZQ, Varani J, Kang S, Voorhees JJ (1996) Molecular basis of sun-induced premature skin ageing and retinoid antagonism. Nature 379:335–339
55. Gill SE, Parks WC (2008) Metalloproteinases and their inhibitors: regulators of wound healing. Int J Biochem Cell Biol 40:1334–1347
56. Quan T, Fisher GJ (2015) Role of age-associated alterations of the dermal extracellular matrix microenvironment in human skin aging: a mini-review. Gerontology 61:427–434
57. Quan T, Qin Z, Xia W, Shao Y, Voorhees JJ, Fisher GJ (2009) Matrix-degrading metalloproteinases in photoaging. J Investig Dermatol Symp Proc 14:20–24
58. Varani J, Dame MK, Rittie L, Fligiel SE, Kang S, Fisher GJ, Voorhees JJ (2006) Decreased collagen production in chronologically aged skin: roles of age-dependent alteration in fibroblast function and defective mechanical stimulation. Am J Pathol 168:1861–1868
59. Tilling T, Moll I (2012) Which are the cells of origin in merkel cell carcinoma? J Skin Cancer 2012:680410
60. Moll I, Zieger W, Schmelz M (1996) Proliferative Merkel cells were not detected in human skin. Arch Dermatol Res 288:184–187

61. Vaigot P, Pisani A, Darmon YM, Ortonne JP (1987) The majority of epidermal Merkel cells are non-proliferative: a quantitative immunofluorescence analysis. Acta Derm Venereol 67:517–520
62. Calder KB, Smoller BR (2010) New insights into merkel cell carcinoma. Adv Anat Pathol 17:155–161
63. Morrison KM, Miesegaes GR, Lumpkin EA, Maricich SM (2009) Mammalian Merkel cells are descended from the epidermal lineage. Dev Biol 336:76–83
64. Van Keymeulen A, Mascre G, Youseff KK, Harel I, Michaux C, De Geest N, Szpalski C, Achouri Y, Bloch W, Hassan BA, Blanpain C (2009) Epidermal progenitors give rise to Merkel cells during embryonic development and adult homeostasis. J Cell Biol 187:91–100
65. Zur Hausen A, Rennspiess D, Winnepenninckx V, Speel EJ, Kurz AK (2013) Early B-cell differentiation in Merkel cell carcinomas: clues to cellular ancestry. Cancer Res 73:4982–4987
66. Brummer GC, Bowen AR, Bowen GM (2016) Merkel cell carcinoma: current issues regarding diagnosis, management, and emerging treatment strategies. Am J Clin Dermatol 17:49–62
67. Rush Z, Fields RC, Lee N, Brownell I (2011) Radiation therapy in the management of Merkel cell carcinoma: current perspectives. Expert Rev Dermatol 6:395–404
68. Saini AT, Miles BA (2015) Merkel cell carcinoma of the head and neck: pathogenesis, current and emerging treatment options. Onco Targets Ther 8:2157–2167
69. Cassler NM, Merrill D, Bichakjian CK, Brownell I (2016) Merkel cell carcinoma therapeutic update. Curr Treat Options in Oncol 17:36
70. Paulson KG, Iyer JG, Tegeder AR, Thibodeau R, Schelter J, Koba S, Schrama D, Simonson WT, Lemos BD, Byrd DR, Koelle DM, Galloway DA, Leonard JH, Madeleine MM, Argenyi ZB, Disis ML, Becker JC, Cleary MA, Nghiem P (2011) Transcriptome-wide studies of merkel cell carcinoma and validation of intratumoral CD8+ lymphocyte invasion as an independent predictor of survival. J Clin Oncol 29:1539–1546
71. Inoue T, Yoneda K, Manabe M, Demitsu T (2000) Spontaneous regression of merkel cell carcinoma: a comparative study of TUNEL index and tumor-infiltrating lymphocytes between spontaneous regression and non-regression group. J Dermatol Sci 24:203–211
72. Paulson KG, Iyer JG, Blom A, Warton EM, Sokil M, Yelistratova L, Schuman L, Nagase K, Bhatia S, Asgari MM, Nghiem P (2013) Systemic immune suppression predicts diminished Merkel cell carcinoma-specific survival independent of stage. J Invest Dermatol 133:642–646
73. Aldabagh B, Joo J, Yu SS (2014) Merkel cell carcinoma: current status of targeted and future potential for immunotherapies. Semin Cutan Med Surg 33:76–82
74. Sihto H, Bohling T, Kavola H, Koljonen V, Salmi M, Jalkanen S, Joensuu H (2012) Tumor infiltrating immune cells and outcome of Merkel cell carcinoma: a population-based study. Clin Cancer Res 18:2872–2881
75. Schadendorf D, Lebbe C, Zur Hausen A, Avril MF, Hariharan S, Bharmal M, Becker JC (2017) Merkel cell carcinoma: epidemiology, prognosis, therapy and unmet medical needs. Eur J Cancer 71:53–69
76. Afanasiev OK, Yelistratova L, Miller N, Nagase K, Paulson K, Iyer JG, Ibrani D, Koelle DM, Nghiem P (2013) Merkel polyomavirus-specific T cells fluctuate with merkel cell carcinoma burden and express therapeutically targetable PD-1 and Tim-3 exhaustion markers. Clin Cancer Res 19:5351–5360
77. Nghiem PT, Bhatia S, Lipson EJ, Kudchadkar RR, Miller NJ, Annamalai L, Berry S, Chartash EK, Daud A, Fling SP, Friedlander PA, Kluger HM, Kohrt HE, Lundgren L, Margolin K, Mitchell A, Olencki T, Pardoll DM, Reddy SA, Shantha EM, Sharfman WH, Sharon E, Shemanski LR, Shinohara MM, Sunshine JC, Taube JM, Thompson JA, Townson SM, Yearley JH, Topalian SL, Cheever MA (2016) PD-1 blockade with pembrolizumab in advanced Merkel-cell carcinoma. N Engl J Med 374:2542–2552
78. Keir ME, Butte MJ, Freeman GJ, Sharpe AH (2008) PD-1 and its ligands in tolerance and immunity. Annu Rev Immunol 26:677–704
79. Mantripragada K, Birnbaum A (2015) Response to anti-PD-1 therapy in metastatic Merkel cell carcinoma metastatic to the heart and pancreas. Cureus 7:e403

80. Zeng Q, Gomez BP, Viscidi RP, Peng S, He L, Ma B, Wu TC, Hung CF (2012) Development of a DNA vaccine targeting Merkel cell polyomavirus. Vaccine 30:1322–1329
81. Gomez BP, Wang C, Viscidi RP, Peng S, He L, Wu TC, Hung CF (2012) Strategy for eliciting antigen-specific CD8+ T cell-mediated immune response against a cryptic CTL epitope of merkel cell polyomavirus large T antigen. Cell Biosci 2:36
82. Wendzicki JA, Moore PS, Chang Y (2015) Large T and small T antigens of Merkel cell polyomavirus. Curr Opin Virol 11:38–43
83. Gomez B, He L, Tsai YC, Wu TC, Viscidi RP, Hung CF (2013) Creation of a Merkel cell polyomavirus small T antigen-expressing murine tumor model and a DNA vaccine targeting small T antigen. Cell Biosci 3:29
84. Ma JE, Brewer JD (2014) Merkel cell carcinoma in immunosuppressed patients. Cancers (Basel) 6:1328–1350
85. Engels EA, Frisch M, Goedert JJ, Biggar RJ, Miller RW (2002) Merkel cell carcinoma and HIV infection. Lancet 359:497–498
86. Wieland U, Silling S, Scola N, Potthoff A, Gambichler T, Brockmeyer NH, Pfister H, Kreuter A (2011) Merkel cell polyomavirus infection in HIV-positive men. Arch Dermatol 147:401–406
87. Kolia-Diafouka P, Foulongne V, Boulle N, Ngou J, Kelly H, Sawadogo B, Delany-Moretlwe S, Mayaud P, Segondy M, Group HS (2016) Detection of four human polyomaviruses (MCPyV, HPyV6, HPyV7 and TSPyV) in cervical specimens from HIV-infected and HIV-uninfected women. Sex Transm Infect 88:695–702
88. Vahabpour R, Nasimi M, Naderi N, Salehi-Vaziri M, Mohajel N, Sadeghi F, Keyvani H, Monavari SH (2017) Merkel cell polyomavirus IgG antibody levels are associated with progression to AIDS among HIV-infected individuals. Arch Virol 162:963–969
89. Fukumoto H, Sato Y, Hasegawa H, Katano H (2013) Frequent detection of Merkel cell polyomavirus DNA in sera of HIV-1-positive patients. Virol J 10:84
90. Li M, Saghafi N, Freymiller E, Basile JR, Lin YL (2013) Metastatic Merkel cell carcinoma of the oral cavity in a human immunodeficiency virus-positive patient and the detection of Merkel cell polyomavirus. Oral Surg Oral Med Oral Pathol Oral Radiol 115:e66–e71
91. Wieland U, Kreuter A (2011) Merkel cell polyomavirus infection and Merkel cell carcinoma in HIV-positive individuals. Curr Opin Oncol 23:488–493
92. Hourdequin KC, Lefferts JA, Brennick JB, Ernstoff MS, Tsongalis GJ, Pipas JM (2013) Merkel cell polyomavirus and extrapulmonary small cell carcinoma. Oncol Lett 6:1049–1052
93. Kim GJ, Lee JH, Lee DH (2017) Clinical and prognostic significance of Merkel cell polyomavirus in nonsmall cell lung cancer. Medicine (Baltimore) 96:e5413
94. Hashida Y, Imajoh M, Nemoto Y, Kamioka M, Taniguchi A, Taguchi T, Kume M, Orihashi K, Daibata M (2013) Detection of Merkel cell polyomavirus with a tumour-specific signature in non-small cell lung cancer. Br J Cancer 108:629–637
95. Gheit T, Munoz JP, Levican J, Gonzalez C, Ampuero S, Parra B, Gaggero A, Corvalan AH, Meneses M, Tommasino M, Aguayo F (2012) Merkel cell polyomavirus in non-small cell lung carcinomas from Chile. Exp Mol Pathol 93:162–166
96. Colombara DV, Manhart LE, Carter JJ, Hawes SE, Weiss NS, Hughes JP, Qiao YL, Taylor PR, Smith JS, Galloway DA (2016) Absence of an association of human polyomavirus and papillomavirus infection with lung cancer in China: a nested case-control study. BMC Cancer 16:342
97. Malhotra J, Waterboer T, Pawlita M, Michel A, Cai Q, Zheng W, Gao YT, Lan Q, Rothman N, Langseth H, Grimsrud TK, Yuan JM, Koh WP, Wang R, Arslan AA, Zeleniuch-Jacquotte A, Boffetta P (2016) Serum biomarkers of polyomavirus infection and risk of lung cancer in never smokers. Br J Cancer 115:1131–1139
98. Dworkin AM, Tseng SY, Allain DC, Iwenofu OH, Peters SB, Toland AE (2009) Merkel cell polyomavirus in cutaneous squamous cell carcinoma of immunocompetent individuals. J Invest Dermatol 129:2868–2874

99. Murakami M, Imajoh M, Ikawa T, Nakajima H, Kamioka M, Nemoto Y, Ujihara T, Uchiyama J, Matsuzaki S, Sano S, Daibata M (2011) Presence of Merkel cell polyomavirus in Japanese cutaneous squamous cell carcinoma. J Clin Virol 50:37–41
100. Kaibuchi-Noda K, Yokota K, Matsumoto T, Sawada M, Sakakibara A, Kono M, Tomita Y, Watanabe D, Fukumoto H, Katano H, Akiyama M (2011) Detection of Merkel cell polyomavirus in cutaneous squamous cell carcinoma before occurrence of Merkel cell carcinoma. J Am Acad Dermatol 65:e152–e154
101. Mertz KD, Paasinen A, Arnold A, Baumann M, Offner F, Willi N, Cathomas G (2013) Merkel cell polyomavirus large T antigen is detected in rare cases of nonmelanoma skin cancer. J Cutan Pathol 40:543–549
102. Reisinger DM, Shiffer JD, Cognetta AB Jr, Chang Y, Moore PS (2010) Lack of evidence for basal or squamous cell carcinoma infection with Merkel cell polyomavirus in immunocompetent patients with Merkel cell carcinoma. J Am Acad Dermatol 63:400–403
103. Mitteldorf C, Mertz KD, Fernandez-Figueras MT, Schmid M, Tronnier M, Kempf W (2012) Detection of Merkel cell polyomavirus and human papillomaviruses in Merkel cell carcinoma combined with squamous cell carcinoma in immunocompetent European patients. Am J Dermatopathol 34:506–510
104. Yahyapour Y, Sadeghi F, Alizadeh A, Rajabnia R, Siadati S (2016) Detection of Merkel cell polyomavirus and human papillomavirus in esophageal squamous cell carcinomas and non-cancerous esophageal samples in northern Iran. Pathol Oncol Res 22:667–672
105. Imajoh M, Hashida Y, Nemoto Y, Oguri H, Maeda N, Furihata M, Fukaya T, Daibata M (2012) Detection of Merkel cell polyomavirus in cervical squamous cell carcinomas and adenocarcinomas from Japanese patients. Virol J 9:154
106. Du-Thanh A, Dereure O, Guillot B, Foulongne V (2014) Merkel cell polyomavirus: its putative involvement in a particular subset of cutaneous lymphoma with possibly unfavorable outcome. J Clin Virol 61:161–165
107. Song Y, Gyarmati P (2017) Identification of Merkel cell polyomavirus from a patient with acute myeloid leukemia. Genome Announc 5:e01241-16
108. Cimino PJ Jr, Bahler DW, Duncavage EJ (2013) Detection of Merkel cell polyomavirus in chronic lymphocytic leukemia T-cells. Exp Mol Pathol 94:40–44
109. Pantulu ND, Pallasch CP, Kurz AK, Kassem A, Frenzel L, Sodenkamp S, Kvasnicka HM, Wendtner CM, Zur HA (2010) Detection of a novel truncating Merkel cell polyomavirus large T antigen deletion in chronic lymphocytic leukemia cells. Blood 116:5280–5284
110. Katano H, Ito H, Suzuki Y, Nakamura T, Sato Y, Tsuji T, Matsuo K, Nakagawa H, Sata T (2009) Detection of Merkel cell polyomavirus in Merkel cell carcinoma and Kaposi's sarcoma. J Med Virol 81:1951–1958
111. Du-Thanh A, Guillot B, Dereure O, Foulongne V (2015) Detection of Merkel cell and other human polyomavirus DNA in lesional and nonlesional skin from patients with Kaposi sarcoma. Br J Dermatol 173:1063–1065
112. Urso C, Pierucci F, Sollai M, Arvia R, Massi D, Zakrzewska K (2016) Detection of Merkel cell polyomavirus and human papillomavirus DNA in porocarcinoma. J Clin Virol 78:71–73
113. Mertz KD, Schmid M, Burger B, Itin P, Palmedo G, Scharer L, Kutzner H, Fernandez Figueras MT, Cribier B, Pfaltz M, Kempf W (2011) Detection of Merkel cell polyomavirus in epidermodysplasia-verruciformis-associated skin neoplasms. Dermatology 222:87–92

Chapter 5
Current Progress in EBV-Associated B-Cell Lymphomas

Yonggang Pei, Alexandria E. Lewis, and Erle S. Robertson

Abstract Epstein-Barr virus (EBV) was the first human tumor virus discovered more than 50 years ago. EBV-associated lymphomagenesis is still a significant viral-associated disease as it involves a diverse range of pathologies, especially B-cell lymphomas. Recent development of high-throughput next-generation sequencing technologies and in vivo mouse models have significantly promoted our understanding of the fundamental molecular mechanisms which drive these cancers and allowed for the development of therapeutic intervention strategies. This review will highlight the current advances in EBV-associated B-cell lymphomas, focusing on transcriptional regulation, chromosome aberrations, in vivo studies of EBV-mediated lymphomagenesis, as well as the treatment strategies to target viral-associated lymphomas.

Keywords Epstein-Barr virus • Latent infection • B-cell lymphomas

5.1 Introduction

Approximately two million new cases of cancer are annually attributed to infectious agents. 12% to 15% of human cancers are associated with oncogenic virus infection and are suspected to be major drivers [1, 2]. Uncovering the roles of infectious agents will help facilitate our understanding of the mechanism of cancer pathogenesis mediated by infectious agents and develop potential methods for therapeutic intervention. Epstein-Barr virus (EBV), also known as herpesvirus 4, was the first human tumor virus to attract significant attention since it was discovered associated with Burkitt's lymphoma in 1964 [3]. EBV infects more than 95% of the world's

Y. Pei • A.E. Lewis • E.S. Robertson (✉)
Department of Otorhinolaryngology-Head and Neck Surgery, and Microbiology, Tumor Virology Program, Abramson Comprehensive Cancer Center, Perelman School of Medicine at the University of Pennsylvania, 3610 Hamilton Walk, 201E Johnson Pavilion, Philadelphia, PA 19104, USA
e-mail: erle@upenn.edu

Q. Cai et al. (eds.), *Infectious Agents Associated Cancers: Epidemiology and Molecular Biology*, Advances in Experimental Medicine and Biology 1018,
DOI 10.1007/978-981-10-5765-6_5

population and sustains lifelong asymptomatic infection. Its ability to induce oncogenesis is likely due to suppression of the immune system or a result of the uncontrolled proliferation. A recent study demonstrated that 1.8% of cancer deaths were related to EBV-attributable malignancies worldwide [4].

Initial infection of EBV is usually asymptomatic or can cause infectious mononucleosis (IM) [5]. The following lytic infection in epithelial cells results in the expression of the complete viral gene program. Previous studies clearly showed that EBV had the ability to transform human primary B lymphocytes into lymphoblastoid cell lines (LCLs) [6, 7]. To date, EBV is still the most efficient transforming virus in culture and can rapidly transform resting B cells in vitro [8, 9]. The persistence of EBV infection is mainly in B cells and leads to EBV associated B-cell lymphoma, typically in individuals with suppressed immune systems. Nasopharyngeal carcinoma (NPC) and EBV-associated gastric carcinoma (GC) are also related with EBV-infected epithelial cells, but whether or not the virus is a major contribution to the pathogenesis of these tumors is still unclear. Therefore, the presence and precise contributions of EBV to numerous human cancers is a challenge to explain. However, it also provides a great opportunity to the development of novel prophylactic or therapeutic methods.

5.2 EBV-Associated B-Cell Lymphomas

5.2.1 Burkitt's Lymphoma (BL)

Burkitt's lymphoma (BL) can be classified into three forms based on the geographic distribution: endemic BL (eBL), sporadic BL (sBL), and HIV-associated BL [10]. The discovery of EBV in BL tumors and the fact that almost 100% of endemic BL are EBV positive support the possibility that BL tumors are driven by EBV as a major contributor. Further sera-epidemiological studies have provided evidence that African BL tumors are positive for EBV [11]. One critical feature of BL tumors is the translocation and activation of *MYC* [10]. MYC overexpression in BL tumors results from a translocation event between the *MYC* gene and immunoglobulin locus which further regulates the downstream network and facilitates tumorigenesis [12, 13]. Most EBV-positive BL tumors consistently express latent antigen EBNA1 as the predominant latent antigen and are termed latency I [14]. Previous studies show that EBNA1 can play antiapoptotic roles which also contributes to increased tumorigenicity [15, 16]. In addition, and different from that observed in Africa, only 15–20% of BL tumors are EBV positive in other parts of the world [12]. The extremely uncommon observation is consistent with the fact that EBV together with malaria can increase the frequency of BL tumors. However, the mechanism of their interaction is not fully understood and needs further investigation [12, 17].

5.2.2 *Diffuse Large B-Cell Lymphoma (DLBCL)*

Diffuse large B-cell lymphoma (DLBCL) is the most common non-Hodgkin lymphoma (NHL), accounting for 40% of adult NHL [18]. Two major subtypes of DLBCL, germinal center B cell (GCB) and activated B cell (ABC), were divided based on genomic signatures [19]. Approximately 10% DLBCL is EBV positive, which has been described in the World Health Organization (WHO) classification system [20]. EBV-positive DLBCL is mainly identified in the elderly because the median age of these patients is 71 years, although in younger patients can also be found [20, 21]. The incidence of EBV among DLBCL patients is less than 5% in the United States and European countries but 10–15% in Asian and Latin American countries [21–24]. EBV-positive DLBCL is associated with activation of NF-κB and JAK/STAT signaling pathways, but the detailed mechanisms of tumorigenesis will need to be further investigated [25].

5.2.3 *Posttransplant Lymphoproliferative Disease (PTLD)*

Posttransplant lymphoproliferative disease (PTLD) is mainly derived from B cells in transplant patients [26, 27]. It is often associated with EBV infection in the context of an impaired immune surveillance system. Furthermore, 60–80% of PTLDs are shown to be EBV positive [28]. EBV is the crucial driver of PTLDs development that is typically early-onset cases of posttransplantation [29]. Early-onset PTLDs that are associated with EBV-infected B cells are usually polyclonal or oligoclonal, while most late-onset PTLDs with or without EBV infection are monoclonal [30]. The transplant-associated immunosuppression in PTLDs leads to expression of EBNA3 family members in addition to all the latent antigens, which are characteristics of latency III-associated EBV infection [9]. The prevention and treatment of EBV-associated PTLDs rely on surgery with irradiation, immunotherapy with monoclonal antibodies (e.g., rituximab), and antiviral drugs [31]. The development of T-cell-based therapies has been very promising to treat EBV-driven PTLDs by transferring patient-derived ex vivo amplified EBV-specific cytotoxic T cells back to patients [32].

5.2.4 *Hodgkin Lymphoma (HL)*

Hodgkin lymphoma (HL) is characterized by the presence of Hodgkin-Reed-Sternberg (HRS) cells [33]. The direct link of EBV and HL is confirmed by the detection of EBER expression in HRS cells using EBER-specific in situ hybridization [34]. In addition, EBNA1, LMP1, and LMP2A are also expressed in EBV-infected HRS cells [35]. HL cells are B-cell originated and derived from the germinal center. They require the necessary signals to escape apoptosis as a result of the lack of functional BCRs [36]. Therefore, in EBV-infected HRS cells, LMP1 mimics the CD40 receptor, recruits TRAF family members, and further activates downstream NF-κB signaling pathways to promote cell survival by inhibiting cell apoptosis [37]. Meanwhile, LMP2A recruits cytoplasmic kinase to activate B-cell Ig receptors or activates the PI3K-AKT pathway in the absence of Ig receptors to promote B-cell survival and growth [9, 38].

5.2.5 *EBV-Associated B-Cell Lymphoma in the Context of HIV*

The increased reports of EBV-associated lymphomas with the onset of acquired immunodeficiency syndrome (AIDS) imply a molecular connection between EBV and HIV in the infected hosts [39]. In HIV-associated lymphomas, EBV infection can be found in 80% of DLBCL and 80–100% of primary central nervous system lymphomas (PCNSL) [40]. BL can occur before HIV infection even if circulating CD4+ T-cell numbers are normal. DLBCL typically occurs only after HIV infection when circulating CD4+ T cells are exhausted [41]. AIDS-BLs involve the typical *MYC* translocation and are less frequently infected by EBV [42, 43]. These observations suggest that HIV may be a potential stimulator which leads to an increase in the risk of EBV-mediated *MYC* translocation and therefore lymphomagenesis. Most AIDS-associated lymphomas that are EBV positive do express broad expression of the latent antigens and are type III latency program. This is likely due to the suppressed immune system and so a loss of control of the EBV-positive cells.

5.3 Molecular Biology of EBV-Mediated B-Cell Lymphomas

EBV is an oncogenic herpesvirus because of its ability to immortalize human primary B lymphocytes in vitro. In general, EBV primary infection is asymptomatic, and the following persistent infection will be established in memory B cells after an early period of virus production [44]. Therefore, two typical EBV infections can be established in the host: lytic infection in epithelial cells and latent infection in memory B cells [45, 46]. The initial events of EBV primary infection are the focus of current studies, but the detailed mechanisms are still not completely understood. In latent infection, specific transcription programs are defined as latency I, II, and III according to the expression of the viral-encoded latent antigens, which are thought to be the critical drivers of EBV-associated lymphomagenesis.

5.3.1 EBV-Associated Transcription Regulatory Network

One hallmark of cancer is the dysregulation of gene expression [47]. The characteristic of effective in vitro transformation by EBV indicates its strong ability to regulate cellular transcriptional programs. With the rapid development of high-throughput sequencing technologies, more and more studies are focused on a complicated regulatory network during EBV-mediated B-cell transformation by utilizing the common database such as NCBI GEO, ENCODE, and TCGA project [48–50].

To determine the molecular mechanisms which drive lymphomagenesis, EBV-transformed lymphoblastoid cell lines (LCLs) are one of the best systems to perform in vitro studies. More recent studies have concentrated on EBV latent protein-mediated regulatory networks using next-generation high-seq analysis, of which the frequently used is ChIP-seq (Table 5.1). ChIP-seq analysis indicated that EBNA2 can convert B lymphocytes to LCLs by targeting H3K4me1 modified sites as well as noncoding regions to regulate cellular gene expression to drive proliferation of LCLs [51]. In addition, EBNA2 induces a new pattern of genome-wide binding through recruitment of RBPJκ and EBF1 to drive LCL survival [52]. EBNA2 recruits the SWI/SNF ATPase BRG1 to bind large-scale *MYC* enhancers activating its expression [53]. EBNA-LP binds with B-cell transcription factors (TFs), which are highly similar to EBNA2 including RBPJκ and EBF [54]. These high-seq data provides evidence to support the explanation that both EBNA2 and EBNA-LP are crucial for LCL outgrowth. EBNA3C, another EBV latent antigen essential for LCLs growth, is associated with cellular transcription factors. It binds to BATF/IRF4 and SPI1/IRF4 sites to repress CDKN2A transcription through the recruitment of Sin3A in LCLs [55]. EBV latent proteins EBNA3A and EBNA3C inhibit BCL2L11 transcription by recruiting the H3K27 methyltransferase EZH2 to silence long-range enhancers [53]. ChIP-seq analysis shows that EBNA2 and EBNA3s (EBNA3A, EBNA3B, and EBNA3C) can target multiple cellular genes through cell-specific regulation of long-range enhancer-promoter interactions [56]. Another study indicated that while these four latent antigens can competitively bind to RBPJκ at its repressive sites to control cellular genes expression, EBNA3s are more likely to interact with other transcription factors [57]. For example, IRF4 is essential for EBNA3C to associate with specific sites on viral and cellular DNA [16, 55, 57, 58]. A recent study identified a number of host dependency factors in BL and LCLs using CRISPR/Cas9 loss-of-function screen [59]. These specific genes, including PI3K/AKT, cFLIP, BATF/IRF4, and IRF2, are likely crucial in regulating downstream transcriptional network to facilitate cell growth and survival.

During EBV primary infection, the correlative latent antigens convert resting B cells to LCLs, and their dependent function may rely on super-enhancers to control B-cell growth [60]. EBV super-enhancers (ESEs) with higher H3K27c signals involve the oncogenes MYC and Bcl2 to promote LCL growth and survival, which provides new insights on EBV-induced lymphoproliferation [60]. EBNA2, EBNA3A, and EBNA3C can enhance RUNX3 expression via RBPJκ to regulate the upstream RUNX3 super-enhancer and meanwhile control the downstream RUNX1

Table 5.1 The transcription factors (TFs) identified in ChIP-seq analysis

Associated TFs[a]	Targets	Cell lines	References
EBNA1	Human genome,	Raji	[4]
	EBV latent promoter		
EBNA2, RBPJκ, CTCF, EBF, RELA, H3K9ac, H3K4me1, Pol II, P300	Human genome	GM12878	[51]
EBNA2, EBF1, RBPJk	Human genome,	LCL,	[52]
	EBV latent promoter	Mutu III	
EBNA2, H3K27Ac	Human genome	GM12878,	[53]
		Mutu III	
EBNA-LP, *EBNA2*	Human genome	GM12878	[54]
EBNA3C,Sin3A, REST, *EBNA2*, RBPJκ, IRF4, BATF, SPI1, RUNX3, p300, Pol II, H3K4me1, H3K4me3, H3K9ac	Human genome	GM12878	[55]
EBNA2, RBPJκ, H3K27ac, H3K4me1, BRD4, P300, Pol II, BATF, EBF1, PAX5, SPI1, Sp1, NFAT, STAT5, ETS1, IRF4, CTCF, RAD21, SMC3, YY1; *EBNA3A*, *EBNA3C*, EBNA-LP, RelA, RelB, cRel, p50, p52	Human genome	GM12878	[60]
Notch1, RBPJκ, ZNF143	Human genome	GM12878	[109]
EBNA3A, *EBNA3C*, *EBNA2*, RBPJκ, BATF, IRF4, SPI1, RUNX3, NF-κB, MEF2A, PAX5, POU2F2, MAX, MYC, POL2, SIN3A, H3K27ac	Human genome	GM12878	[110]
CHD2, CFOS, BRCA, EGR1, PBX3, BCL3, GCN5, p300, TBP, TAF1, CTCF, Pol II, TCF12, EBF1, SP1, PU.1, PAX5, BATF, JUND, SMC3, RAD21, H3K27ac	Human genome,	GM12878	[111]
	EBV latent promoter		
CTCF, RAD21, RPB1	EBV genome	Raji	[112]
Cohesin, RNA Polymerase II			
EBNA3A, *EBNA3B*, *EBNA3C*	Human genome	LCL	[113]

[a]EBV latent antigens are *underlined*

expression [61]. Additionally, abundant enhancers (eRNAs) are also transcribed from ESEs and are regulated by the activity of ESEs [62]. For example, the inactivation of EBNA2 and bromodomain-containing protein 4 (BRD4) in ESEs will significantly reduce the expression of eRNAs and further the MYC protein, therefore affecting LCL growth [62].

About 300 novel EBV transcripts have been predicted by combining multiple platform data from PacBio SMART Iso-Seq, RNA-Seq, and deep-CAGE, which illustrates the complex regulation of viral gene transcription during EBV infection [63]. Studies on miRNA targetome show that EBV miRNAs mainly target cellular transcription factors to manipulate the microenvironment during latent infection, suggesting the importance of EBV-expressed miRNAs in contributing to viral-mediated oncogenesis [64]. Furthermore, EBV miRNAs can modulate immune recognition to protect infected cells from killing by cytotoxic EBV-specific $CD4^+$ T cells through repression of pro-inflammatory cytokine release, naïve $CD4^+$ T-cell differentiation, and peptide presentation, which allow for establishment of latent infection and development of lymphomas [65]. Similarly, EBV miRNAs can use multiple pathways to evade immune surveillance and killing by EBV-specific $CD8^+$ T cells [66].

Although the development and manipulation of high-throughput sequencing technologies provide us a deeper and wider understanding of EBV-mediated transformation or lymphomagenesis (Fig. 5.1), the complicated regulatory network targeted by EBV latent infection is still being explored. Furthermore, systematic proteomic analyses can possibly validate some of the genomic observations and gain additional insights into EBV-host interactions [67]. In the future, more efficient systems and more advanced technologies with higher resolution, specificity, and sensitivity will be helpful in revealing the complex EBV-host interactions in associated lymphomas.

5.3.2 Genomic Instability and Chromosome Aberrations

Genomic instability is a hallmark of cancer that increases the risks of oncogenic chromosome alterations [1, 9]. Previous studies have indicated that EBV persistent infection can result in chromosome aberrations in associated lymphomas [9, 26, 68]. EBV latent antigens play crucial roles in driving genomic instability. To be specific, EBNA1 may function to contribute to genomic instability through activation of the RAG gene or induction of reactive oxygen species (ROS) [17, 29]. EBNA3C can promote genomic instability by inhibiting BubR1 transcription and inactivating the mitotic spindle checkpoint [69]. Additionally, EBNA3C can compromise the mitotic spindle checkpoint and block caspase-mediated cell death, leading to abnormal mitosis and DNA damage accumulation [15, 70]. Although the detailed mechanism of EBNA3C-mediated genetic instability needs further investigation, multiple functions of EBNA3C may contribute to genetic instability directly or indirectly by binding with cell cycle or DNA damage checkpoint proteins, including cyclin A [71], Chk2 [72], cyclin D1 [73], p53 [74, 75], and the E2F family member E2F1/E2F6 [28, 76]. LMP1-associated genomic instability may also result from telomerase activation and DNA damage response (DDR) inhibition [69, 77]. Intriguingly, EBV tegument protein BNRF1 could also induce centrosome amplification and further chromosome instability during lytic infection, suggesting that EBV viral particles may be sufficient to modify host chromosome without the establishment of latent infection [78].

In addition, the EBV genome can frequently integrate into host cell chromosomes in persistently infected B cells [22, 79, 80]. This integration increases the

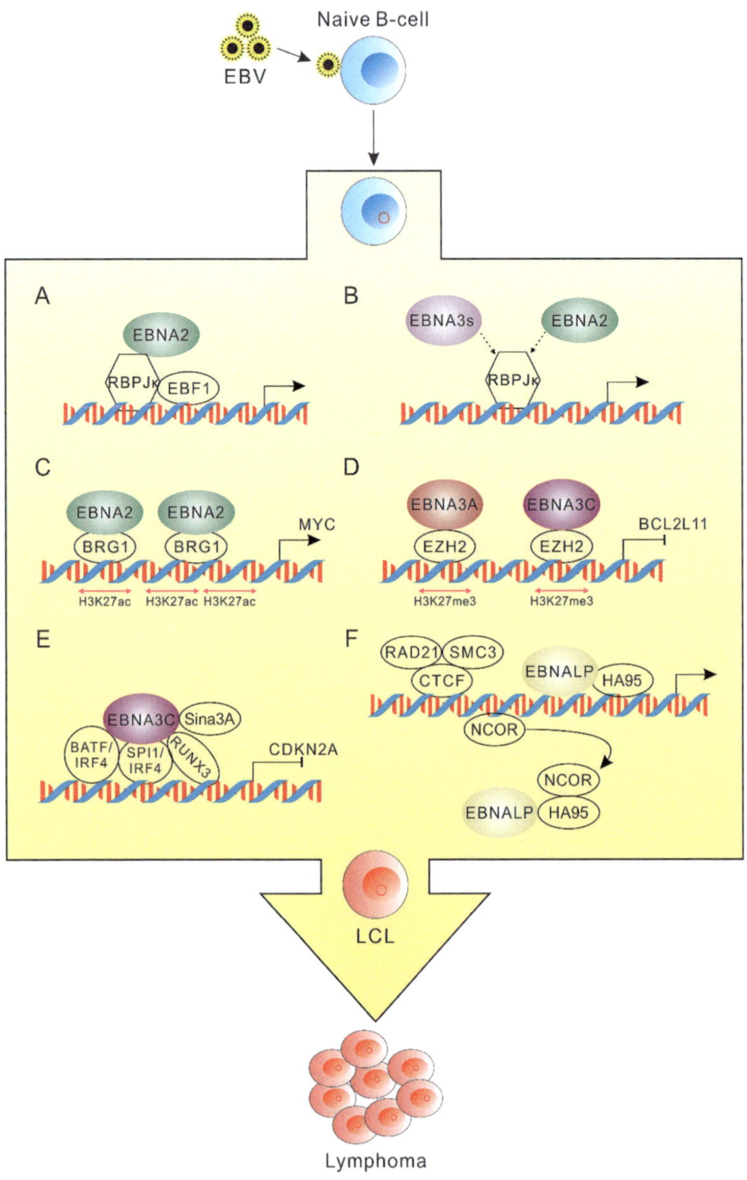

Fig. 5.1 EBV latent antigen-associated cellular signaling pathways from the current high-throughput sequencing data during EBV-mediated lymphomagenesis. (**a**) EBNA2 regulates target genes expression through the recruitment of transcription factors RBPJκ and EBF1. (**b**) EBNA3s and EBNA2 bind with partially the same RBPJκ genomic sites. The interaction between RBPJκ and EBNA3s or EBNA2 will result in different effects of downstream gene expression, which are also associated with other EBNA-interacting cell transcription factors. (**c**) EBNA2 activates the three clusters of upstream enhancers of *MYC* promoter with increased H3K27Ac and BRG1 binding, and then EBNA2 mediates MYC activation through promoting the interaction of *MYC* promoter and the activated upstream enhancers. (**d**) EBNA3A and EBNA3C repress BCL2L11

possibility of lymphomagenesis when the constitutive regions release the viral genome which leads to loss of normal DNA or chromosome instability [81]. For instance, the integration of the EBV genome into chromosome 6q15 blocks the expression of the tumor repressor BACH2 in Burkitt lymphoma cell lines [80]. Using whole genome sequencing technology, a recent study reports that a comprehensive view of integration sites shows that they are randomly distributed across the entire host genome in EBV-positive Raji (Burkitt's lymphoma cells), and C666-1 (nasopharyngeal carcinoma cells) and so may be contributing to lymphomagenesis [25]. The frequent chromosome recombination, involved in chromosome 8 and c-Myc activation, is also noted in Burkitt's lymphoma cells after combined treatment with EBV and purified 4-deoxyphorbol ester [82].

5.3.3 *In Vivo Models of EBV Infection*

Host-range restriction is a major limitation of EBV research because humans are the exclusive natural host for EBV. Therefore, the development of a more efficient in vivo system to support the studies from in vitro results will provide additional information related to the complicated EBV-host interactions. An important achievement on in vivo system began with the development of *scid*-hu PBL mouse through the injection of human peripheral blood leukocytes (PBL) into C.B-17 *scid* mice that lack B and T cells because of the severe combined immunodeficiency (SCID) phenotype [83, 84]. Later, another *scid*-hu thy/liv mouse was generated by implanting fetal thymus, liver cells, and fetal lymph nodes into C.B-17 *scid* mice [85]. However, these mice have obvious shortcomings of generated graft versus host disease and transient immune responses [86]. Subsequently, a new series of mice models were generated to overcome the preceding disadvantages by transplanting human hematopoietic stem cells (HSCs) into various mice such as NOD/Shi-*scid* $Il2rg^{null}$ (NOG) [87], BALB/c $Rag2^{-/-}Il2rg^{-/-}$ (BRG) [88], and NOD/LtSz-*scid* $Il2rg^{-/-}$ (NSG) [89]. These transplanted HSCs reconstituted the human immune system by differentiating into diversified cells, including B cells, T cells, natural killer (NK) cells, dendritic cells (DCs), monocytes, and macrophages [86].

Given the great improvement in mouse models, it is possible to further study the mechanisms of EBV-associated lymphomagenesis in vivo using humanized mice. Previous studies have shown that EBV could infect humanized BALB/c

Fig. 5.1 (continued) expression by inactivating the upstream enhancers of its promoter. The inactivation is associated with increased H3K27me3 and EZH2 binding as well as the inhibition of interactions between BCL2L11 promoter and its enhancers. (**e**) EBNA3C binds to the promoters through BATF/IRF4, SPI1/IRF4, and RUNX and further recruits Sin3A to inhibit CDKN2A expression. (**f**) EBNA-LP regulates the derepression of target genes by removing NCOR repression complex from the promoters with the help of HA95 and further promotes the long-distance enhancer-promoter interaction through CTCF, RAD21, and SMC3 proteins. EBV latent antigens are *highlighted* by colorful patterns, while cellular factors are labeled with colorless patterns

$Rag2^{-/-}Il2rg^{-/-}$ mice and induce specific T-cell response [88]. A mouse model for EBV infection was established by transplanting only $CD34^+$-depleted human cord blood mononuclear cells into NOD/LtSz-*scid* $Il2rg^{-/-}$ (NSG) mice [90]. The results from this EBV-infected mouse model indicated that the PD-1/CTLA-4 blockade will induce strong specific T-cell responses and inhibit the outgrowth of EBV-associated lymphomas [90].

To further support human T cells which demonstrate HLA-restricted cytotoxic functions in mouse models, an immunodeficient NSG-HLA-A2/HHD mouse was created through the introduction of HLA-A2 allele into $CD34^+CD38^-$ HSC-transplanted NSG mice [91]. The new mouse model showed a relatively complete immune system that expresses HLA class I heavy and light chains, promotes human T-cell development, and produces functional $CD4^+$ and $CD8^+$ T cells. In this mouse model, EBV infection will result in B-cell-associated lymphoproliferative diseases, which can be inhibited by HLA-restricted CTL cytotoxicity [91]. What's more, NK cells are necessary to control infectious mononucleosis (IM) symptoms by targeting EBV lytic antigens and so control lytic infection [92]. Furthermore, NOD/SCID-hu BLT mice (or BLT mice) are developed by transplanting *scid*-hu thy/liv mice with autologous $CD34^+$ cells which combines the advantages of *scid*-hu thy/liv mice model and $CD34^+$ cell-transplanted NOD/SCID mice model [36]. BLT mice were shown to have a more complete human immune system, of which the T cells generate long-term, specific adaptive immune responses after EBV infection via human major histocompatibility complex (MHC) class I and II [36].

In 2011, an improved humanized mouse model was developed through the transplantation of human fetal $CD34^+$ hematopoietic stem cells and thymus/liver tissue into NOD/LtSz-*scid* $Il2rg^{-/-}$ (NSG) mice [93]. The mouse model supports long-term EBV latent infection and lymphoma development. Further experiments showed that EBV lytic infection was critical for B-cell lymphomagenesis with limited help of the immune system [93]. The following application of this mouse model with wild-type EBV or LMP1-deficient EBV infection demonstrated that LMP1 may not be essential for EBV-mediated lymphomagenesis but that T cells may substitute LMP1 function for development of B-cell lymphomas [94].

Different from the application of humanized mouse model, a recent study reported establishment of a transgenic mouse model with conditional LMP1/2A coexpression in germinal center (GC) B cells [95]. In this mouse model, LMP1/2A showed very limited function in immunocompetent mice, while they promote B-cell lymphoproliferative diseases in the context of T-cell or NK-cell deficiency [95].

5.4 Treatment of B-Cell Lymphomas

Diffuse large B-cell lymphoma (DLBCL) continues to be one of the few lymphomas that remain curable due to advancements made over the last decade. More than half of the patients can be cured using treatments that include chemo-, radio-, or immunotherapeutic regimens [96]. However, approximately 30–40% of patients diagnosed will develop relapsed or refractory disease after being treated for DLBCL [97, 98]. Treatment of these patients has become extremely difficult due to the resistance that has grown with the disease [99]. The improved outcome in patients with DLBCL and relapsed-refractory DLBCL (RR-DLBCL) is largely attributed to the incorporation of rituximab into standard regimens [99, 100]. With further findings and introduction of novel specific anticancer agents and therapeutic approaches, treatment and survival of affected patients are likely to improve tremendously [101].

DLBCL is commonly treated with R-CHOP, a combination of rituximab, cyclophosphamide, doxorubicin, vincristine, and prednisone, and it has shown great benefits for patients [102]. Tolerance in patients of all ages has been demonstrated, and survival rates have increased, specifically in patients diagnosed with non-Hodgkin's lymphoma [103]. Recent findings indicate that in combination with rituximab or R-CHOP, drugs lenalidomide and epratuzumab could be effective in not only first-line treatment of DLBCL but also RR-DLBCL [96]. Other novel agents such as ibrutinib, bortezomib, CC-122, and pidilizumab have been shown to be successful in the first-line treatment of DLBCL as both single agents or in combination with rituximab-based chemotherapy [96]. Studies have also investigated the role of the NF-κB/Rel family, specifically nuclear factor kappa-B (NF-κB) and RelA (p65) in DLBCL. High p65 nuclear expression is a significant adverse biomarker in patients with early-stage (I/II) DLBCL [104]. Findings have shown that with p65 inactivation, cell growth and survival can be effectively inhibited. Furthermore, activation of the JAK-STAT and NF-κB pathways is characteristic of EBV-positive DLBCL [25]. Therefore, development of therapies targeting these pathways would be of potential benefit for these patients and lead to an improvement in their post-therapy outcomes.

Another major development in the treatment of DLBCL is CAR T-cell therapy. This therapy utilizes chimeric antigen receptor (CAR)-engineered T cells specifically engineered to recognize their target antigen through the scFv-binding domain [105]. This recognition results in the activation of T cells in a major histocompatibility complex (MHC)-independent manner [106]. Investigation of this therapy has demonstrated promising outcomes by targeting CD19, CD20, or CD30 which is significant for B-cell malignancies such as B-cell non-Hodgkin's lymphoma (B-NHL) and Hodgkin's lymphoma (HL) [106]. Though still in development, success has been shown in treatment of patients, and with a deeper understanding of its functional role, the future of this novel therapy will likely prove to be promising for many diseases.

Research has led to the discovery that B-aggressive lymphoma-1 protein and ADP-ribosyltransferase BAL1/ARTD9 may serve as a novel potential drug target for treatment [96, 107]. Combining a drug(s) targeting STAT1 or the macrodomains of BAL1/ARTD9 with common day therapeutic treatments might be a successful strategy toward increasing the sensitivity of HR-DLBCL to classic therapy [107]. Several other potential therapies have been identified through other ongoing investigations including the targeting of Deltex-3-like E3 ubiquitin ligase (DTX3L) and the BET Bromodomain Protein BRD4 [1, 96, 108]. Preliminary studies indicate that DTX3L controls CXCR4, a chemokine receptor [108]. Further studies would need to be done to identify the link, if any, of DTX3L via CRCX4 with DLBCL. However, a therapy involving this control mechanism shows great potential [108]. Regarding BRD4, studies have shown that the BET inhibitors have the ability to inhibit oncogenic NF-κB activity through decreased expression of the NF-κB target genes IL6 and IL10 [1]. These findings, along with the developments in understanding the functions of NF-κB and RelA (p65), highly support the need for further research into developing a therapeutic drug targeting NF-κB complex.

Further investigation on these therapies, with or without standard immunochemotherapy, would provide major insights and pave the way to developing successful treatments for patients suffering from more aggressive types of DLBCL or RR-DLBCL or even different types of lymphomas. It is also believed that acquired drug resistance is mediated by a finite set of pathways. If these pathways can be identified and the targets that need to be suppressed or activated can be determined, sensitivity could be restored to drugs that were used successfully in a prior line of therapy or optimize the efficiency of the available therapeutic personalized regimens [13, 96].

5.5 Conclusions

EBV was discovered more than 50 years ago, but a large body of questions remain unanswered. Although EBV infects more than 90 % of the world's population, only a subset of the related infections results in lymphomagenesis. The lifelong relationships between host and EBV suggest the importance of the immune system in normal individuals. For many immunodeficient patients, EBV-induced lymphomagenesis is a frequent occurrence. Although EBV-associated lymphomas have been studied for many years, the precise roles of EBV in these processes are still unclear. EBV can infect B cells and establish latent infection, further inducing them toward lymphomagenesis under specific conditions in the microenvironment. Although the in vitro model of EBV infection has been established for many years, the detailed strategies of EBV infection, which includes latent and lytic infection, are not completely understood. The complex regulatory network is associated with regulation of numerous transcription factors, viral lytic/latent antigens, and their associated relationships. In addition, the development of NPC or GC after EBV infection has not been completely investigated because of the limitation of an efficient in vitro and in vivo model system. It is anticipated that the combined application of high-throughput next-generation sequencing technologies and in vivo mouse models will significantly improve our understanding of EBV biology in the near future and the development of potential therapeutic intervention strategies.

Acknowledgments This work was supported by National Cancer Institute at the National Institutes of Health public health service Grants P30-CA016520, P30-DK050306, R01-CA171979, P01-CA174439, and R01-CA177423 to ESR. The funders had no role in study design, data collection and analysis, decision to publish, or preparation of the manuscript. We sincerely apologize to the authors whose important work could not be cited here due to space limitations.

References

1. Ceribelli M, Kelly PN, Shaffer AL, Wright GW, Xiao W et al (2014) Blockade of oncogenic IkappaB kinase activity in diffuse large B-cell lymphoma by bromodomain and extraterminal domain protein inhibitors. Proc Natl Acad Sci U S A 111:11365–11370
2. Morini E, Dietrich P, Salani M, Downs HM, Wojtkiewicz GR et al (2016) Sensory and autonomic deficits in a new humanized mouse model of familial dysautonomia. Hum Mol Genet 25:1116–1128
3. Epstein MA, Achong BG, Barr YM (1964) Virus particles in cultured Lymphoblasts from Burkitt's lymphoma. Lancet 1:702–703
4. Lu F, Wikramasinghe P, Norseen J, Tsai K, Wang P et al (2010) Genome-wide analysis of host-chromosome binding sites for Epstein-Barr Virus Nuclear Antigen 1 (EBNA1). Virol J 7:262
5. Henle G, Henle W, Diehl V (1968) Relation of Burkitt's tumor-associated herpes-ytpe virus to infectious mononucleosis. Proc Natl Acad Sci U S A 59:94–101
6. Pope JH (1967) Establishment of cell lines from peripheral leucocytes in infectious mononucleosis. Nature 216:810–811
7. Henle W, Diehl V, Kohn G, Zur Hausen H, Henle G (1967) Herpes-type virus and chromosome marker in normal leukocytes after growth with irradiated Burkitt cells. Science 157:1064–1065
8. Cerchietti L, Damm-Welk C, Vater I, Klapper W, Harder L et al (2011) Inhibition of anaplastic lymphoma kinase (ALK) activity provides a therapeutic approach for CLTC-ALK-positive human diffuse large B cell lymphomas. PLoS One 6:e18436
9. Mesri EA, Feitelson MA, Munger K (2014) Human viral oncogenesis: a cancer hallmarks analysis. Cell Host Microbe 15:266–282
10. Rund D (2014) Targeting drug resistance to close the gap in diffuse large B-cell lymphoma. Leuk Lymphoma 55:1966–1967
11. Lv X, Feng L, Ge X, Lu K, Wang X (2016) Interleukin-9 promotes cell survival and drug resistance in diffuse large B-cell lymphoma. J Exp Clin Cancer Res 35:106
12. Yang C, Lu P, Lee FY, Chadburn A, Barrientos JC et al (2008) Tyrosine kinase inhibition in diffuse large B-cell lymphoma: molecular basis for antitumor activity and drug resistance of dasatinib. Leukemia 22:1755–1766
13. Wilson WH (2006) Drug resistance in diffuse large B-cell lymphoma. Semin Hematol 43:230–239
14. Yu X, Li Z (2015) New insights into MicroRNAs involves in drug resistance in diffuse large B cell lymphoma. Am J Transl Res 7:2536–2542
15. Parker GA, Touitou R, Allday MJ (2000) Epstein-Barr virus EBNA3C can disrupt multiple cell cycle checkpoints and induce nuclear division divorced from cytokinesis. Oncogene 19:700–709
16. Banerjee S, Lu J, Cai Q, Saha A, Jha HC et al (2013) The EBV latent antigen 3C inhibits apoptosis through targeted regulation of interferon regulatory factors 4 and 8. PLoS Pathog 9:e1003314
17. Tsimbouri P, Drotar ME, Coy JL, Wilson JB (2002) bcl-xL and RAG genes are induced and the response to IL-2 enhanced in EmuEBNA-1 transgenic mouse lymphocytes. Oncogene 21:5182–5187
18. De Paepe P, De Wolf-Peeters C (2007) Diffuse large B-cell lymphoma: a heterogeneous group of non-Hodgkin lymphomas comprising several distinct clinicopathological entities. Leukemia 21:37–43
19. Rosenwald A, Wright G, Chan WC, Connors JM, Campo E et al (2002) The use of molecular profiling to predict survival after chemotherapy for diffuse large-B-cell lymphoma. N Engl J Med 346:1937–1947
20. Heslop HE (2005) Biology and treatment of Epstein-Barr virus-associated non-Hodgkin lymphomas. Hematol Am Soc Hematol Educ Program: 260–266

21. Oyama T, Yamamoto K, Asano N, Oshiro A, Suzuki R et al (2007) Age-related EBV-associated B-cell lymphoproliferative disorders constitute a distinct clinicopathologic group: a study of 96 patients. Clin Cancer Res 13:5124–5132
22. Gibson SE, Hsi ED (2009) Epstein-Barr virus-positive B-cell lymphoma of the elderly at a United States tertiary medical center: an uncommon aggressive lymphoma with a nongerminal center B-cell phenotype. Hum Pathol 40:653–661
23. Hoeller S, Tzankov A, Pileri SA, Went P, Dirnhofer S (2010) Epstein-Barr virus-positive diffuse large B-cell lymphoma in elderly patients is rare in western populations. Hum Pathol 41:352–357
24. Park S, Lee J, Ko YH, Han A, Jun HJ et al (2007) The impact of Epstein-Barr virus status on clinical outcome in diffuse large B-cell lymphoma. Blood 110:972–978
25. Kato H, Karube K, Yamamoto K, Takizawa J, Tsuzuki S et al (2014) Gene expression profiling of Epstein-Barr virus-positive diffuse large B-cell lymphoma of the elderly reveals alterations of characteristic oncogenetic pathways. Cancer Sci 105:537–544
26. Sabik JF 3rd (2016) Why coronary artery bypass grafting remains the standard of care for patients with complex, multivessel coronary artery disease. J Thorac Cardiovasc Surg 152:1227–1228
27. Del Sorbo L, Ranieri VM, Keshavjee S (2012) Extracorporeal membrane oxygenation as "bridge" to lung transplantation: what remains in order to make it standard of care? Am J Respir Crit Care Med 185:699–701
28. Schurko B, Oh WK (2008) Docetaxel chemotherapy remains the standard of care in castration-resistant prostate cancer. Nat Clin Pract Oncol 5:506–507
29. Gruhne B, Sompallae R, Marescotti D, Kamranvar SA, Gastaldello S et al (2009) The Epstein-Barr virus nuclear antigen-1 promotes genomic instability via induction of reactive oxygen species. Proc Natl Acad Sci U S A 106:2313–2318
30. Henderson MA (2006) Completion lymphadenectomy for melanoma patients with a positive sentinel node biopsy remains standard of care. Ann Surg Oncol 13:761–763
31. Boules TN, Proctor MC, Aref A, Upchurch GR Jr, Stanley JC et al (2005) Carotid endarterectomy remains the standard of care, even in high-risk surgical patients. Ann Surg 241:356–363
32. Bollard CM, Rooney CM, Heslop HE (2012) T-cell therapy in the treatment of post-transplant lymphoproliferative disease. Nat Rev Clin Oncol 9:510–519
33. Kuppers R, Rajewsky K (1998) The origin of Hodgkin and Reed/Sternberg cells in Hodgkin's disease. Annu Rev Immunol 16:471–493
34. Choudhary SK, Rezk NL, Ince WL, Cheema M, Zhang L et al (2009) Suppression of human immunodeficiency virus type 1 (HIV-1) viremia with reverse transcriptase and integrase inhibitors, CD4+ T-cell recovery, and viral rebound upon interruption of therapy in a new model for HIV treatment in the humanized Rag2−/−{gamma}c−/− mouse. J Virol 83:8254–8258
35. Yajima M, Imadome K, Nakagawa A, Watanabe S, Terashima K et al (2008) A new humanized mouse model of Epstein-Barr virus infection that reproduces persistent infection, lymphoproliferative disorder, and cell-mediated and humoral immune responses. J Infect Dis 198:673–682
36. Melkus MW, Estes JD, Padgett-Thomas A, Gatlin J, Denton PW et al (2006) Humanized mice mount specific adaptive and innate immune responses to EBV and TSST-1. Nat Med 12:1316–1322
37. Mosialos G, Birkenbach M, Yalamanchili R, VanArsdale T, Ware C et al (1995) The Epstein-Barr virus transforming protein LMP1 engages signaling proteins for the tumor necrosis factor receptor family. Cell 80:389–399
38. Merchant M, Caldwell RG, Longnecker R (2000) The LMP2A ITAM is essential for providing B cells with development and survival signals in vivo. J Virol 74:9115–9124
39. Doll DC, List AF (1982) Burkitt's lymphoma in a homosexual. Lancet 1:1026–1027
40. Gloghini A, Dolcetti R, Carbone A (2013) Lymphomas occurring specifically in HIV-infected patients: from pathogenesis to pathology. Semin Cancer Biol 23:457–467

41. Moir S, Fauci AS (2013) Insights into B cells and HIV-specific B-cell responses in HIV-infected individuals. Immunol Rev 254:207–224
42. Klein G (1983) Specific chromosomal translocations and the genesis of B-cell-derived tumors in mice and men. Cell 32:311–315
43. Ziegler JL, Drew WL, Miner RC, Mintz L, Rosenbaum E et al (1982) Outbreak of Burkitt's-like lymphoma in homosexual men. Lancet 2:631–633
44. Jha HC, Pei Y, Robertson ES (2016) Epstein-Barr virus: diseases linked to infection and transformation. Front Microbiol 7:1602
45. Tsurumi T, Fujita M, Kudoh A (2005) Latent and lytic Epstein-Barr virus replication strategies. Rev Med Virol 15:3–15
46. Amon W, Farrell PJ (2005) Reactivation of Epstein-Barr virus from latency. Rev Med Virol 15:149–156
47. Bradner JE, Hnisz D, Young RA (2017) Transcriptional addiction in cancer. Cell 168:629–643
48. Barrett T, Wilhite SE, Ledoux P, Evangelista C, Kim IF et al (2013) NCBI GEO: archive for functional genomics data sets--update. Nucleic Acids Res 41:D991–D995
49. Consortium EP (2004) The ENCODE (ENCyclopedia of DNA elements) project. Science 306:636–640
50. Cancer Genome Atlas Research N, Weinstein JN, Collisson EA, Mills GB, Shaw KR et al (2013) The cancer genome atlas pan-cancer analysis project. Nat Genet 45:1113–1120
51. Zhao B, Zou J, Wang H, Johannsen E, Peng CW et al (2011) Epstein-Barr virus exploits intrinsic B-lymphocyte transcription programs to achieve immortal cell growth. Proc Natl Acad Sci U S A 108:14902–14907
52. Lu F, Chen HS, Kossenkov AV, DeWispeleare K, Won KJ et al (2016) EBNA2 drives formation of new chromosome binding sites and target genes for B-cell master regulatory transcription factors RBP-jkappa and EBF1. PLoS Pathog 12:e1005339
53. Wood CD, Veenstra H, Khasnis S, Gunnell A, Webb HM, et al. (2016) MYC activation and BCL2L11 silencing by a tumour virus through the large-scale reconfiguration of enhancer-promoter hubs. Elife 5
54. Portal D, Zhou H, Zhao B, Kharchenko PV, Lowry E et al (2013) Epstein-Barr virus nuclear antigen leader protein localizes to promoters and enhancers with cell transcription factors and EBNA2. Proc Natl Acad Sci U S A 110:18537–18542
55. Jiang S, Willox B, Zhou H, Holthaus AM, Wang A et al (2014) Epstein-Barr virus nuclear antigen 3C binds to BATF/IRF4 or SPI1/IRF4 composite sites and recruits Sin3A to repress CDKN2A. Proc Natl Acad Sci U S A 111:421–426
56. McClellan MJ, Wood CD, Ojeniyi O, Cooper TJ, Kanhere A et al (2013) Modulation of enhancer looping and differential gene targeting by Epstein-Barr virus transcription factors directs cellular reprogramming. PLoS Pathog 9:e1003636
57. Wang A, Welch R, Zhao B, Ta T, Keles S et al (2015) Epstein-Barr virus nuclear antigen 3 (EBNA3) proteins regulate EBNA2 binding to distinct RBPJ genomic sites. J Virol 90:2906–2919
58. Pei Y, Banerjee S, Jha HC, Sun Z, Robertson ES (2017) An essential EBV latent antigen 3C binds Bcl6 for targeted degradation and cell proliferation. PLoS Pathog 13:e1006500
59. Ma Y, Walsh MJ, Bernhardt K, Ashbaugh CW, Trudeau SJ et al (2017) CRISPR/Cas9 screens reveal Epstein-Barr virus-transformed B cell host dependency factors. Cell Host Microbe 21(580–591):e587
60. Zhou H, Schmidt SC, Jiang S, Willox B, Bernhardt K et al (2015) Epstein-Barr virus oncoprotein super-enhancers control B cell growth. Cell Host Microbe 17:205–216
61. Gunnell A, Webb HM, Wood CD, McClellan MJ, Wichaidit B et al (2016) RUNX super-enhancer control through the notch pathway by Epstein-Barr virus transcription factors regulates B cell growth. Nucleic Acids Res 44:4636–4650
62. Liang J, Zhou H, Gerdt C, Tan M, Colson T et al (2016) Epstein-Barr virus super-enhancer eRNAs are essential for MYC oncogene expression and lymphoblast proliferation. Proc Natl Acad Sci U S A 113(49):14121–14126

63. O'Grady T, Wang X, Honer Zu Bentrup K, Baddoo M, Concha M et al (2016) Global transcript structure resolution of high gene density genomes through multi-platform data integration. Nucleic Acids Res 44:e145
64. Skalsky RL, Corcoran DL, Gottwein E, Frank CL, Kang D et al (2012) The viral and cellular microRNA targetome in lymphoblastoid cell lines. PLoS Pathog 8:e1002484
65. Tagawa T, Albanese M, Bouvet M, Moosmann A, Mautner J et al (2016) Epstein-Barr viral miRNAs inhibit antiviral CD4+ T cell responses targeting IL-12 and peptide processing. J Exp Med 213:2065–2080
66. Albanese M, Tagawa T, Bouvet M, Maliqi L, Lutter D et al (2016) Epstein-Barr virus microRNAs reduce immune surveillance by virus-specific CD8+ T cells. Proc Natl Acad Sci U S A 113:E6467–e6475
67. Ersing I, Nobre L, Wang LW, Soday L, Ma Y et al (2017) A temporal proteomic map of Epstein-Barr virus lytic replication in B cells. Cell Rep 19:1479–1493
68. Etoh T, Baba H, Taketomi A, Nakashima H, Kohnoe S et al (1998) Sequential methothrextate and 5-fruororacil therapy for diffuse bone metastasis from gastric cancer. Anticancer Res 18:2085–2088
69. Gray NA, Kapojos JJ, Burke MT, Sammartino C, Clark CJ (2016) Patient kidney disease knowledge remains inadequate with standard nephrology outpatient care. Clin Kidney J 9:113–118
70. Meissner MH (2010) Conventional anticoagulant therapy remains the current standard of care for the treatment of iliofemoral deep venous thrombosis. Dis Mon 56:642–652
71. Knight JS, Sharma N, Kalman DE, Robertson ES (2004) A cyclin-binding motif within the amino-terminal homology domain of EBNA3C binds cyclin A and modulates cyclin A-dependent kinase activity in Epstein-Barr virus-infected cells. J Virol 78:12857–12867
72. Choudhuri T, Verma SC, Lan K, Murakami M, Robertson ES (2007) The ATM/ATR signaling effector Chk2 is targeted by Epstein-Barr virus nuclear antigen 3C to release the G2/M cell cycle block. J Virol 81:6718–6730
73. Saha A, Halder S, Upadhyay SK, Lu J, Kumar P et al (2011) Epstein-Barr virus nuclear antigen 3C facilitates G1-S transition by stabilizing and enhancing the function of cyclin D1. PLoS Pathog 7:e1001275
74. Yi F, Saha A, Murakami M, Kumar P, Knight JS et al (2009) Epstein-Barr virus nuclear antigen 3C targets p53 and modulates its transcriptional and apoptotic activities. Virology 388:236–247
75. Saha A, Bamidele A, Murakami M, Robertson ES (2011) EBNA3C attenuates the function of p53 through interaction with inhibitor of growth family proteins 4 and 5. J Virol 85:2079–2088
76. Pei Y, Banerjee S, Sun Z, Jha HC, Saha A et al (2016) EBV nuclear antigen 3C mediates regulation of E2F6 to inhibit E2F1 transcription and promote cell proliferation. PLoS Pathog 12:e1005844
77. DeLong WB, Polissar NL, Neradilek MB, Laam LA (2015) To the Editor: Re: Controversy: acute cauda equina syndrome caused by a disk lesion: is emergent surgery the correct option? In: Mahadevappa et al. (ed) Surgical decompression remains the standard of care, by McLain et al. Spine (Phila Pa 1976) 40: E1120
78. Shumilov A, Tsai MH, Schlosser YT, Kratz AS, Bernhardt K et al (2017) Epstein-Barr virus particles induce centrosome amplification and chromosomal instability. Nat Commun 8:14257
79. Crowther M, Donadini MP (2010) Hematology/oncology clinics of North America. Hypercoagulable states and new anticoagulants. Preface. Hematol Oncol Clin North Am 24:xiii–xxiv
80. Brown JR (2013) Hematology/oncology clinics of North America. Chronic lymphocytic leukemia. Preface. Hematol Oncol Clin North Am 27:xiii–xxiv
81. Jox A, Rohen C, Belge G, Bartnitzke S, Pawlita M et al (1997) Integration of Epstein-Barr virus in Burkitt's lymphoma cells leads to a region of enhanced chromosome instability. Ann Oncol 8(Suppl 2):131–135

82. O'Donnell MT, Greer LT, Nelson J, Shriver C, Vertrees A (2014) Diversion remains the standard of care for modern management of war-related rectal injuries. Mil Med 179:778–782
83. Kessinger A (1998) Consensus conference on high-dose therapy with hematopoietic stem cell transplantation in diffuse large-cell lymphoma. Type of cells, optimal mobilization of stem cells – positive and negative selection. Ann Oncol 9(Suppl 1):S23–S30
84. Masip J, Vecilla F, Paez J (1998) Diffuse pulmonary hemorrhage after fibrinolytic therapy for acute myocardial infarction. Int J Cardiol 63:95–97
85. McCune JM, Namikawa R, Kaneshima H, Shultz LD, Lieberman M et al (1988) The SCID-hu mouse: murine model for the analysis of human hematolymphoid differentiation and function. Science 241:1632–1639
86. Fujiwara S, Matsuda G, Imadome K (2013) Humanized mouse models of epstein-barr virus infection and associated diseases. Pathogens 2:153–176
87. Ito M, Hiramatsu H, Kobayashi K, Suzue K, Kawahata M et al (2002) NOD/SCID/gamma(c) (null) mouse: an excellent recipient mouse model for engraftment of human cells. Blood 100:3175–3182
88. Traggiai E, Chicha L, Mazzucchelli L, Bronz L, Piffaretti JC et al (2004) Development of a human adaptive immune system in cord blood cell-transplanted mice. Science 304:104–107
89. Ishikawa F, Yasukawa M, Lyons B, Yoshida S, Miyamoto T et al (2005) Development of functional human blood and immune systems in NOD/SCID/IL2 receptor {gamma} chain(null) mice. Blood 106:1565–1573
90. Ma SD, Xu X, Jones R, Delecluse HJ, Zumwalde NA et al (2016) PD-1/CTLA-4 blockade inhibits Epstein-Barr virus-induced lymphoma growth in a cord blood humanized-mouse model. PLoS Pathog 12:e1005642
91. Shultz LD, Saito Y, Najima Y, Tanaka S, Ochi T et al (2010) Generation of functional human T-cell subsets with HLA-restricted immune responses in HLA class I expressing NOD/SCID/IL2r gamma (null) humanized mice. Proc Natl Acad Sci U S A 107:13022–13027
92. Chijioke O, Muller A, Feederle R, Barros MH, Krieg C et al (2013) Human natural killer cells prevent infectious mononucleosis features by targeting lytic Epstein-Barr virus infection. Cell Rep 5:1489–1498
93. Ma SD, Hegde S, Young KH, Sullivan R, Rajesh D et al (2011) A new model of Epstein-Barr virus infection reveals an important role for early lytic viral protein expression in the development of lymphomas. J Virol 85:165–177
94. Ma SD, Xu X, Plowshay J, Ranheim EA, Burlingham WJ et al (2015) LMP1-deficient Epstein-Barr virus mutant requires T cells for lymphomagenesis. J Clin Invest 125:304–315
95. Minamitani T, Ma Y, Zhou H, Kida H, Tsai CY et al (2017) Mouse model of Epstein-Barr virus LMP1- and LMP2A-driven germinal center B-cell lymphoproliferative disease. Proc Natl Acad Sci U S A 114:4751–4756
96. Camicia R, Winkler HC, Hassa PO (2015) Novel drug targets for personalized precision medicine in relapsed/refractory diffuse large B-cell lymphoma: a comprehensive review. Mol Cancer 14:207
97. Perry AR, Goldstone AH (1998) High-dose therapy for diffuse large-cell lymphoma in first remission. Ann Oncol 9(Suppl 1):S9–14
98. Fisher RI, Gaynor ER, Dahlberg S, Oken MM, Grogan TM et al (1993) Comparison of a standard regimen (CHOP) with three intensive chemotherapy regimens for advanced non-Hodgkin's lymphoma. N Engl J Med 328:1002–1006
99. Friedberg JW, Fisher RI (2008) Diffuse large B-cell lymphoma. Hematol Oncol Clin North Am 22(941–952):ix
100. Raut LS, Chakrabarti PP (2014) Management of relapsed-refractory diffuse large B cell lymphoma. South Asian J Cancer 3:66–70
101. Zinzani PL, Pellegrini C, Argnani L, Broccoli A (2016) Prolonged disease-free survival in elderly relapsed diffuse large B-cell lymphoma patients treated with lenalidomide plus rituximab. Haematologica 101:e385–e386
102. Feugier P, Van Hoof A, Sebban C, Solal-Celigny P, Bouabdallah R et al (2005) Long-term results of the R-CHOP study in the treatment of elderly patients with diffuse large B-cell

lymphoma: a study by the Groupe d'Etude des Lymphomes de l'Adulte. J Clin Oncol 23:4117–4126
103. Dotan E, Aggarwal C, Smith MR (2010) Impact of rituximab (Rituxan) on the treatment of B-cell non-Hodgkin's lymphoma. P T 35:148–157
104. Zhang M, Xu-Monette ZY, Li L, Manyam GC, Visco C et al (2016) RelA NF-kappaB subunit activation as a therapeutic target in diffuse large B-cell lymphoma. Aging (Albany NY) 8:3321–3340
105. Ramos CA, Heslop HE, Brenner MK (2016) CAR-T cell therapy for lymphoma. Annu Rev Med 67:165–183
106. Wang Z, Wu Z, Liu Y, Han W (2017) New development in CAR-T cell therapy. J Hematol Oncol 10:53
107. Camicia R, Bachmann SB, Winkler HC, Beer M, Tinguely M et al (2013) BAL1/ARTD9 represses the anti-proliferative and pro-apoptotic IFNgamma-STAT1-IRF1-p53 axis in diffuse large B-cell lymphoma. J Cell Sci 126:1969–1980
108. Holleman J, Marchese A (2014) The ubiquitin ligase deltex-3l regulates endosomal sorting of the G protein-coupled receptor CXCR4. Mol Biol Cell 25:1892–1904
109. Wang H, Zou J, Zhao B, Johannsen E, Ashworth T et al (2011) Genome-wide analysis reveals conserved and divergent features of Notch1/RBPJ binding in human and murine T-lymphoblastic leukemia cells. Proc Natl Acad Sci U S A 108:14908–14913
110. Schmidt SC, Jiang S, Zhou H, Willox B, Holthaus AM et al (2015) Epstein-Barr virus nuclear antigen 3A partially coincides with EBNA3C genome-wide and is tethered to DNA through BATF complexes. Proc Natl Acad Sci U S A 112:554–559
111. Arvey A, Tempera I, Tsai K, Chen HS, Tikhmyanova N et al (2012) An atlas of the Epstein-Barr virus transcriptome and epigenome reveals host-virus regulatory interactions. Cell Host Microbe 12:233–245
112. Holdorf MM, Cooper SB, Yamamoto KR, Miranda JJ (2011) Occupancy of chromatin organizers in the Epstein-Barr virus genome. Virology 415:1–5
113. Paschos K, Bazot Q, Ho G, Parker GA, Lees J et al (2016) Core binding factor (CBF) is required for Epstein-Barr virus EBNA3 proteins to regulate target gene expression. Nucleic Acids Res 45:2368–2383

Chapter 6
EBV Infection and Glucose Metabolism in Nasopharyngeal Carcinoma

Jun Zhang, Lin Jia, Chi Man Tsang, and Sai Wah Tsao

Abstract To establish persistent infection in cells, viruses evolve strategies to alter host cellular pathways to regulate cell proliferation and energy metabolism which support viral infection. Epstein-Barr virus (EBV) undergoes both lytic and latent infection to achieve persistent and lifelong infection in human. EBV readily infects human B cells, driving their transformation to proliferative lymphoblastoid cell lines (LCL), and eventually establishes lifelong latent infection in memory B cells. In contrary, EBV undergoes lytic replication upon infection into normal epithelial cells which is essential for the replication of EBV genome and production of infectious viral particles for transmission through saliva. EBV shuttles between B cells and epithelial cells to complete its infection cycle. EBV infection is closely associated with nasopharyngeal carcinoma (NPC) and is present in practically 100% of undifferentiated NPC. In contrast to undergo lytic infection of normal pharyngeal epithelium, EBV establishes latent infection in NPC. The switch from lytic infection to latent infection may represent an early and essential step in the development of NPC. Recent studies in both B cells and NPC cells latently infected with EBV reveal alterations in cell metabolism to support persistent and latent EBV infection. Events underlying the switching of lytic to latent EBV infection in NPC cells are largely undefined. Molecular events and alterations of cell metabolism are likely to play crucial roles in switching EBV infection from lytic to latent in NPC cells. Latent EBV infection and expression of viral genes, including LMP1, LMP2, and possibly EBV-encoded micro RNAs, may play essential roles in alterations of cell metabolism to support NPC pathogenesis. Alteration of energy metabolism is an essential hallmark of cancer. The role of altered energy metabolism in host cells in modulating latent and lytic EBV infection in NPC cells is unclear. In this review, we will discuss the impact of genetic alterations in NPC to module cellular metabolism and its influence on latent infection and lytic reactivation of EBV infection in NPC cells. In particular, the role of EBV-encoded genes in driving glucose metabolism and their contribution to NPC pathogenesis will be discussed. This new perspective

J. Zhang • L. Jia • C.M. Tsang • S.W. Tsao (✉)
School of Biomedical Sciences, Li Ka Shing Faculty of Medicine,
The University of Hong Kong, 21 Sassoon Road, Pokfulam, Hong Kong
e-mail: gswtsao@hku.hk

Q. Cai et al. (eds.), *Infectious Agents Associated Cancers: Epidemiology and Molecular Biology*, Advances in Experimental Medicine and Biology 1018,
DOI 10.1007/978-981-10-5765-6_6

on the interplay between EBV infection and altered host metabolic pathways in NPC pathogenesis may offer novel and effective therapeutic strategies in the treatment of NPC and other EBV-associated malignancies.

Keywords Epstein-Barr virus • Latent infection • Nasopharyngeal carcinoma • Glucose metabolism

6.1 Introduction

The EBV is a type of human γ-herpesvirus infecting over 90% of the populations worldwide. EBV infection is associated with human malignancies of both lymphoid and epithelial origins [1, 2]. Upon infection, EBV effectively transforms and immortalizes B cells into lymphoblastoid cell lines with unlimited proliferative potential, a hallmark property of EBV contributing to B-cell malignancies [3]. Nearly 100% of undifferentiated NPC and a subtype of gastric cancer (about 10%) are latently infected with EBV infection [4]. Compared to B-cell malignancies, the pathogenic roles of EBV in human epithelial cancers are much less understood. In complete contrast to EBV infection of primary B cells, infection of primary epithelial cells by EBV does not induce cell proliferation. In contrast, the EBV-infected primary epithelial cells readily undergo growth arrest. As EBV infection is practically present in all undifferentiated NPC cells, presumably, EBV infection should have selective growth advantages to NPC cells in patients. The nature of growth advantages of EBV infection in NPC remains to be determined but have been postulated to involve immune evasion and survival advantage of infected NPC cells. EBV readily undergoes lytic infection in normal epithelial cells. However, in NPC cells, EBV infection is predominantly latent with expression of restricted number of latent genes including EBER, EBNA1, LMP1, LMP2, and miR-BARTs. An earlier study has shown that LMP2 induces mTOR activation to upregulate c-myc protein translation [5]. Recent studies showed that LMP1 drives mTORC1 signaling and glucose metabolism [6, 7] supporting the important role of EBV infection in energy metabolism in NPC pathogenesis. Recent genomic profiling of NPC also revealed frequent mutations leading to activation of NF-κB [8, 9] and other cell signaling pathways such as ERBB/PI3K signaling [10], which are upstream events involved in mTOR signaling. Activation of mTOR signaling and enhanced glucose metabolism may play a crucial role to support latent infection of EBV and contribute to NPC pathogenesis.

6.2 EBV Establishes Latent Infection in NPC

Both lytic and latent infection of EBV are involved to maintain persistent and life-long infection in human. EBV devises specific strategies to switch its cellular tropism to facilitate the shuttling of virus between epithelial cells and B cells during its infection cycle [11]. EBV establishes latent infection in human memory B cells and is believed to be the reservoir for persistent EBV infection in human. Differentiation of infected B cells to plasma cells will trigger lytic reactivation of EBV resulting in production of infectious viruses to infect epithelial cells. The EBV episomes in infected epithelial cells replicate efficiently and packaged into infectious viruses for transmission. Lytic replication of EBV in infected oropharyngeal epithelial cells has been postulated to be the continuous source of infectious virus shedding into saliva which is the major route of EBV transmission [12]. As mentioned, EBV infection in NPC is predominantly latent [2]. The switching of EBV infection from lytic to latent mode may represent an early and essential step in NPC pathogenesis.

Three types of latent infection program of EBV in human cells have been observed. In NPC, EBV undergoes a specific type of latency infection program (referred as latency type II) where expression of latent EBV genes is limited to EBERs, EBNA1, LMP1, LMP2, and BART transcripts. The BART transcripts are expressed with high abundance in NPC cells, which are further processed to EBV-encoded microRNAs (miR-BARTs). Interestingly, the miR-BARTs are expressed at exceptionally high levels in epithelial cancers, including the NPC and EBV-associated gastric cancer, but at reduced levels in lymphoid malignancies (type I latency) and very low levels in EBV-transformed lymphoblastoid cell lines (type III latency). The high expression of miR-BARTs in NPC as well as EBV-associated gastric cancer suggests the pathogenic roles of miR-BARTs in NPC [13]. Events regulating latent infection of EBV in NPC are largely undefined [11]. Interestingly, overexpression of Cyclin D1 and inactivation of p16, which are common events in NPC, support stable and latent EBV infection in immortalized nasopharyngeal epithelial cells [14]. Additional genetic mutations and alterations of host cell signaling in NPC cells are likely to contribute to the establishment of latent EBV infection.

6.3 Glucose Metabolism and Viral Infection

Alteration in cell signaling pathways to rewire energy metabolism is an essential hallmark of human cancer. In the presence of oxygen, differentiated tissues and normal cells metabolize glucose mainly rely on oxidative phosphorylation to generate energy which is a highly efficient process, resulting in generation of up to 38 ATP molecules per molecule of glucose metabolized. Under low oxygen condition (hypoxia), normal cells switch to anaerobic glycolysis which results in 2 ATP per glucose molecule and generation of lactate. Enhanced glucose consumption, via glycolysis despite the presence of oxygen (referred as aerobic glycolysis), is common in cancer cells and is referred to as the "Warburg's effect" [15, 16]. Besides ATP, enhanced aerobic glycolysis in cancer cells also generates substantial number of biosynthetic metabolites, which are essential for biomass production associated with cell growth. Enhanced glycolysis also results in high accumulation of lactate which has important influence on the tumor microenvironment. Adaptions of host cell metabolism to viral infection have been observed in cells infected with EBV, KSHV, HPV, and HCV [17–19]. Virus-encoded oncoproteins, such as E6 and E7 of the high-risk HPVs, drive cell proliferation which requires high demand for energy and biosynthetic metabolites to support cell growth [20]. The key pathway involved in balancing cell metabolisms to meet requirement of cell growth is the mTOR signaling pathway, which could sense the levels of energy and nutrients in cells. The viral oncoproteins also regulate metabolic pathways to control the nutrient assimilation required by proliferating cells [21]. In EBV-infected NPC cells, the expression of EBV-encoded LMP1 has been shown to enhance glycolysis by modulating multiple signaling pathways including FGF1 [22], AMPK [23], and mTORC1 [6] to drive up glucose metabolism. The viral oncoproteins may also act as transcription factors to upregulate the enzyme activities directly involved in glucose metabolism [24].

There are much remains to be determined on how EBV infection in NPC alters the glucose metabolism and contributes to the malignant transformation of nasopharyngeal epithelial cells. Recently, the landscape of somatic mutations in NPC has been reported [8–10]. Their involvement in glucose metabolism in NPC and latent EBV infection will be discussed. Altered energy metabolism in NPC may stabilize latent EBV infection. Expression of latent EBV products may further enhance and stabilize energy reprogramming in EBV-infected NPC cells. A close interplay between genetic alterations and EBV infection to enhance energy metabolism may support NPC development.

6.4 Genomic Profiling of NPC Reveals Alteration of Multiple Signaling Pathways Involved in Glucose Metabolism

Several comprehensive genomic profiling studies in NPC have revealed unique genetic landscapes relevance to NPC pathogenesis [8–10]. The most common genetic mutations observed in NPC are those involved in NF-κB activation [8, 9]. NF-κB signaling activation could be linked to mTOR activation which is a key driver in glucose metabolism in cancer cells [25]. Other common mutations in NPC alterations that are implicated in glucose metabolism including *PTEN/AKT/PIK3CA* mutations, which activate PI3K/AKT signaling upstream of mTOR activation, were also identified.

6.4.1 Activation of NF-κB Signaling

Two recent genomic studies in NPC employing whole exome sequencing (WES) have identified frequent mutations in multiple upstream negative regulators of NF-κB signaling including *CYLD*, *TRAF3*, *NFKB1A*, *TNFAIP3*, and *NLRC5* [8, 9]. The CYLD cleaves the lysine 63-linked polyubiquitin chains from target proteins, including NEMO (IKKγ) which is involved in phosphorylation and degradation of IκBs, which are inhibitors of NF-κB signaling. The CYLD also deubiquitinates TRAF2 which is an activator of NF-κB signaling. Furthermore, the CYLD inhibits activation of bcl3, which was reported in an early study to be involved in atypical activation of NF-κB in NPC associated with p50 dimmers [26]. The TRAF3 is involved in the suppression of NIK-activating NF-κB signaling. Most mutations of *TRAF3* in NPC are in the domain regions involved in this suppression of NIK activation. The *NFKB1A* encodes IκBα, which belongs to the NF-κB inhibitor family (IκBs). The NLRC5 is a potent inhibitor of NF-κB activation and competes with NEMO for binding to IKKα and IKKβ. Collectively, mutations of all these negative regulators upstream of NF-κB signaling may contribute to the constitutive activation of NF-κB commonly detected in NPC [26, 27]. Interestingly, an exclusive relationship between mutation of these negative regulators of NF-κB and high expression levels of LMP1 in NPC was observed [9]. The potent function of NF-κB activation of the EBV-encoded LMP1 is well documented. Together, LMP1 expression and mutations of upstream negative regulators of NF-κB account for the majority of NPC samples examined. Hence, the genomic landscape of NPC reveals an essential role of NF-κB signaling in NPC pathogenesis either by genomic mutation or expression of EBV-encoded LMP1 [9]. The role of NF-κB signaling contributing to establishment of latent EBV infection in NPC is unclear.

Earlier study has showed that activation of NF-κB signaling by overexpressing p65 in EBV-infected cells inhibits activation of lytic promoter of EBV [28]. In lymphocytes and epithelial NPC latently infected with EBV, treatment with an NF-κB inhibitor (Bay 11–7082) resulted in expression of lytic viral protein [28, 29]. Using a more specific inhibitor of NF-κB, NBD peptide, we also observed lytic reactivation of EBV in infected NPC cells (Tsao SW. unpublished observations). The NBD peptide specifically inhibits NF-κB activation by binding to the NEMO-binding domain of IKK to inactivate the kinase activity of IKK complex upstream of NF-κB signaling [30]. Furthermore, expression of LMP1 or activated CD40 domain, which effectively activates NF-κB signaling in EBV-infected lymphocytes, also suppressed lytic reactivation of EBV [31]. These studies support a role of NF-κB activation in establishment of latent infection of EBV in infected cells. The underlying mechanisms are unclear.

As a key inflammatory modulator, aberrantly activation of NF-κB signaling contributes multiple hallmarks of cancer, including energy metabolism [32–34]. An early study reported that NF-κB activation is directly involved in activation of mTORC1 signaling [25]. The IKK kinase complex is upstream of NF-κB signaling and composed of IKKα, IKKβ, and IKKγ (NEMO). The IKKβ could directly interact and phosphorylate the TSC1, which is the negative regulator of the mTORC1 complex [25]. Phosphorylation of TSC1 at Ser^{487} and Ser^{522} by IKKβ results in the release of TSC1 suppression on mTORC1 signaling. Furthermore, the IKKα, another key component of the kinase complex in NF-κB activation, could also phosphorylate mTOR at Ser^{1415} to activate mTORC1 signaling [35]. Activation of mTORC1 signaling enhances uptake of glucose through activation of HIF1α to upregulate *Glut-1* transcription which increases the glucose influx to support aerobic glycolysis [21]. We showed that NF-κB activation by LMP1 could directly enhance transcription of *Glut-1*, and another study also showed that LMP1 induced *Glut-1* translocation to plasma membrane, both of which increase the glucose influx into cells to support aerobic glycolysis [6, 7].

Here, we postulated that metabolic adaption through NF-κB activation to activate mTOR and glucose metabolism may minimize metabolic stress associated with viral infection and play an essential role in establishment of latent EBV infection in NPC cells. In support of this hypothesis, we have also observed enhanced lytic replication in EBV-infected NPC cells upon treatment with rapamycin (a specific inhibitor of mTORC1) (Tsao SW, unpublished observation). Similarly, inhibition of lytic reactivation in BX-1 EBV-infected AGS (gastric cancer cells) by rapamycin has also been reported [36].

6.4.2 Activation of PTEN/PI3K/AKT Signaling Pathway

The PTEN/PI3K/AKT signaling pathway, which is a key signaling pathway upstream of mTOR activation, is frequently activated in cancer cells. Mutation of their negative regulator, *PTEN*, and activation mutations in *PIK3CA* could activate PI3K/AKT signaling [21, 37]. Recent genomic profiling studies of NPC revealed deletion of *PTEN*, amplification, and hot spot mutations of *PIK3CA* [8–10]. The mTORC1, a serine/threonine protein kinase, is the key sensor in cells to regulate cell metabolism by balancing the energy status and controlling the synthesis of essential metabolic resources in cells including proteins, nucleotides, and lipids [38]. The constitutive activation of mTORC1 signaling is commonly observed in human cancers including NPC [39]. Activation of mTORC1 signaling enhances glucose uptake and consumption to ensure sufficient energy and generation of biosynthetic metabolites for growth and proliferation of cancer cells [21]. Interestingly, this adaptive process also confers metastatic potential to cancer cells and their resistance to chemotherapy [40]. PTEN is an inhibitor of AKT, which suppresses PI3K activity, the common cell signaling pathway in mTOR activation. Mutation and deletion of *PTEN* were detected in NPC [9] which may lead to activation of PI3K/AKT and mTOR. Interestingly, loss of function mutation of *PTEN* in mammalian cells may also regulate aerobic glycolysis via a PI3K-independent manner through the E3 ubiquitin ligase activity of APC/C-Cdh1 to enhance aerobic glycolysis [41]. Other mutations detected in NPC including the ERBB2/ERBB3 may also converge to PI3K/MAPK signaling leading to mTOR activation [9, 10].

6.5 Latent EBV Genes Drive Glucose Metabolism in NPC

6.5.1 LMP1

The LMP1 is a well-documented EBV-encoded oncoprotein expressed during latent EBV infection in NPC. LMP1 is a potent activator of multiple signaling pathways. Its role in activation of NF-κB signaling is well documented [42]. The LMP1 has been shown to enhance aerobic glycolysis in NPC cells through alteration of metabolism-associated pathways [6, 7, 22, 23, 43, 44]. A recent study showed that LMP1 induces expression of hexokinase 2 (HK2), a key enzyme to control and enhance the glycolysis process [43]. LMP1 expression is positively correlated with HK2 expression in NPC tissue and poor overall survival of NPC patients following radiation therapy. Enhanced aerobic glycolysis also conferred insensitivity to radiation therapy in LMP1-expressing cells. Suppression of HK2 expression induces apoptosis in NPC cells. The transport of glucose over the plasma membrane by its glucose transporters (Gluts) is the first rate-limiting step of glucose metabolism. We recently reported that LMP1 upregulates *Glut-1* transcription in NPC cells to enhance aerobic glycolysis, a process dependent on activation of mTORC1 and NF-κB signaling [6]. Blocking aerobic glycolysis by specific chemical and genetic inhibitors also suppressed multiple LMP1-mediated malignant phenotypes. An earlier study also showed that LMP1 mediated the relocation of Glut-1 to plasma membrane in EBV-infected B cells involving activation of IKKβ/NF-κB signaling [7]. Localization of the Glut-1 to cell membrane enhances glucose uptake to support cell proliferation and also confers resistance to apoptosis. Hence expression of LMP1 serves as an important driver in EBV-infected cells to accelerate aerobic glycolysis by targeting glycolysis-associated events, particularly activation of Glut-1 and HK2.

Additional signaling pathways or upstream modulators have been identified in driving glucose metabolism in cancer cells. The LMP1 is also involved in modulating these pathways. The LMP1 was reported to induce FGF expression and secretion to promote FGFR1 signaling which drive aerobic glycolysis [22]. Blockade of FGFR signaling by small molecules suppressed the LMP1-induced transformed phenotypes. HoxC8 is a negative regulator of aerobic glycolysis commonly downregulated in NPC. LMP1 could suppress the expression of HoxC8 via stalling the activity of RNA polymerase II (RNA Pol II) [44]. The AMPK-mTOR axis activity is well known for its involvement in energy metabolism. LMP1 was reported to inhibit AMPK activity and signaling in immortalized nasopharyngeal epithelial cells [45]. Inhibition of AMPK suppressed LMP1-induced proliferation and transformation of immortalized nasopharyngeal epithelial cells. The inhibition of AMPK signaling also accelerated glucose uptake and lactate production and conferred resistance of NPC cells to apoptosis induced by irradiation [46]. In B cells, infection with EBV induced cell proliferation which demands increased supply of energy and metabolites. These observations are concordant with the close association of proliferation of EBV-infected B cells with AMPK inhibition and mTOR activity [47]. A role of latent EBV infection and expression of the EBV-encoded LMP1 in driving glucose metabolism to meet the increased demand of energy and metabolites for cell proliferation is emerging.

6.5.2 LMP2

LMP2A is another latent EBV-encoded protein expressed in NPC that could activate mTORC1 signaling. Expression of LMP2A in HONE1 cells leads to mTOR activation as revealed by the phosphorylation of 4E–BP1 which is the downstream effector of mTOR [5]. Activation of PI3K/AKT signaling pathway is the upstream in mediating mTOR activation and may provide growth advantage to LMP2A-expressing cells. Activation of mTOR by LMP2A may also contribute to the enhanced glucose metabolism in latent EBV-infected NPC cells.

6.5.3 miR-BARTs

EBV-encoded miRNAs have been identified to be highly expressed in EBV-associated cancers and contribute to viral latency, cell survival, as well as immune escape. High expression of miR-BARTs is found in epithelial malignant cells with latency type II EBV infection, including NPC and gastric carcinoma, but low in EBV-infected B cells which strongly indicates a specific role of miR-BARTs in the pathogenesis of EBV-associated epithelial malignancies [13].

The role of NF-κB in driving latent EBV infection and expression of miR-BARTs has been reported [28, 48]. Expression of miR-BARTs is known to be deregulated in EBV-associated tumors and associated with NPC pathogenesis [49, 50]. Interestingly, constitutively activation of NF-κB upregulates LMP1 and miR-BART expression in EBV-infected NPC cells [48]. Multiple miR-BARTs can downregulate LMP1 expression by targeting its 3′-UTR and thereby form a negative feedback loop to modulate the level of LMP1 in NPC. The miR-BARTs have also been demonstrated to suppress EBV lytic replication in both B cell and epithelial cell lines induced by TPA. miR-BART6 was reported to target DICER, and miR-BART20-5p has the ability to stabilize EBV latency by directly targeting BZLF1 and BRLF1; both of them can govern the EBV entry into the lytic replication phase [51, 52]. These studies are in agreement with the view that miR-BARTs can support latency infection.

At present, the evidence for EBV microRNA in glucose metabolism is limited. A recent report showed that the miR-BART-1 may be involved in enhancing aerobic glycolysis through upregulation of a panel of genes involved in cell metabolism [53]. Based on RNA deep sequencing analysis, overexpression of miR-BART-1 leads to the upregulation of *G6PD*, *PHGDH*, *PAST1*, *IDH2*, *and PISD* and downregulation of *UGT8*, *LDHB*, *SGPL1*, *and DHRS3* [53]. It remains to be determined whether the effects of miR-BART in glucose metabolism are direct or indirect through modulation of metabolism-related pathways.

6.6 The Impact of Glucose Metabolism to EBV Infection in NPC

As aforementioned, EBV infection readily immortalizes and transforms primary B cells both in vitro and in vivo but not in primary epithelial cells. Primary epithelial cells have finite life span in culture and readily undergo senescence upon passages. EBV infection may induce cellular stress and proliferation arrest in primary epithelial cells. EBV could infect and undergo lytic replication in oral keratinocyte grown as three-dimensional organotypic culture [54]. However, neither latent infection of EBV nor proliferation of EBV-infected epithelial cells was observed in EBV-infected stratified keratinocytes in the organotypic culture. This supports the postulation that EBV undergoes lytic infection in normal epithelium to amplify the EBV genomes and generate infectious viral particles for transmission.

Establishment of latent EBV infection requires modification of host cell signaling. Latent EBV infection could be established in telomerase-immortalized nasopharyngeal epithelial cells [14, 55]. Immortalization is a prerequisite property of cancer cells and is regarded as an early event in human carcinogenesis [15]. Metabolic stress is a major barrier for cell immortalization. The high demand for energy and biosynthetic metabolites to sustain continuous proliferation requires metabolic adaptation in both immortalized and cancer cells. Our recent study showed that mTORC1, as well as NF-κB signaling, is commonly activated during the immortalization of nasopharyngeal epithelial cells mediated by telomerase [56]. Another barrier to immortalization is cellular senescence induced by reactive oxygen species (ROS), which are generated during cell proliferation. Primary cells propagated for extended period of culture will accumulate ROS to induce cellular senescence and apoptosis [57]. Enhanced glycolysis could protect cells from apoptosis due to ROS induced oxidative stress and facilitate immortalization [58].

Metabolic stress has been characterized as a major barrier for immortalization and latency establishment of B cells mediated by EBV infection [47]. Only a small population of EBV-infected B cells could be immortalized by EBV. Analysis of these EBV-immortalized B cells revealed activation of aerobic glycolysis with high glucose metabolism which is associated with suppression of AMPK and activation of mTOR signaling. Accordingly, activation of AMPK and a decrease of mTOR activity were detected in the growth-arrested B cells that further failed to be immortalized upon EBV infection [47]. Furthermore, inhibition of mTORC1 in the EBV-infected epithelial cells with specific inhibitor, rapamycin, effectively elevated the lytic EBV replication in a dose- and time-dependent manner as evidenced by the increase of *Zta* and *Rta* transcripts and their translated proteins in rapamycin-treated cells [36]. As aforementioned, blockade of mTORC1 by rapamycin induces lytic reactivation in NPC and gastric cancer cells. Blockage of mTORC1 activation by rapamycin may induce a starvation status by slowing down the glucose uptake, which may be a physiological signaling for EBV to switch into lytic replication. Enhanced glucose metabolism may represent essential cellular properties to support latent EBV infection which warrants further investigations.

6.7 Enhanced Glucose Metabolism Alters the Tumor Microenvironment

The tumor microenvironment is known to play an important function for modulation of tumor growth, progression and metastasis to distant sites, and development of acquired treatment resistance and accounts for poor patient prognosis [59]. The tumor microenvironment contains multiple elements, including immune cells, stromal fibroblasts, and tumor-associated endothelial cells, all of which are known to be involved in modulating malignant behaviors of cancer cells. A typical feature of tumor microenvironment is high acidity, which plays key roles for tumor progression. As a result of enhanced aerobic glycolysis, accumulation of lactate is a major contributor for the high acidity of tumor microenvironment. Lactate has been shown to promote angiogenesis, cell migration, metastasis, and growth sufficiency in cancer cells [15, 60–62]. High concentrations of lactate are associated with development of distant cancer metastasis [62–65]. Expression of the EBV-encoded LMP1 in NPC has been postulated to enhance malignant properties of NPC by inducing angiogenesis [66–68], cell motility, and immune escape [69–71]. Some of these malignant properties of NPC cells may be accounted for by the enhanced aerobic glycolysis and accumulation of elevated level of lactate in the tumor microenvironment. Interestingly, treatment of LMP1-expressing nasopharyngeal epithelial cells with aerobic glycolysis inhibitors, STF-31 and 2-DG, suppressed the LMP1-induced cell migration and invasion supporting a role of EBV gene-driven aerobic glycolysis in NPC metastasis (Tsao's unpublished data).

Lactate may also contribute to immune escape of tumor cells through suppressing monocyte migration and release of cytokines [72, 73], inhibiting activation of T cells and natural killer cells [74–76]. NPC is characterized by substantial infiltration of immune cells in the tumor microenvironment including dendritic cells, monocytes, T cells, and B cells. The contribution of aerobic glycolysis driven by EBV-infected NPC cells to modulate host immune responses remains to be determined.

6.8 Conclusions

The role of EBV in NPC pathogenesis has been enigmatic. The underlying mechanism supporting latent EBV infection and growth advantage in NPC are not well defined. Genomic profiling revealed that mutations involved in activation of NF-κB and PTEN/PI3K/AKT may drive mTOR signaling to support latent infection of

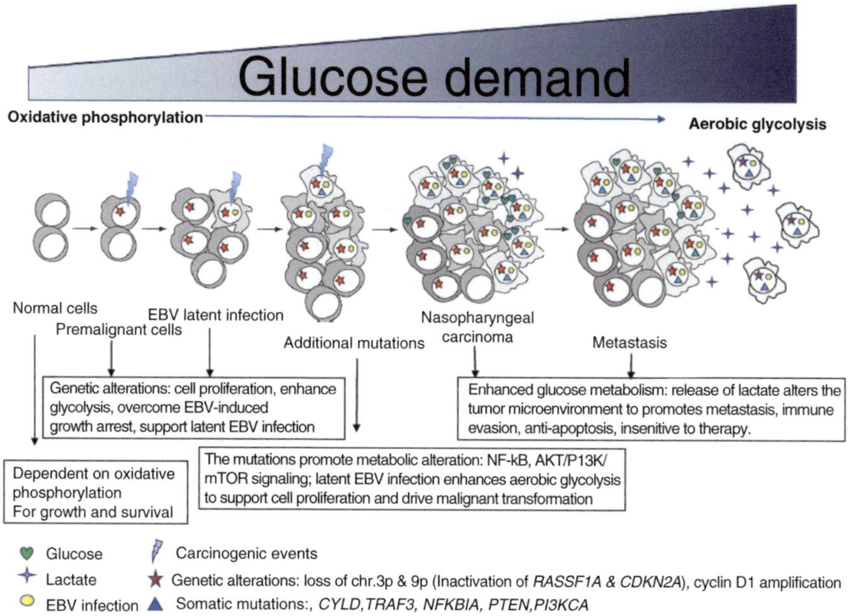

Fig. 6.1 Schematic diagram showing how EBV infection and glucose metabolism may contribute to pathogenesis and progression of nasopharyngeal carcinoma (NPC). The glucose demand increases during malignant transformation of nasopharyngeal epithelial cells which serves as a selective force for latent EBV infection and NPC development. EBV initially establishes latent infection in premalignant nasopharyngeal epithelial cells harboring genetic alterations, e.g., cyclin D1 amplification and p16 deletion. Expression of latent EBV genes, e.g., LMP1 and LMP2, drives glycose metabolism and supports clonal expansion of EBV-infected premalignant nasopharyngeal epithelial cells harboring additional mutations. Genetic alterations involved in activation of NF-κB signaling have selective advantage growth in EBV-infected nasopharyngeal epithelial cells for its ability to activate mTOR and glucose metabolism which support latent EBV infection. Positive feedback of latent EBV genes further enhances glucose metabolism to support malignant transformation of premalignant nasopharyngeal epithelial cells. Enhanced glucose metabolism and accumulation of metabolites of aerobic glycolysis, e.g., lactate, further modulate the tumor microenvironment to promote NPC progression including immune evasion, angiogenesis, anti-apoptosis, resistance to treatment, invasion, and metastasis

EBV in NPC cells. Latent infection of NPC cells and expression of latent EBV genes further drive glucose metabolism and modify the tumor microenvironment to enhance malignant properties of NPC including immune evasion and antiapoptosis. Enhanced glucose metabolism is a common hallmark of human cancer. In NPC, enhanced glucose metabolism has been demonstrated by the EBV-encoded LMP1. A schematic diagram illustrating how latent EBV infection and enhance glucose metabolism may drive the development of NPC and its progression is shown in Fig. 6.1. An interactive interplay between glucose metabolism and EBV gene-driven glucose metabolism may modulate malignant properties of NPC cells including angiogenesis, invasion, metastasis, and resistance to therapy. Understanding the key events involved in altered glucose metabolism and the role of latent EBV infection in NPC pathogenesis may reveal novel therapeutic targets to suppress NPC metastasis and potentially reverse resistance to treatment therapy in NPC patients.

Acknowledgment This project was supported by funding from the Research Grants Council, Hong Kong: General Research Fund (17120814, 17161116, 17111516, 171110315), Area of Excellence (AoE/M-06/08), Theme-Based Research Scheme (T12-401/13-R), Collaborative Research Fund (Project reference: C7027-16G); Health and Medical Research Fund of Hong Kong (12110782, 13120872, 04151726); and CRCG, University of Hong Kong, Hong Kong.

References

1. Vockerodt M, Yap LF, Shannon-Lowe C, Curley H, Wei W, Vrzalikova K, Murray PG (2015) The Epstein-Barr virus and the pathogenesis of lymphoma. J Pathol 235:312–322
2. Tsao SW, Tsang CM, To KF, Lo KW (2015) The role of Epstein-Barr virus in epithelial malignancies. J Pathol 235:323–333
3. Minamitani T, Ma Y, Zhou H, Kida H, Tsai CY, Obana M, Okuzaki D, Fujio Y, Kumanogoh A, Zhao B, Kikutani H, Kieff E, Gewurz BE, Yasui T (2017) Mouse model of Epstein-Barr virus LMP1- and LMP2A-driven germinal center B-cell lymphoproliferative disease. Proc Natl Acad Sci U S A 114:4751–4756
4. Young LS, Yap LF, Murray PG (2016) Epstein-Barr virus: more than 50 years old and still providing surprises. Nat Rev Cancer 16:789–802
5. Moody CA, Scott RS, Amirghahari N, Nathan CO, Young LS, Dawson CW, Sixbey JW (2005) Modulation of the cell growth regulator mTOR by Epstein-Barr virus-encoded LMP2A. J Virol 79:5499–5506
6. Zhang J, Jia L, Lin W, Yip YL, Lo KW, Lau VM, Zhu D, Tsang CM, Zhou Y, Deng W, Lung HL, Lung ML, Cheung LM, Tsao SW (2017) Epstein-Barr virus encoded latent membrane protein-1 upregulates glucose transporter-1 transcription via the mTORC1/NF-kappaB signaling pathways. J Virol 91:e02168
7. Sommermann TG, O'Neill K, Plas DR, Cahir-McFarland E (2011) IKKbeta and NF-kappaB transcription govern lymphoma cell survival through AKT-induced plasma membrane trafficking of GLUT1. Cancer Res 71:7291–7300
8. Zheng H, Dai W, Cheung AK, Ko JM, Kan R, Wong BW, Leong MM, Deng M, Kwok TC, Chan JY, Kwong DL, Lee AW, Ng WT, Ngan RK, Yau CC, Tung S, Lee VH, Lam KO, Kwan CK, Li WS, Yau S, Chan KW, Lung ML (2016) Whole-exome sequencing identifies multiple loss-of-function mutations of NF-kappaB pathway regulators in nasopharyngeal carcinoma. Proc Natl Acad Sci U S A 113:11283–11288
9. Li YY, Chung GT, Lui VW, To KF, Ma BB, Chow C, Woo JK, Yip KY, Seo J, Hui EP, Mak MK, Rusan M, Chau NG, Or YY, Law MH, Law PP, Liu ZW, Ngan HL, Hau PM, Verhoeft KR, Poon PH, Yoo SK, Shin JY, Lee SD, Lun SW, Jia L, Chan AW, Chan JY, Lai PB, Fung CY, Hung ST, Wang L, Chang AM, Chiosea SI, Hedberg ML, Tsao SW, van Hasselt AC, Chan AT, Grandis JR, Hammerman PS, Lo KW (2017) Exome and genome sequencing of nasopharynx cancer identifies NF-kappaB pathway activating mutations. Nat Commun 8:14121
10. Lin DC, Meng X, Hazawa M, Nagata Y, Varela AM, Xu L, Sato Y, Liu LZ, Ding LW, Sharma A, Goh BC, Lee SC, Petersson BF, Yu FG, Macary P, Oo MZ, Ha CS, Yang H, Ogawa S, Loh KS, Koeffler HP (2014) The genomic landscape of nasopharyngeal carcinoma. Nat Genet 46:866–871
11. Hutt-Fletcher LM (2017) The long and complicated relationship between Epstein-Barr virus and epithelial cells. J Virol 91:e01677

12. Hadinoto V, Shapiro M, Sun CC, Thorley-Lawson DA (2009) The dynamics of EBV shedding implicate a central role for epithelial cells in amplifying viral output. PLoS Pathog 5:e1000496
13. Kang D, Skalsky RL, Cullen BR (2015) EBV BART MicroRNAs target multiple pro-apoptotic cellular genes to promote epithelial cell survival. PLoS Pathog 11:e1004979
14. Tsang CM, Yip YL, Lo KW, Deng W, To KF, Hau PM, Lau VM, Takada K, Lui VW, Lung ML, Chen H, Zeng M, Middeldorp JM, Cheung AL, Tsao SW (2012) Cyclin D1 overexpression supports stable EBV infection in nasopharyngeal epithelial cells. Proc Natl Acad Sci U S A 109:E3473–E3482
15. Hanahan D, Weinberg RA (2011) Hallmarks of cancer: the next generation. Cell 144:646–674
16. Warburg O (1956) On the origin of cancer cells. Science 123:309–314
17. Claus C, Liebert UG (2014) A renewed focus on the interplay between viruses and mitochondrial metabolism. Arch Virol 159:1267–1277
18. Sanchez EL, Lagunoff M (2015) Viral activation of cellular metabolism. Virology 479-480:609–618
19. Rosenwasser S, Ziv C, Creveld SG, Vardi A (2016) Virocell metabolism: metabolic innovations during host-virus interactions in the ocean. Trends Microbiol 24:821–832
20. Noch E, Khalili K (2012) Oncogenic viruses and tumor glucose metabolism: like kids in a candy store. Mol Cancer Ther 11:14–23
21. Laplante M, Sabatini DM (2012) mTOR signaling in growth control and disease. Cell 149:274–293
22. Lo AKF, Dawson CW, Young LS, Ko CW, Hau PM, Lo KW (2015) Activation of the FGFR1 signalling pathway by the Epstein-Barr viurs-encoded LMP1 rpomtes aerobi glyolysis and transformation of human nasopharyngeal epithelial cells. J Pathol 237:238–248
23. Lu JC, Tang M, Li HD, Xu ZJ, Weng XX, Li JJ, Yu XF, Zhao LQ, Liu HW, Hu YB, Tan ZQ, Yang LF, Zhong MZ, Zhou J, Fan J, Bode AM, Yi W, Gao JH, Sun LQ, Cao Y (2016) EBV-LMP1 suppresses the DNA damage response through DNA-PK/AMPK signaling to promote radioresistance in nasopharyngeal carcinoma. Cancer Lett 380:191–200
24. Thai M, Graham NA, Braas D, Nehil M, Komisopoulou E, Kurdistani SK, McCormick F, Graeber TG, Christofk HR (2014) Adenovirus E4ORF1-induced MYC activation promotes host cell anabolic glucose metabolism and virus replication. Cell Metab 19:694–701
25. Lee DF, Kuo HP, Chen CT, Hsu JM, Chou CK, Wei Y, Sun HL, Li LY, Ping B, Huang WC, He X, Hung JY, Lai CC, Ding Q, Su JL, Yang JY, Sahin AA, Hortobagyi GN, Tsai FJ, Tsai CH, Hung MC (2007) IKK beta suppression of TSC1 links inflammation and tumor angiogenesis via the mTOR pathway. Cell 130:440–455
26. Thornburg NJ, Pathmanathan R, Raab-Traub N (2003) Activation of nuclear factor-kappaB p50 homodimer/Bcl-3 complexes in nasopharyngeal carcinoma. Cancer Res 63:8293–8301
27. Chung GT, Lou WP, Chow C, To KF, Choy KW, Leung AW, Tong CY, Yuen JW, Ko CW, Yip TT, Busson P, Lo KW (2013) Constitutive activation of distinct NF-kappaB signals in EBV-associated nasopharyngeal carcinoma. J Pathol 231:311–322
28. Brown HJ, Song MJ, Deng H, Wu TT, Cheng G, Sun R (2003) NF-kappaB inhibits gammaherpesvirus lytic replication. J Virol 77:8532–8540
29. Liu SF, Wang H, Lin XC, Xiang H, Deng XY, Li W, Tang M, Cao Y (2008) NF-kappaB inhibitors induce lytic cytotoxicity in Epstein-Barr virus-positive nasopharyngeal carcinoma cells. Cell Biol Int 32:1006–1013
30. May MJ, D'Acquisto F, Madge LA, Glockner J, Pober JS, Ghosh S (2000) Selective inhibition of NF-kappa B activation by a peptide that blocks the interaction of NEMO with the I kappa B kinase complex. Science 289:1550–1554
31. Adler B, Schaadt E, Kempkes B, Zimber-Strobl U, Baier B, Bornkamm GW (2002) Control of Epstein-Barr virus reactivation by activated CD40 and viral latent membrane protein 1. Proc Natl Acad Sci U S A 99:437–442
32. Ben-Neriah Y, Karin M (2011) Inflammation meets cancer, with NF-kappa B as the matchmaker. Nat Immunol 12:715–723

33. Mauro C, Leow SC, Anso E, Rocha S, Thotakura AK, Tornatore L, Moretti M, De Smaele E, Beg AA, Tergaonkar V, Chandel NS, Franzoso G (2011) NF-kappaB controls energy homeostasis and metabolic adaptation by upregulating mitochondrial respiration. Nat Cell Biol 13:1272–1279
34. Johnson RF, Perkins ND (2012) Nuclear factor-kappaB, p53, and mitochondria: regulation of cellular metabolism and the Warburg effect. Trends Biochem Sci 37:317–324
35. Dan HC, Cooper MJ, Cogswell PC, Duncan JA, Ting JP, Baldwin AS (2008) Akt-dependent regulation of NF-{kappa} B is controlled by mTOR and raptor in association with IKK. Genes Dev 22:1490–1500
36. Adamson AL, Le BT, Siedenburg BD (2014) Inhibition of mTORC1 inhibits lytic replication of Epstein-Barr virus in a cell-type specific manner. Virol J 11:110
37. Hay N, Sonenberg N (2004) Upstream and downstream of mTOR. Genes Dev 18:1926–1945
38. Shimobayashi M, Hall MN (2014) Making new contacts: the mTOR network in metabolism and signalling crosstalk. Nat Rev Mol Cell Biol 15:155–162
39. Chen J, Hu CF, Hou JH, Shao Q, Yan LX, Zhu XF, Zeng YX, Shao JY (2010) Epstein-Barr virus encoded latent membrane protein 1 regulates mTOR signaling pathway genes which predict poor prognosis of nasopharyngeal carcinoma. J Transl Med 8:30
40. Pusapati RV, Daemen A, Wilson C, Sandoval W, Gao M, Haley B, Baudy AR, Hatzivassiliou G, Evangelista M, Settleman J (2016) mTORC1-dependent metabolic reprogramming underlies escape from glycolysis addiction in cancer cells. Cancer Cell 29:548–562
41. Garcia-Cao I, Song MS, Hobbs RM, Laurent G, Giorgi C, de Boer VC, Anastasiou D, Ito K, Sasaki AT, Rameh L, Carracedo A, Vander Heiden MG, Cantley LC, Pinton P, Haigis MC, Pandolfi PP (2012) Systemic elevation of PTEN induces a tumor-suppressive metabolic state. Cell 149:49–62
42. Edwards RH, Marquitz AR, Raab-Traub N (2015) Changes in expression induced by Epstein-Barr virus LMP1-CTAR1: potential role of bcl3. MBio 6:e00441
43. Xiao L, Hu ZY, Dong X, Tan Z, Li W, Tang M, Chen L, Yang L, Tao Y, Jiang Y, Li J, Yi B, Li B, Fan S, You S, Deng X, Hu F, Feng L, Bode AM, Dong Z, Sun LQ, Cao Y (2014) Targeting Epstein-Barr virus oncoprotein LMP1-mediated glycolysis sensitizes nasopharyngeal carcinoma to radiation therapy. Oncogene 33:4568–4578
44. Cai L, Ye Y, Jiang Q, Chen Y, Lyu X, Li J, Wang S, Liu T, Cai H, Yao K, Li JL, Li X (2015) Epstein-Barr virus-encoded microRNA BART1 induces tumour metastasis by regulating PTEN-dependent pathways in nasopharyngeal carcinoma. Nat Commun 6:7353
45. Lo AK, Lo KW, Ko CW, Young LS, Dawson CW (2013) Inhibition of the LKB1-AMPK pathway by the Epstein-Barr virus-encoded LMP1 promotes proliferation and transformation of human nasopharyngeal epithelial cells. J Pathol 230:336–346
46. Lu J, Tang M, Li H, Xu Z, Weng X, Li J, Yu X, Zhao L, Liu H, Hu Y, Tan Z, Yang L, Zhong M, Zhou J, Fan J, Bode AM, Yi W, Gao J, Sun L, Cao Y (2016) EBV-LMP1 suppresses the DNA damage response through DNA-PK/AMPK signaling to promote radioresistance in nasopharyngeal carcinoma. Cancer Lett 380:191–200
47. McFadden K, Hafez AY, Kishton R, Messinger JE, Nikitin PA, Rathmell JC, Luftig MA (2016) Metabolic stress is a barrier to Epstein–Barr virus-mediated B-cell immortalization. Proc Natl Acad Sci 113:E782–E790
48. Verhoeven RJ, Tong S, Zhang G, Zong J, Chen Y, Jin DY, Chen MR, Pan J, Chen H (2016) NF-kappaB signaling regulates expression of Epstein-Barr virus BART MicroRNAs and long noncoding RNAs in nasopharyngeal carcinoma. J Virol 90:6475–6488
49. Marquitz AR, Raab-Traub N (2012) The role of miRNAs and EBV BARTs in NPC. Semin Cancer Biol 22:166–172
50. Klinke O, Feederle R, Delecluse HJ (2014) Genetics of Epstein-Barr virus microRNAs. Semin Cancer Biol 26:52–59
51. Jung YJ, Choi H, Kim H, Lee SK (2014) MicroRNA miR-BART20-5p stabilizes Epstein-Barr virus latency by directly targeting BZLF1 and BRLF1. J Virol 88:9027–9037

52. Iizasa H, Wulff BE, Alla NR, Maragkakis M, Megraw M, Hatzigeorgiou A, Iwakiri D, Takada K, Wiedmer A, Showe L, Lieberman P, Nishikura K (2010) Editing of Epstein-Barr virus-encoded BART6 MicroRNAs controls their Dicer targeting and consequently affects viral latency. J Biol Chem 285:33358–33370
53. Ye YF, Zhou Y, Zhang L, Chen YX, Lyu XM, Cai LM, Lu YY, Deng Y, Wang JG, Yao KT, Fang WY, Cai HB, Li X (2013) EBV-miR-BART1 is involved in regulating metabolism-associated genes in nasopharyngeal carcinoma. Biochem Bioph Res Co 436:19–24
54. Temple RM, Zhu J, Budgeon L, Christensen ND, Meyers C, Sample CE (2014) Efficient replication of Epstein-Barr virus in stratified epithelium in vitro. Proc Natl Acad Sci U S A 111:16544–16549
55. Tsang CM, Zhang G, Seto E, Takada K, Deng W, Yip YL, Man C, Hau PM, Chen H, Cao Y, Lo KW, Middeldorp JM, Cheung AL, Tsao SW (2010) Epstein-Barr virus infection in immortalized nasopharyngeal epithelial cells: regulation of infection and phenotypic characterization. Int J Cancer 127:1570–1583
56. Zhu DD, Zhang J, Deng W, Yip YL, Lung HL, Tsang CM, Law WT, Yang J, Lau VMY, Shuen WH, Lung ML, Cheung ALM, Tsao SW (2016) Significance of NF-kappa B activation in immortalization of nasopharyngeal epithelial cells. Int J Cancer 138:1175–1185
57. Shay JW, Roninson IB (2004) Hallmarks of senescence in carcinogenesis and cancer therapy. Oncogene 23:2919–2933
58. Kondoh H, Lleonart ME, Bernard D, Gil J (2007) Protection from oxidative stress by enhanced glycolysis; a possible mechanism of cellular immortalization. Histol Histopathol 22:85–90
59. Albini A, Sporn MB (2007) The tumour microenvironment as a target for chemoprevention. Nat Rev Cancer 7:139–147
60. Justus CR, Dong L, Yang LV (2013) Acidic tumor microenvironment and pH-sensing G protein-coupled receptors. Front Physiol 4:354
61. Hirschhaeuser F, Sattler UG, Mueller-Klieser W (2011) Lactate: a metabolic key player in cancer. Cancer Res 71:6921–6925
62. Dhup S, Dadhich RK, Porporato PE, Sonveaux P (2012) Multiple biological activities of lactic acid in cancer: influences on tumor growth, angiogenesis and metastasis. Curr Pharm Design 18:1319–1330
63. Schwickert G, Walenta S, Sundfor K, Rofstad EK, Mueller-Klieser W (1995) Correlation of high lactate levels in human cervical cancer with incidence of metastasis. Cancer Res 55:4757–4759
64. Walenta S, Wetterling M, Lehrke M, Schwickert G, Sundfor K, Rofstad EK, Mueller-Klieser W (2000) High lactate levels predict likelihood of metastases, tumor recurrence, and restricted patient survival in human cervical cancers. Cancer Res 60:916–921
65. Walenta S, Mueller-Klieser WF (2004) Lactate: mirror and motor of tumor malignancy. Semin Radiat Oncol 14:267–274
66. Yang L, Liu L, Xu Z, Liao W, Feng D, Dong X, Xu S, Xiao L, Lu J, Luo X, Tang M, Bode AM, Dong Z, Sun L, Cao Y (2015) EBV-LMP1 targeted DNAzyme enhances radiosensitivity by inhibiting tumor angiogenesis via the JNKs/HIF-1 pathway in nasopharyngeal carcinoma. Oncotarget 6:5804–5817
67. Sakakibara S, Tosato G (2009) Regulation of angiogenesis in malignancies associated with Epstein-Barr virus and Kaposi's sarcoma-associated herpes virus. Future Microbiol 4:903–917
68. Jonigk D, Izykowski N, Maegel L, Schormann E, Ludewig B, Kreipe H, Hussein K (2014) Tumour angiogenesis in Epstein-Barr virus-associated post-transplant smooth muscle tumours. Clin Sarcoma Res 4:1
69. Ning S (2011) Innate immune modulation in EBV infection. Herpesviridae 2:1
70. Munz C, Moormann A (2008) Immune escape by Epstein-Barr virus associated malignancies. Semin Cancer Biol 18:381–387
71. Merlo A, Turrini R, Dolcetti R, Martorelli D, Muraro E, Comoli P, Rosato A (2010) The interplay between Epstein-Barr virus and the immune system: a rationale for adoptive cell therapy of EBV-related disorders. Haematol-Hematol J 95:1769–1777

72. Goetze K, Walenta S, Ksiazkiewicz M, Kunz-Schughart LA, Mueller-Klieser W (2011) Lactate enhances motility of tumor cells and inhibits monocyte migration and cytokine release. Int J Oncol 39:453–463
73. Dart A (2016) Tumour metabolism: lactic acid: not just a waste product? Nat Rev Cancer 16:676–677
74. Husain Z, Huang YN, Seth P, Sukhatme VP (2013) Tumor-derived lactate modifies antitumor immune response: effect on myeloid-derived suppressor cells and NK cells. J Immunol 191:1486–1495
75. Romero-Garcia S, Moreno-Altamirano MM, Prado-Garcia H, Sanchez-Garcia FJ (2016) Lactate contribution to the tumor microenvironment: mechanisms, effects on immune cells and therapeutic relevance. Front Immunol 7:52
76. Brand A, Singer K, Koehl GE, Kolitzus M, Schoenhammer G, Thiel A, Matos C, Bruss C, Klobuch S, Peter K, Kastenberger M, Bogdan C, Schleicher U, Mackensen A, Ullrich E, Fichtner-Feigl S, Kesselring R, Mack M, Ritter U, Schmid M, Blank C, Dettmer K, Oefner PJ, Hoffmann P, Walenta S, Geissler EK, Pouyssegur J, Villunger A, Steven A, Seliger B, Schreml S, Haferkamp S, Kohl E, Karrer S, Berneburg M, Herr W, Mueller-Klieser W, Renner K, Kreutz M (2016) LDHA-associated lactic acid production blunts tumor immunosurveillance by T and NK cells. Cell Metab 24:657–671

Chapter 7
Kaposi's Sarcoma-Associated Herpesvirus: Epidemiology and Molecular Biology

Shasha Li, Lei Bai, Jiazhen Dong, Rui Sun, and Ke Lan

Abstract Kaposi's sarcoma-associated herpesvirus (KSHV), also known as *Human herpesvirus 8* (HHV-8), is a member of the lymphotropic gammaherpesvirus subfamily and a human oncogenic virus. Since its discovery in AIDS-associated KS tissues by Drs. Yuan Chang and Patrick Moore, much progress has been made in the past two decades. There are four types of KS including classic KS, endemic KS, immunosuppressive therapy-related KS, and AIDS-associated KS. In addition to KS, KSHV is also involved in the development of primary effusion lymphoma (PEL) and certain types of multicentric Castleman's disease. KSHV manipulates numerous viral proteins to promote the progression of angiogenesis and tumorigenesis. In this chapter, we review the epidemiology and molecular biology of KSHV and the mechanisms underlying KSHV-induced diseases.

Keywords Kaposi's sarcoma-associated herpesvirus • KSHV life cycle • KSHV epidemiology • KSHV pathogenesis

7.1 Discovery, Definition, and Classification of Kaposi's Sarcoma-Associated Herpesvirus (KSHV)

7.1.1 Discovery of KSHV

Kaposi's sarcoma (KS) was initially described by the Hungarian-born dermatologist Moritz Kaposi in 1872. He reported this new disease as "idiopathic purplish pigmented sarcoma of the skin," which is characterized by multiple purple patches

S. Li • K. Lan (✉)
State Key Laboratory of Virology, College of Life Sciences, Wuhan University, Wuhan 430072, People's Republic of China
e-mail: klan@whu.edu.cn

L. Bai • J. Dong • R. Sun
Key Laboratory of Molecular Virology and Immunology, Institute Pasteur of Shanghai, Chinese Academy of Sciences, Shanghai 200031, People's Republic of China

Q. Cai et al. (eds.), *Infectious Agents Associated Cancers: Epidemiology and Molecular Biology*, Advances in Experimental Medicine and Biology 1018,
DOI 10.1007/978-981-10-5765-6_7

or nodular lesions located in the lower extremity. In his clinical report, this sarcoma was first observed on the feet of five men aged 40–68 years old; later evidence showed that similar lesions were present in the trachea, esophagus, stomach, liver, and bowel. He declared that this disease was incurable and patients will eventually die within 5 years [1]. KS was considered rare for a long time since its description. Until the 1980s, the annual incidence of KS was reported to be as low as 0.021–0.061 per 100,000 in the United States, and it often occurred in elderly men [1]. Case reports indicated that the high-incidence areas were equatorial tropical Africa or sub-Saharan Africa, Eastern Central Europe [2], Xinjiang of China [3, 4], etc.

In 1981, Alvin Friedman-Kien firstly reported KS syndrome in 41 young homosexual men aged between 26 and 51 [2]. The discovery of acquired immunodeficiency syndrome (AIDS) resulted in an increase in the number of AIDS-associated KS cases. The outbreak of the AIDS epidemic in 1980s shifted attention to this neoplastic complication of AIDS and its association with human pathogens. Much effort has been made to identify the pathogens involved in KS. In 1972, typical herpes-type virus particles were found in five of eight selected tissue culture lines derived from different cases of KS [5], although one of these viruses was determined to be cytomegalovirus (CMV). Another study confirmed the presence of "intranuclear herpes-type viral inclusions" in KS specimens [6]. The association between AIDS and KS and the discovery of human immunodeficiency virus (HIV) led to the assumption that HIV may be the causal agent of KS. However, this hypothesis cannot explain the distinct distribution of KS among HIV patients with different transmission pathways. For example, the incidence of KS is lower in patients infected with HIV through blood products than in homosexual and bisexual AIDS patients [7]. In addition, KS was initially reported in HIV-negative patients. This phenomenon indicated that other environmental factors or infectious agents may be responsible for the emergence of KS.

The breakthrough occurred in 1994 when Yuan Chang and Patrick Moore et al. discovered new sequences in KS lesions that were absent in control tissue using the representational difference analysis (RDA) technique [8]. These new sequences were similar but not the same as those of the *Gammaherpesvirinae*, such as *Herpesvirus saimiri* and Epstein-Barr virus. It was later determined that they belonged to a new human herpesvirus termed Kaposi's sarcoma-associated herpesvirus (KSHV). Further research demonstrated that KSHV is the etiological agent of Kaposi's sarcoma.

Herpesviridae is a large family of DNA viruses that cause diseases in various animals and humans. The formal name of KSHV is *Human herpesvirus 8* (HHV-8) according to the International Committee on Taxonomy of Viruses (ICTV). It belongs to the *Gammaherpesvirinae* subfamily, which is characterized by lymphocytic tropism. The *Gammaherpesvirinae* subfamily can be further divided into five genera according to the 2015 ICTV report of virus taxonomy, such as the *Lymphocryptoviridae*, containing the human Epstein-Barr virus (EBV), and *Rhadinoviridae*, containing human KSHV.

7.1.2 Virion Structure

As a typical herpesviral particle, KSHV consists of three major morphologically distinct parts: a highly ordered nucleocapsid encasing a linear double-strand viral DNA, an outer envelope bilayer containing large amounts of viral glycoprotein on the surface, and an electron-dense layer called the tegument layer between the capsid and the outer envelope (Table 7.1).

The three-dimensional (3D) structure of the KSHV capsid was analyzed in detail using cryo-electron microscopy reconstruction. Similar to the structure of herpes simplex virus type 1 (HSV-1) and human cytomegalovirus (HCMV), the KSHV capsid possesses a shell composed of 12 pentons, 150 hexons, and 320 triplexes arranged on a T = 16 icosahedral lattice [9]. Four viral structural proteins constitute these units. The major capsid protein (MCP) encoded by open reading frame (ORF) 25 forms hexameric and pentameric capsomers; the triplexes are heterotrimeric structures encoded by one molecule of ORF62 and two molecules of ORF26 [10]. These proteins show significant amino acid sequence homology to the capsid proteins of alpha- and betaherpesviruses. ORF65, as the smallest and highly antigenic capsid protein, is attached to the upper domain of the major capsid protein in hexons but not to that in pentons [11]. ORF56 also plays a pivotal in the KSHV capsid assembly process [12, 13]. There are at least three capsid species: A (empty capsid), B (containing structural protein), and C (containing the viral genome) [10]. B-capsids contain various inner scaffolding proteins that probably

Table 7.1 Major KSHV virion structural proteins

Category	Gene	Function
Capsid proteins	ORF17.5	Scaffold protein
	ORF25	Major capsid protein
	ORF26	Capsid triplex homologs
	ORF62	Capsid triplex homologs
	ORF65	Capsid-associated structural protein
Envelope proteins	ORF8	gB
	K8.1	Glycoprotein
	ORF22	gH
	ORF39	gM
	ORF47	gL
	ORF53	gN
	ORF68	Glycoprotein
Tegument proteins	ORF21	TK
	ORF33	Tegumentation and viral assembly
	ORF45	Assembled capsid-tegument complexes transportation
	ORF52	Tegumentation and viral assembly
	ORF63	Tegument protein
	ORF64	Scaffold protein interact with capsid and tegument protein
	ORF75	Viral particle transportation

represent different intermediates rather than a structurally homogenous group during the process of KSHV capsid assembly and maturation [14]. Cryo-electron tomography revealed that an internally localized, umbilicated portal complex exists in the KSHV capsid [15].

Seven viral glycoproteins are interspersed in the envelope, including ORF8 (gB), ORF22 (gH), ORF39 (gM), ORF47 (gL), ORF53 (gN), K8.1, and ORF68 [16–21]. These glycoproteins function in mediating KSHV interactions with cell surface receptors and virion entry into host cells. The tegument layer remains largely undefined and is thought to be an amorphous structure. Several possible tegument proteins have been characterized as tegument components based on the following criteria: (1) resistance to protease digestion in the absence of detergent and (2) susceptibility to protease digestion with envelope dissolved [21–23]. Under these conditions, ORF11, ORF21, ORF33, ORF45, ORF50, ORF52, ORF63, ORF64, and ORF75 are considered tegument proteins with more to be characterized in the future [21–23]. Studies on the structure of tegument in HSV-1 [24, 25], HCMV [26], and MHV-68 [27] suggest a two-layered organization of the tegument. The inner layer may interact with the capsid, and the outer layer is loosely organized to conform to the shape of the envelope [23]. A recent study of the KSHV virion showed that five tegument densities cap each penton vertex. Each KSHV tegument density includes a penton-binding globular region, a helix-bundle stalk region, and a β-sheet-rich triplex-binding region [28]. The role of tegument proteins in mediating the interaction of inner capsid and outer envelope suggests their function of tegumentation and viral assembly and egress processes. Major KSHV virion structural proteins are listed in Table 7.1.

7.1.3 Viral Genome

The KSHV genome is a double-stranded DNA consisting of a 140.5-kb-long unique coding region flanked by a GC-rich terminal repeat region. About 90 ORFs are present in this long unique region, including 66 *Herpesvirus saimiri*-conserved ORFs [29, 30]. The repeat regions are 803 bp in length and are 85% G+C. The number of repeats varies at each end, but the total number remains at 35–45 [31].

Similar to other rhadinoviruses, KSHV has a large number of regulatory genes with similarity to known cellular genes. This is probably due to the virus pirating these genes from the host cell genome during evolution. They include IL-6, BCL-2, several different interferon regulatory factor homologs, cyclin D, and other genes whose function resembles that of their cellular homologs [32]. In addition, KSHV has a significant number of unique genes that are not found in other herpesviruses. These specific genes are named K1-K15 based on the locations from left to right in the genome [33]. In addition, several noncoding RNAs, such as polyadenylated nuclear RNA (PAN RNA), and 13 pre-miRNAs, encoding 25 mature microRNAs (miRNAs), are encoded by KSHV [34–37].

7.2 Epidemiology of KSHV

7.2.1 Prevalence and Genetic Diversity

The prevalence of KSHV infection varies globally. The infection rates in Western Europe and America are low, whereas the rates in Africa are higher [38]. The Mediterranean region (e.g., Italy and Spain) is an exception in that it shows a relatively high prevalence of KSHV infection [39, 40]. KSHV is also prevalent in the Xinjiang province of China, where high rates are found in the Uyghur and Kazakh ethnic groups [3, 4]. The seroepidemiological approach is the major method to determine the prevalence of KSHV infection. Seropositivity not only reflects past exposure to KSHV but also the presence of an ongoing viral infection. However, in most cases, KSHV infection is not associated with any symptoms, and the antibody titers in these individuals are lower than those in individuals with KSHV-related diseases. Thus, the reported seropositivity may be underestimated. To determine seroprevalence, testing of LANA (the major latently expressed antigen) and at least one lytic antigen (e.g., K8.1, ORF65) is recommended for variable seroreactivity [38].

Unlike the rates of KSHV infection, which may be difficult to determine accurately in epidemiological studies, the incidence of KSHV infection-related diseases such as KS is clear. The incidence of KS is around 1 in 100,000 in the general population, whereas it is approximately 1 in 20 in HIV-infected individuals, increasing to almost 1 in 3 in HIV-infected gay men prior to the introduction of highly active antiretroviral treatment (HAART) [7, 41]. The overall incidence of KS in the United States is as follows (age-standardized incidence in males per 100,000): non-Hispanic whites, 0.8; white Hispanics, 1.4; and blacks, 2.4. In Italy, the incidence ranges from 0.2 to 2.0. The incidence of KS in Africa is much higher and can be above 22 in endemic or HIV pandemic regions [38]. KS is one of the most common cancers in many subequatorial African countries. Almost all KS patients are seropositive for KSHV infection [42]. The geographic variation in KS incidence corresponds to the imbalanced worldwide seropositivity of KSHV infection. Therefore, the association of KS with KSHV infection strongly supports that KSHV is the etiologic agent of KS. Moreover, KSHV genomic DNA and viral antigens can be detected in all cases of KS [43].

Although KSHV is a DNA virus with a highly conserved genome, remarkable sequence variability can be found in the regions surrounding the terminal repeats, which are the ORFs of K1 and K15 [44, 45]. The genetic diversity in the K1 and K15 regions is used as a marker of strain diversity, which can help track the spread of KSHV and study the evolution of KSHV, as well as the relationship with its human host. The K1 gene encodes a transmembrane protein. Its N-terminal domain is highly variable, and the hypervariable regions are named by V1 and V2 [46]. The sequence variability may be related to the host immune pressure or the recognition of highly polymorphic cell surface proteins [47]. KSHV isolates can be classified into four major subtypes, A, B, C, and D, according to the K1 sequence variation.

The genetic diversity can only be found among different individuals, which means the viral sequences are stable in a given individual [46]. KSHV-infected individuals in Europe, the United States, the Middle East, and Asia are majorly A and C subtypes, whereas viral isolates from sub-Saharan Africa belong to the B strain. The type D strain is found primarily in Australia, South Asia, and the Pacific Islands [38]. The A and C subgroups (and only rare B isolates) have two allelic variants, termed P (for prototype) and M (for minority) in the K15 coding sequence. The P and M K15 isolates share 33% amino acid identity [45]. Additional alleles (N and Q) were reported in a recent study [48]. The diversity of the KSHV genome is also related to its pathogenesis, as suggested by a few studies. Subtype A is the predominant KSHV subtype in HIV-positive patients in Brazil, whereas subtype C is found mostly in the HIV-negative population [49, 50]. KSHV subtype A is also more frequent than subtype C in China [51]. Patients with rapid progression of KS are infected by KSHV subtype A, whereas KSHV subtype C is related to slow progression [52]. Additional evidence is necessary to support the relationship between genetic diversity and KSHV pathogenesis.

7.2.2 Transmission Routes and Susceptibility Factors

The most obvious transmission route of KSHV is sexual or blood-borne transmission, reflecting the high prevalence of KS in the AIDS population. Another significant fact is that the seroprevalence of KSHV infection can range from 25% to 60% in the homosexual population [53]. The seroprevalence rates are lower among women than men, but heterosexual transmission occurs. Blood-borne transmission (blood transfusion) of KSHV was documented in a prospective observational cohort study [54]. However, a historical cohort study showed that there is no statistically significant difference in KSHV seroconversion between the transfusion and non-transfusion groups [38]. The incidence of KS is also high in organ transplant recipients because of immunosuppressive therapy, and this is known as iatrogenic KS [55]. Iatrogenic KS may originate from the donor, although the evidence of this is weak, and there are no formal guidelines for the clinical screening of organ transplantation patients [38].

In addition to the studies of AIDS-KS and iatrogenic KS regarding the routes of transmission, important epidemiological studies have been performed in the endemic KS population in Africa [56–58]. Reports of KSHV-infected children in endemic regions suggest that KSHV infection can occur at a very early age and the transmission routes of KSHV are not limited to sexual or blood-borne transmission. A strong familial association (a child is positive if the mother or a sibling is positive for KSHV infection) in endemic regions has been reported [59, 60]. Studies in rural Uganda further support that KSHV infection in endemic regions is transmitted mostly through nonsexual routes (contacts with family members in childhood and continued into adulthood) [61]. Moreover, KSHV viral particles can be found in the maternal saliva and breast milk in endemic areas, which suggests an oral/salivary transmission route for KSHV [62]. Studies in the AIDS-KS population showed that KSHV is consistently detected in patients' saliva, which is in a line with the study in endemic KS [63]. However, further mechanistic studies are needed to explain the epidemiological observations.

The susceptibility factors involved in KSHV infection and those involved in the development of KSHV-related diseases are two different but tightly related issues. It is difficult to evaluate the susceptibility factors of KSHV infection using seroprevalence data without specific surveys of the behavioral and biological background. However, the existence of prevalence regions for endemic and classic KS suggests that the host genetic and behavioral factors and environmental factors (certain volcanic soils, arthropod bites, and living in rural areas) may be involved in KSHV infection [64]. For example, Uyghur and Kazakh ethnic groups, which live in the Xinjiang province of China, have significantly higher rates of KSHV infection than the Han population [65]. Although KSHV infection is necessary for the development of KS, it is not sufficient for its pathogenesis. Susceptibility factors have been identified for the oncogenic outcome of KSHV infection. Cases of AIDS-KS and iatrogenic KS indicate that the immune status of the host is critical for the pathogenesis of KSHV infection [38]. However, a more direct role of HIV infection alone in the development of the disease cannot be excluded [65]. Genetic polymorphisms of inflammatory and immune-response genes have been associated with the classic KS risk [38]. The susceptibility factors for the development of KSHV-related diseases should be studied in the same background as that of KSHV infection.

7.2.3 KSHV-Related Diseases

7.2.3.1 Kaposi's Sarcoma

KS can be classified into four subtypes according to geographical distribution and clinical origins, namely, classic KS, endemic KS, iatrogenic KS, and AIDS-related KS [42]. Dr. Moritz Kaposi first described the rare, frequently indolent tumor of the skin in older men of Mediterranean and Eastern European origins currently known as classic KS. Endemic KS, which is more aggressive than classic KS, was first described in the sub-Saharan region in the 1960s. Iatrogenic KS was identified among immunosuppressive patients, such as those undergoing transplantation surgery [66]. AIDS-related KS, the most common subtype, robustly appeared along with the HIV pandemic. These subtypes are histologically indistinguishable regarding their clinical detection. KS lesions are characterized by poorly formed and dilated vascular spaces, where the spindle-shaped cell proliferates. These spindle cells are thought to be the KS tumor cells. The infiltration of inflammatory mononuclear cells including lymphocytes, plasma cells, and some macrophages is consistently observed in KS lesions [67]. The symptoms of KS vary, ranging from indolence to aggressive tumors leading to significant morbidity and mortality. Cutaneous lesions are mostly found in the lower extremities, face, and genitalia and are typically multifocal, with the appearance of papules, patches, plaques, or nodules. Patch lesions correspond to the earliest stage, whereas other forms are characteristic of more advanced disease. However, these clinical manifestations do not necessarily reflect the progression of KS because different lesion forms can appear simultaneously. KS lesions are also found in the oral cavity and internal organs (gastrointestinal tract, lungs, and lymph nodes) [68]. Diagnostic confirmation of KS is done through immunohistochemistry for LANA detection in biopsies.

7.2.3.2 Primary Effusion Lymphoma and Multicentric Castleman's Disease

Primary effusion lymphoma (PEL) is a rare tumor associated with KSHV infection, and it is also known as body cavity-based lymphoma (BCBL). The incidence of this tumor is very low, even in populations with high KSHV seroprevalence (3% of AIDS-related lymphomas and 0.4% of all AIDS-unrelated diffuse large cell NHLs) [38]. PEL is commonly found as lymphomatous effusions in body cavities. Although it is more common in HIV-positive males, HIV-negative men and women can also develop PEL [69]. Diagnostic criteria for PEL have been proposed, including immunoblastic anaplastic large cell morphology, null cell phenotype, and B-cell genotype [70]. The PEL cell is assumed and confirmed at a preterminal stage of B-cell differentiation. Multiple copies of the KSHV genome can be found in PEL cells, ranging from 40 to 80 copies per cell, and the viral particles of KSHV can be isolated from cells derived from PEL, which makes it a useful tool to study the virology of KSHV. It also boosts the serologic assays in the clinical diagnosis. Similar to KS, KSHV infection is necessary but not sufficient for the development of PEL. EBV infection may be also involved in the development of this disease because both viral genomes are found in many types of PEL [38].

Patients with Castleman's disease usually have multiple enlarged lymph nodes, hence the name multicentric Castleman's disease (MCD). Approximately 90% of patients with MCD have the plasma cell-type morphology. The association between KSHV infection and MCD was established shortly after the discovery of KSHV [71]. Both PEL and MCD can originate from KSHV-infected preterminal B cells, although cells in MCD do not undergo the GC reaction, whereas those in PEL do [38]. The symptoms of MCD include fever, malaise, wasting, hypoalbuminemia, cytopenia, and hyponatremia. The systemic symptoms in MCD are related to the excess production of cytokines (both IL-6 and vIL-6). Lytic antigens can be detected in MCD, indicating that abundant lytic viral replication occurs; this makes it a feature of MCD by comparison with other KSHV-related diseases [72].

7.3 KSHV Life Cycle

KSHV can successfully infect multiple cells including endothelial cells, B cells, monocytes, epithelial cells, and keratinocytes to establish latency [71]. KSHV can also infect human and mouse fibroblast cells, owl monkey kidney cells, BHK-21 (baby hamster kidney) cells, and CHO (Chinese hamster ovary) cells [73, 74]. However, KSHV does not infect rodents in vivo, and there are no small or primate animal models that can imitate human KSHV infection and pathogenesis. KSHV infection includes three steps: (1) manipulation of the host signaling pathways to enter and traffic in the cytoplasm of target cells; (2) delivery of the viral genome into the nucleus; and (3) initiation of viral gene expression for successful de novo infection. KSHV entry is the fundamental as well as key process for de novo infection.

7.3.1 Virus Entry

KSHV entry is a complex multistep process involving viral envelope glycoproteins as well as a variety of cell surface molecules that are utilized by KSHV for its binding and entry.

Because KSHV displays a broad cell tropism, KSHV uses a cell type-specific approach for entry [75]. For example, KSHV enters human B cells, fibroblasts, epithelial cells, and endothelial cells by endocytosis. Specifically, KSHV enters human dermal microvascular endothelial cells (HMVEC-D) and human foreskin fibroblasts (HFFs) by macropinocytosis and clathrin-mediated endocytosis, respectively [76–80]. Macropinocytosis, a specific form of endocytosis, is the major route of KSHV infection in endothelial HMVEC and HUVEC cells. Clathrin-mediated endocytosis is another major route involving the uptake of KSHV into the cell from the surface via clathrin-coated vesicles.

KSHV entry is initiated by the binding of KSHV-encoded glycoproteins to receptors in the host cell membrane. KSHV encodes several glycoproteins including gB (ORF8), gH (ORF22), gL (ORF47), gM (ORF39), and gN (ORF53), which are conserved among herpesviruses [81, 82]. KSHV also encodes certain unique lytic cycle-associated glycoproteins such as ORF4, gpK8.1A, gpK8.1B, K1, K14, and K15, among which ORF4 and gpK8.1A are associated with the KSHV envelope [16, 17, 83]. These glycoproteins are important for the following processes: virus-host cell initial attachment, virus entry, viral particle assembly, and virus egress. Among these glycoproteins, KSHV gB is the most vital envelope glycoprotein involved in the initiation of entry [84].

Similar to other herpesviruses, there are two categories of host cellular receptors that can be recognized by KSHV glycoproteins. One is the host cellular binding receptor heparan sulfate (HS), which promotes KSHV attachment and concentration in target cells [74]. The other one is the entry receptor that is highly specific according to cellular tropism as well as the entry pathways utilized by KSHV. Integrins, DC-SIGN, xCT, and ephrin receptor A2 (EphA2) are all host cellular entry receptors for KSHV. Integrins are entry receptors utilized by KSHV in adherent cells such as endothelial cells, fibroblasts, and monocytes [85]. DC-SIGN is a dendritic cell specific entry receptor that is expressed on the DC cell surface and is exploited by KSHV in human myeloid dendritic cells, macrophages, and B cells during infection [86]. xCT is a fusion-entry receptor for KSHV infection in HMVEC cells [87]. EphA2 is the entry receptor utilized by KSHV in HFFs [88]. The interaction between KSHV glycoproteins and host cellular receptors not only promotes viral entry but also triggers host cell signaling pathways that may modulate the cellular microenvironment to enhance virus entry and infection.

During early infection, focal adhesion kinase (FAK) and Src are induced to facilitate KSHV entry and infection. FAK is an important tyrosine kinase activated by the KSHV-integrin interaction, inducing multiple biological processes including cell adhesion, proliferation, migration, apoptosis, and endocytosis [89]. KSHV glycoprotein gB can phosphorylate FAK, leading to the assembly of a membrane-bound signaling complex and linking other kinases to downstream signaling events, facilitating KSHV entry [90]. Phosphorylated FAK associates with other kinases such as Src and RhoA to facilitate KSHV entry and infection. Moreover, Src kinases are also critical for the endocytosis of KSHV [91].

During early infection, phosphoinositide 3-kinase (PI3K) interacts with RhoA-GTP to assist KSHV entry and infection. PI3K is involved in KSHV entry as a signal mediator, which can be activated via the interaction between KSHV and the cellular receptor EphA2. This results in signaling to downstream RhoA GTPases and other effectors to promote endosome formation and endosome trafficking during KSHV entry [92]. The RhoA GTPase pathway, which is induced by KSHV gB through the FAK-Src-PI3K pathway, is a vital signaling pathway regulating endocytosis of KSHV. Recent reports showed that reactive oxygen species (ROS) generated by KSHV have a significant impact on the entry of KSHV by amplifying the initial host signal including EphA2, FAK, Src, and Rho GTPase [93].

In conclusion, the coordinated activities of these proteins play an important role in regulating the mechanism of KSHV entry and infection.

7.3.2 Intracellular Trafficking

KSHV enters the host cell cytosol and delivers its genome in infected cell nuclei as early as 15 min postinfection, and trafficking of KSHV DNA to the nucleus is maximal at 90 min postinfection, suggesting that KSHV trafficking is a very rapid process [94]. During KSHV infection, Rho GTPase is involved in microtubule acetylation and aggregation, which can increase the nuclear delivery of the KSHV genome. Moreover, Rho GTPase utilizes dynein proteins (Dia-2) to rearrange the cytoskeleton, leading to acetylation and aggregation of microtubules. Rho GTPase is activated by KSHV infection targeting the FAK-Src-PI3K signaling pathway [94]. Therefore, KSHV infection-induced Rho GTPase plays an important role in facilitating not only virus entry but also the nuclear delivery of viral DNA.

The endosomal sorting complexes required for transport (ESCRT) proteins, including ESCRT-0, -I, -II, and -III, function in a sequential manner to mediate endosomal trafficking with the VPS4 ATPase. A recent study showed that the ESCRT-I complex protein Tsg101 plays an important role during KSHV trafficking. Small interfering RNA assays showed that Tsg101 does not affect KSHV entry but dramatically reduces the delivery of the KSHV genome to the nucleus [95, 96].

7.3.3 Viral and Host Gene Expression

To overcome the host cell restrictions on virus entry and viral gene expression, KSHV induces robust expression of ERK1/2, nuclear factor κB (NF-κB), and nuclear factor E2-related factor 2 (Nrf2) transcriptional factors early during de novo infection by interacting with the cell surface receptors [97]. For example, Nrf2, as a transcription factor involved in the establishment of de novo KSHV infection, is an important host factor and plays a crucial role in viral gene expression. The inhibition or knockdown of Nrf2 with the chemical brusatol blocks viral gene expression [96, 98].

In conclusion, KSHV has evolved with a prominent survival strategy that reflects the biological complexity of the virus and host interactions [97, 99]. KSHV cell entry involves a sequence of events: (1) macropinocytosis and clathrin-mediated endocytosis facilitate the rapid entry of viral particles into different cell types, (2) modulation of the various host cell functions to enable KSHV trafficking from the cytoplasm into the nucleus, and (3) induction of cytoplasmic ERK1/2, NF-κB, and Nrf2 transcription factors early during infection to initiate viral gene expression soon after the entry of the viral genome into the nucleus (Fig. 7.1). These data provide crucial information for the design of future drugs that can efficiently inhibit the entry of KSHV.

7.4 Two Different Phases of the KSHV Life Cycle

Similar to other human herpesviruses, KSHV has two different phases of infection referred to as persistent latent infection and transient lytic reactivation, which are distinguished by their viral gene expression patterns [100].

7.4.1 Latent Phase and Latent-Associated Proteins

During latent infection, viral gene expression is highly restricted to limit host immune responses while promoting cell survival. The latent viral genome is replicated as a circular episome within the nucleus through host cellular DNA

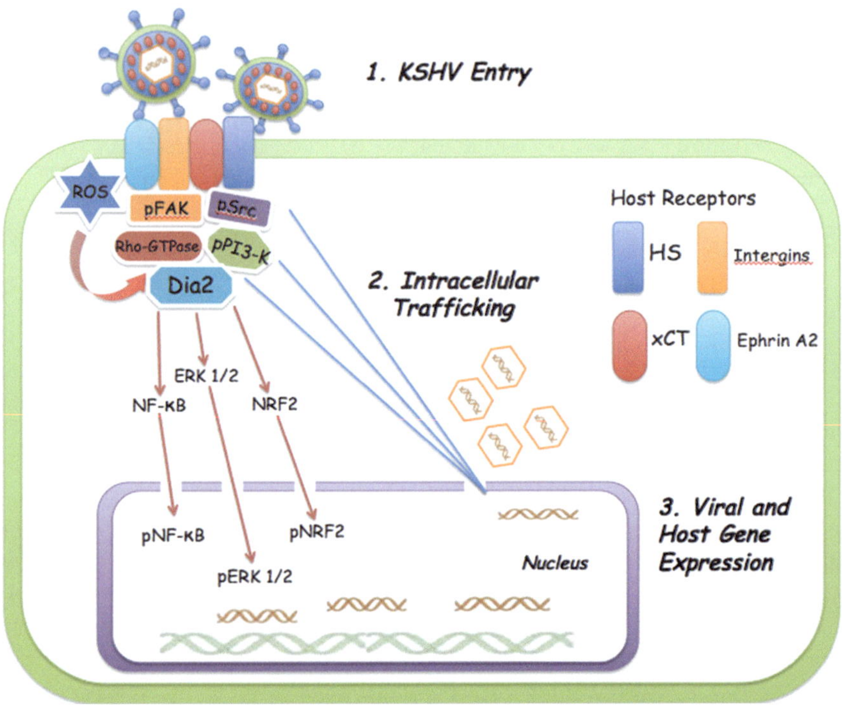

Fig. 7.1 Diagram demonstrating the sequence of events in the processes of KSHV entry and intracellular trafficking. (*1*) KSHV utilizes macropinocytosis and clathrin-mediated endocytosis to facilitate the rapid entry of viral particles into different cell types; (*2*) KSHV modulates the diverse host cellular proteins for its trafficking from the cytoplasm to the nucleus; (*3*) KSHV induces the cytoplasmic ERK1/2, NF-κB, and Nrf2 transcription factors very early during infection to initiate viral gene expression soon after the entry of the viral genome into the nucleus

polymerase, and the replicated genomes are evenly distributed to the daughter cells. Latency is the default pathway of KSHV infection, and although no infectious virions are produced during this stage, it retains the potential for virion production [22, 100]. It is commonly accepted that ORF73/LANA, ORF72/v-cyclin, ORF71/vFLIP, and the K12/kaposin family of proteins (kaposin A, B and C) are latent genes, as are the 13 pre-miRNAs that can yield approximately 25 miRNAs [101, 102].

ORF73/LANA, a major KSHV-encoded latent protein, is located in the nucleus of latently infected cells and has been detected in PEL and KS tumor cells in vivo, in B cells in MCD, and in every latent cell type infected in vitro [101, 103, 104]. LANA is a multifunctional protein consisting of 1,162 amino acids, and it is approximately 220–230 kDa in size. LANA has three major domains. The C-terminal domain binds directly to the conserved TR sequences of the KSHV genome, the N-terminal domain docks onto the host chromosome to hitch a ride during host mitosis to maintain a stable copy number in the latently infected cells, and the central region includes highly acidic amino acid repeats. One of the most critical functions of LANA is maintaining efficient segregation of the viral genomes as a circular episome from generation to generation [105, 106]. The C-terminal DNA-binding domain of LANA interacts with the host TR sequences to initiate semiconservative replication by recruiting the host cell origin recognition complex (ORC) and mini-chromosome maintenance (MCM) proteins, whereas the N-terminal chromosomal binding domain enhances this latent replication process [107]. LANA has also been reported to bind to several viral promoters and suppress viral lytic gene transcription to maintain the latent process. For example, LANA can inactivate the intracellular domain of Notch (ICN), which mediates transactivation of ORF50/replication and transcription activator (RTA) and interacts with RBP-Jk, located in the promoter of RTA, to repress the function of RTA, thereby preventing lytic reactivation [108]. In addition to maintaining the KSHV latent genome, LANA also binds and interacts with multiple cellular proteins, such as the tumor suppressors Rb and p53 to partly inactivate their functions [109]. Meanwhile, LANA can impact host gene expression by interacting with certain transcription factors including ATF4/CREB2 and CPB [110]. In conclusion, LANA is a multifunctional protein that plays a central role in the establishment and maintenance of viral latency.

v-Cyclin, the product of ORF72, shares 54% homology and 32% identity with cellular cyclin D2, which binds and activates cellular cyclin-dependent kinase 6 (CDK6) to regulate cell cycle progression and proliferation [111, 112]. v-Cyclin forms a complex with CDK6 to mediate Rb phosphorylation and activation of p27 and histone H1 [113]. Although the exact role of this viral protein in the regulation of KSHV latency is not fully understood, studies indicate that v-cyclin can interact with CDK6, and the v-cyclin-CDK6 complex participates in mediating the phosphorylation of nucleophosmin (NPM), promoting the interaction between NPM and LANA and the recruitment of HDAC1 to maintain KSHV latency [114]. In addition, v-cyclin might induce apoptosis through the inactivation of the anti-apoptotic factor BCL2, and the expression of v-cyclin is low during latency, which prevents KSHV-triggered apoptosis [115, 116]. In a similar functional relationship to that of murine gammaherpesvirus 68 (MHV68) v-cyclin, KSHV v-cyclin modulates the latent-lytic switch [117].

ORF71/vFLIP, also referred to as K13, is homologous with cellular FLICE [Fas-associated death domain (FADD)-like interleukin-1 beta-converting enzyme, now called caspase-8] [118, 119]. vFLIP activates the NF-κB pathway through two approaches: direct upregulation of the antiapoptotic transcription factor NF-κB and binding to the inhibitor of NEMO (also referred to as IKK-gamma) [120, 121]. NF-κB activation can hinder lytic gene expression, whereas NF-κB inactivation can induce lytic reactivation. Therefore, vFLIP plays a critical role in maintaining KSHV latency and promoting cell proliferation and survival [122]. Moreover, vFLIP can also activate the JNK signal pathway by binding to RIP and TRAF2 upstream of IKK [123].

Kaposin, also known as K12, is composed of at least three proteins named kaposin A, B, and C, which show differential translation initiation [104]. Kaposin A is located in intracellular membranes and the plasma membrane and has the potential to transform rodent fibroblasts, and the resulting cell lines form tumors in nude mice. Kaposin B is a small soluble nuclear protein that can bind and activate

MAP kinase-associated protein kinase 2 (MK2) and the upstream kinase, p38 MAP kinase [124].

miRNAs are 21–23 nucleotide long RNAs that regulate gene expression through base pairing to their mRNA targets. KSHV encodes 13 pre-miRNAs designated as miR-K1 to miR-K12, which generate 25 mature miRNAs that are highly conserved and expressed in all latently infected cells of KS and PEL [37, 125, 126]. Several putative functions for KSHV miRNAs have been proposed since their discovery [127–129]. Deletion of most KSHV miRNAs stimulates KSHV reactivation, suggesting that these miRNAs play an important role in maintaining KSHV latent infection. miR-K1 represses the activation of lytic viral promoters, miR-K12-7-5p target viral ORF50 to stabilize latency [130], and miR-K10 inhibits cell apoptosis by hindering pro-inflammatory responses [131]. Similar to cellular miRNAs, these KSHV-encoded miRNAs have an impact on the development of viral malignancies by affecting the differentiation status of the KSHV-infected cell [132]. Because miRNAs are expressed in latency, they may promote cell survival and regulate oncogenesis and the aberrant angiogenesis phenotype of KS [133, 134].

7.4.2 Lytic Phase and Lytic-Associated Proteins

At 12 h postinfection, KSHV-infected cells display both latent and lytic gene expression. At approximately 24–48 h postinfection, KSHV-infected cells predominantly express viral latent genes, except for a small fraction (1–3%) of cells that enter the lytic replication phase [135]. However, latently infected cells have the potential to undergo lytic reactivation, a process that can turn the virus from latency back into lytic replication mode under different exogenous stimuli. The process of lytic reactivation predominantly involves lytic DNA replication and infectious virion particle production. Similar to other herpesviruses, during lytic reactivation, the entire viral genome is expressed in a temporally regulated mode, resulting in the transcriptional activation of three classes of lytic genes named immediate early (IE) genes, early (E) genes, and late (L) genes [136, 137]. Proteins encoded by the three classes of lytic genes assist in the assembly and release of infectious as well as mature virion particles that egress out of the cell by destroying the cellular membrane [100].

The IE genes consist of ORF50/RTA, ORF45, K8alpha, K8.2, K4.2, K4.1, K4, ORF48, ORF29b, K3, and ORF70, and they are primarily expressed within 10 h of induction. The proteins encoded by the IE genes are involved in gene transcription and cellular modifications for the purpose of viral replication [138]. The E genes, including K8, K5, K2, K12, ORF6, ORF57, ORF74, K9, ORF59, ORF37, K1, K8.1A, ORF21, K2/vIL6, PAN RNA, vIRF1, and ORF65, are expressed between 10 and 24 h post-induction and encode proteins primarily required for DNA replication and gene expression [139]. The L genes, which are expressed after 48 h post-induction, encode viral structural proteins such as membrane glycoproteins (gB and K8.1) and a small viral capsid protein involved in the process of assembly and maturation of the virion particles [140].

The RTA protein (691 amino acids and 110 kDa) has an N-terminal DNA-binding domain and a C-terminal activation domain that can be phosphorylated during lytic reactivation, and it is regarded as both an initiator and a controller of KSHV lytic DNA replication [141]. RTA acts as the latent-lytic switch that induces KSHV lytic reactivation by transcriptionally activating its own promoter. Moreover, RTA transcriptionally activates at least 19 genes through direct binding to their promoters, including ORF8, K4.1, K5, PAN, ORF16, ORF29, ORF45, ORF50, K8, K10.1, ORF59, K12, LANA, K14/vGPCR, K15, the two origins of lytic replication OriLyt-L and OriLyt-R, and the miRNA cluster, inducing latent cells to disrupt latency and complete the lytic cascade. Among these 19 genes, the highest activation by ORF50 is observed in the PAN promoter, which drives the transcription of an abundant noncoding PAN RNA [142]. Direct DNA binding by RTA is not the only approach by which RTA transactivates promoters. RTA can also target other promoters by interacting with RBP-Jk (known as a transcriptional repressor), which can be converted to an activator via the activation domain of RTA [143]. Recent research reported a new mechanism by which KSHV RTA activates the Notch signaling pathway in neighboring cells to inhibit lytic gene expression, which maintains these cells in the latent phase [144].

In conclusion, KSHV lytic reactivation is a complex process that involves a combination of both viral and cellular factors. RTA plays a pivotal role during this process. Lytic reactivation is not only important for infectious virus production but also fundamental for tumor growth, as shown in animal models [145].

7.4.3 Switch Between Latent and Lytic Proteins

LANA and RTA, the two major proteins of KSHV, interact with each other and control the switch between latency and lytic reactivation by targeting the RBP-Jκ effector protein, which is a major transcriptional repressor of the Notch signaling pathway. RTA can transactivate several viral genes, inducing lytic reactivation of latent cells. Therefore, regulation of the expression and function of RTA is vital for the latent-lytic balance and the fate of infected cells. LANA interacts with the RBP-Jκ effector protein to repress the promoter of RTA. LANA competes with RTA for binding to RBP-Jκ, inhibiting RTA self-activation and maintaining the cells in latency. In addition, the RTA protein can activate the expression of LANA by interacting with the RBP-Jκ effector protein on the promoter of LANA during de novo infection, contributing to the establishment of KSHV latency [108, 146]. Therefore, the interaction between LANA and RTA proteins in KSHV-infected cells controls the molecular transition between latency and lytic reactivation (Fig. 7.2).

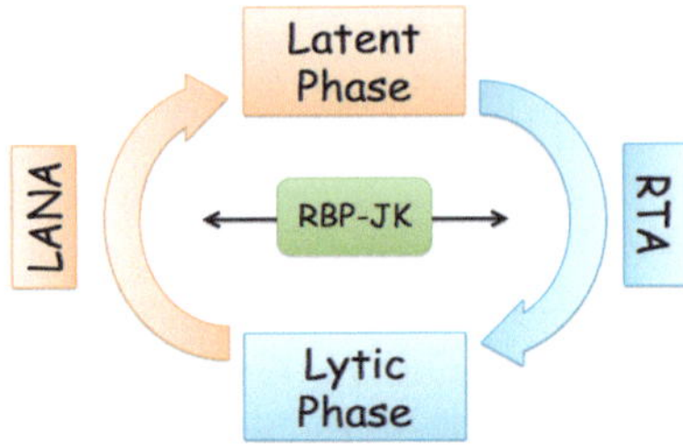

Fig. 7.2 Model for the switch of the latent phase and lytic phase. The model for the switch of the latent phase and lytic phase: LANA and RTA control the switch between latency and lytic reactivation through targeting of the RBP-Jκ effector protein

7.5 KSHV-Related Pathogenesis

7.5.1 An Overview of KSHV-Related Pathogenesis

In established KS lesions, spindle cells account for most of the cell mass and are considered as the driving force of KS; however, KS also contains heterogeneous cell types including B cells, plasma cells, T cells, and monocytes, which orchestrate an inflammatory microenvironment. Interestingly, KS is not formed by a monotonous clonal outgrowth of mesenchymal cells as traditional cancers. In fact, KS lesions progress in a stepwise manner with unique features. Considerable angiogenesis and inflammatory cell infiltration are present at the initial stages. The patched lesions are reddish but flat. Despite lack of in vivo data of de novo KSHV infection, it is speculated that KSHV infects endothelial cells from unknown sources, either the blood vascular system or lymphatic vascular system. Initial infection of endothelial cells promotes endothelial cell proliferation and differentiation, which leads to neovascularization [147]. Whether KSHV infects lymphocytes and endothelial cells simultaneously remains unclear. It is also possible that newly formed vessels recruit lymphocytes to the infection sites and KSHV subsequently infects the recruited cells. The subsequent stage is the plaque stage, as the infected cells constantly undergo excessive proliferation as well as endothelial-to-mesenchymal transition [148] and are transformed into spindle cells. The lesions become more edematous and continue to progress toward the next stage, the nodular stage, which is featured by visible tumor masses. In this stage, the tumor consists of mainly spindle cells. The spindle cells express multiple endothelial markers such as CD31, CD34, and CD36, indicative of their endothelial origin [149, 150]. Although a lymphatic endothelial origin of KS spindle cells has been suggested [151–154], it remains controversial which endothelial cell type, the lymphatic endothelial cell (LEC) or blood vascular endothelial cell (BEC), is primarily infected by KSHV. However, we may not be able to tease apart one from the other, as HSHV infection tends to alter the transcription patterns of the terminally differentiated cells to a mixed expression of both LEC and BEC markers [152, 155, 156].

KSHV infection induces spindle cell morphology in endothelial cells, as evidenced by extensive in vitro data and in vivo observations in the past two decades [152, 155, 157–159]. These spindle cells harbor an incompletely transformed phenotype characterized by excessive proliferation [160–162] and loss of contact inhibition [158, 163, 164], although they do not necessarily induce malignancy. This differs dramatically from real cancers. Instead, KSHV-infected spindle cells may promote the onset of KS, PEL, and MCD in the context of an appropriate environment. Over 95% of healthy people who are infected with KSHV have no symptoms during their lifetime, whereas those who become immunosuppressed display severe symptoms or even death, suggesting that the suppressed immune system is one of the most important factors synergizing with KSHV to induce tumorigenesis. Another essential element is inflammation, as KS lesions occur at sites of injury, to which many inflammatory cells and cytokines are recruited [165]. As discussed above, angiogenesis starts early during KS pathogenesis and persist throughout KS progression, indicating the presence of a third key element. Therefore, immune escape, inflammation, and angiogenesis act together to establish a permissive condition that allows KSHV to function as an oncogenic virus.

7.5.2 KSHV-Induced Angiogenesis and Lymphangiogenesis

The transportation system in the human body is supported by two tubular networks, blood vessels and lymphatic vessels, which are interconnected but relatively independent [166]. The vessels are formed by endothelial cells. KS is highly vascularized. KSHV directly infects endothelial cells, which may facilitate the neovascularization process. On the other hand, proangiogenic cytokines secreted by KSHV-infected cells induce vessel formation in a paracrine manner.

The first key factor shown to induce angiogenesis in KS is basic fibroblast growth factor (bFGF) [167–170], which is highly expressed in KS tissues. Since then, increasing evidence supports that KSHV plays an essential role in altering the phenotype of infected cells. The tight junctions between endothelial cells are also disrupted by KSHV through the degradation of VE-cadherin, which initiates angiogenesis [171, 172]. KSHV-infected endothelial cells form vascular tubes on Matrigel, induce capillary morphogenesis in 3D cultures, and display an enhanced invasive capability [148, 173–176]. The vascular endothelial growth factor (VEGF) family and related signaling pathways are required for angiogenesis, which can also be usurped by KSHV for its own benefit.

Many KSHV-encoded proteins participate in these processes. KSHV LANA, which is encoded by ORF73, increases the proliferation and life span of primary human umbilical vein endothelial cells (HUVEC) and maintains its ability to form vessel-like structures in vitro [177]. LANA also interacts with Daxx and interferes with the Daxx/Ets-1 complex, which antagonizes the inhibitory activity of Daxx on VEGF receptor expression, thereby contributing to its high expression in infected cells [178]. Another protein complex that is targeted by LANA is EZH2 via the

NF-κB pathway, and upregulated EZH2 leads to the induction of Ephrin-B2, a well-known proangiogenic factor [179]. In addition, LANA stabilizes Notch signaling and upregulates the Notch downstream factor Hey1 to achieve pathologic angiogenesis [180]. vGPCR (ORF74) activates JNK/SAPK and p38MAPK signaling cascades and induces an angiogenic phenotype in infected cells [181]. VEGF expression can also be enhanced by vGPCR through the stimulation of the activity of hypoxia-inducible factor (HIF)-1 alpha [182]. vIL6 (ORF-K2) cooperates with vGPCR to upregulate angiopoietin-2 (Ang2) through the mitogen-activated protein kinase (MAPK) pathway [183–185], resulting in RAC1 activation, migration, and sprouting angiogenesis [186]. The KSHV K1 protein induces the expression of VEGF and matrix metalloproteinase-9 [187]. K15 recruits and activates PLCγ1 and downstream calcineurin and NFAT1 to induce RCAN1/DSCR1 expression and capillary tube formation [188]. In addition, vFLIP and vGPCR induce the reprogramming of lymphatic endothelial to mesenchymal transition (EndMT), which requires Notch signaling and leads to the morphological differentiation of cells into capillary structures [148]. Therefore, KSHV directly induces the differentiation of infected cells into an angiogenic phenotype (shown in Fig. 7.3).

During KSHV infection, viral or host cytokines are secreted and promote angiogenesis in a paracrine manner. The most important cytokine family is VEGF, the expression of which is highly regulated by KSHV. The VEGF family has five members: VEGF-A, placental growth factor (PGF), VEGF-B, VEGF-C, and VEGF-D. KSHV-conditioned media contain VEGF-A and VEGF-B, which are sufficient to induce the formation of angiogenic capillaries from cultured cells [189]. The mechanism by which KSHV induces VEGF may involve its ability to stabilize HIF-1α [190], which binds the hypoxia-response elements (HRE) in the enhancers of VEGF [191]. Ang1 and Ang2 are ligands that bind to the tyrosine kinase Tie2, and the interaction between the ligands and receptors regulates angiogenesis. In KS, Ang2 is upregulated and plays a role in vascular permeability and angiogenesis [192]. Many other cytokines including interleukin 6 (IL-6) [193], CCL-2 [194], and prostaglandin E2 [195] are induced and secreted by KSHV-infected cells to promote neovascularization. In addition, KSHV encodes several unique cytokine-like factors to stimulate angiogenesis. A viral homolog of interleukin 6 (vIL-6) assists to promote VEGF-A secretion [183]. Viral macrophage inflammatory proteins (vMIPs) encoded by KSHV include three members. vMIP-II upregulates multiple proangiogenic factors including VEGF in vivo [196]. vMIP-III functions as a CCR4 agonist and stimulates angiogenesis [197]. These proangiogenic cytokines act on both infected cells and neighboring uninfected cells to facilitate new vessel formation in a paracrine manner.

In summary, proangiogenic cytokines, which are either host factors induced by viral infection or virus-encoded proteins, and the activation of the corresponding signaling pathways create a permissive milieu that helps induce the formation of new vessels, contributing to the highly vascularized feature of KS.

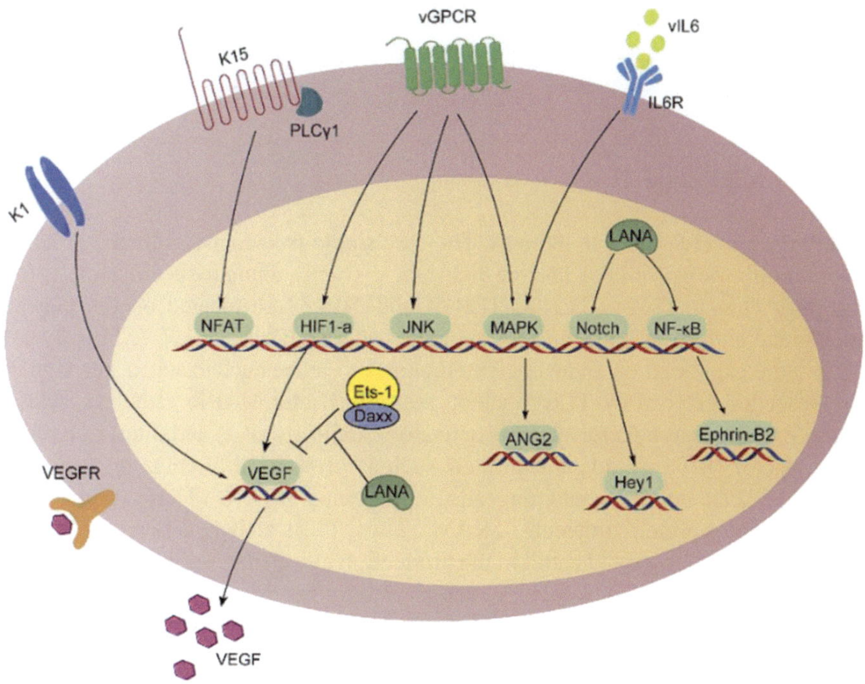

Fig. 7.3 Schematic representation of KSHV protein-induced angiogenesis-related signaling pathway. The KSHV K1 protein induces VEGF expression, and secreted VEGF binds to VEGFR to promote angiogenesis. K15 recruits and activates PLCγ1, which in turn activates downstream NFAT1 to induce RCAN1/DSCR1 expression and capillary tube formation. vGPCR activates JNK/SAPK, p38MAPK, and HIF-1α (which activates VEGF expression), which induce an angiogenic phenotype in infected cells. vIL6 (ORF-K2) cooperates with vGPCR to upregulate angiopoietin-2 (Ang2) through the mitogen-activated protein kinase (MAPK) pathway. LANA promotes angiogenesis in multiple phases: (*1*) LANA stabilizes Notch intracellular domain (ICN) and induces Hey1; (*2*) LANA activates NF-κB and induces proangiogenic factor Ephrin-B2 expression; (*3*) LANA antagonizes the inhibitory activity of Ets-1/Daxx complexes on VEGF and induces VEGF expression

7.5.3 KSHV-Induced Immune Escape and Regulation

During KSHV infection, the host immune system recognizes various pathogen-associated molecular patterns (PAMPs) of KSHV by pattern recognition receptors (PRRs) to clear infection; however, KSHV has developed many ways to counteract host immunosurveillance, which is critical for its survival and related disease progression. PRRs can be classified into four groups based on their localization and function: free receptors in serum, membrane-bound phagocytic receptors, membrane-bound signaling receptors, and cytoplasmic signaling receptors. Among these, four types of PRRs have been extensively studied in the KSHV field: Toll-like receptors (TLRs), nucleotide-binding oligomerization domain (NOD)-like receptors (NLRs), retinoic acid-inducible gene I (RIG-I)-like receptors (RLR), and cytosolic DNA-sensing receptors.

7.5.3.1 KSHV and TLRs

There are ten *TLR* genes in humans. They are single-pass transmembrane proteins containing an extracellular leucine-rich repeat (LRR) domain specific for ligand binding. TLR-1, TLR-2, TLR-4, TLR-5, and TLR-6 are located on the plasma membrane and sense lipids and lipoproteins, whereas TLR-3, TLR-7, TLR-8, and TLR-9 are expressed on endosomal membranes to sense nucleic acids. The signaling cascades induced by TLRs include the NF-κB and MAPK pathways, which activate transcription factors such as activator protein 1 (AP-1) and interferon regulatory factor (IRF) via adaptor proteins such as myeloid differentiation factor 88 (MyD88), TIR domain-containing adaptor-inducing IFN- β (TRIF), and TRIF-related adaptor protein molecule (TRAM). Thus, TLRs activate a diverse range of intracellular responses and result in the production of inflammatory factors, chemotactic factors, and antiviral factors such as interferon (IFN)-α and IFN-β. KSHV manipulates each step of TLR signaling to evade it. KSHV RTA or ORF50 down-regulates TLR2 and TLR4 protein expression and alters their localization on the plasma membrane [198]. The adaptor protein TRIF undergoes RTA-mediated degradation through the ubiquitin-proteasome pathway, which blocks the TLR3 pathway [199]. MyD88 is also targeted by RTA for degradation via direct interaction between RTA and MyD88 [99]. Viral interferon regulatory factors (vIRFs) encoded by KSHV have three members named vIRF-1, vIRF-2, and vIRF-3. vIRF-1 decreases the phosphorylation and subsequent translocation of IRF3 into the nucleus upon TLR3 activation [200].

7.5.3.2 KSHV and NLRs

NLRs consist of another family of PRRs that use LRR scaffold domains to detect pathogen products in the cytoplasm. NLRs contain a nucleotide-binding oligomerization domain (NOD) and an LRR domain close to the carboxyl terminus. Based on the domains near the amino terminus, NLRs are divided into two subfamilies. One subfamily harbors an amino-terminal caspase recruitment domain (CARD) and is represented by NOD1 and NOD2. Upon ligand activation, the NLRs recruit the CARD-containing serine-threonine kinase RIPK2. RIPK2 then activates TAK1 kinase, which activates NF-κB by phosphorylating IKK. Another subfamily has a pyrin domain at its amino terminus and is known as the NLRP family. NLRPs interact with proteins containing other pyrin domains such as PYCARD, which is associated with pro-caspase1 through the CARD domain. The large complex (termed the inflammasome) undergoes autoactivation via proteolytic cleavage to generate active caspase1. Caspase1 cleaves pro-IL1 and pro-IL18 into their mature forms, leading to their secretion to induce inflammation. The pyrin subfamily includes another member known as AIM2. AIM2 replaces the LRR domain with an HIN (H inversion) domain that recognizes double-stranded DNA genomes and triggers caspase1 activation. KSHV blocks NLRs by mimicking the receptor, although it elicits dominant negative activity. KSHV ORF63 functions as a viral homolog of NLRP1 lacking the pyrin domain. ORF63 interacts with NLRP1 and blocks NLRP1/3 to interact with PYCARD, hindering assembly of the inflammasome [201]. In addition to AIM2, IFI16 is a member of the PYHIN subfamily that senses cytosolic dsDNA; instead of escaping recognition, KSHV usurps IFI16 to maintain latency, which is related to viral life cycle regulation [202, 203].

7.5.3.3 KSHV and RLRs

RLRs have an RNA helicase-like domain that can bind viral RNA and two CARD domains at the amino terminus. RLRs interact with the adaptor protein MAVS at the mitochondrial membrane when bound to viral RNA and activate the NF-κB and interferon pathways via IKK and IRF3, respectively. RIG-I/MAVS suppresses both KSHV infection and reactivation, although the underlying mechanism is not completely understood [204]. The KSHV tegument protein ORF64 harbors deubiquitinase (DUB) activity and suppresses K63-linked ubiquitination of RIG-I and RIG-I activation [205], thus counteracting the RLR-induced antiviral interferon pathway.

7.5.3.4 KSHV and the Cytosolic DNA Sensor

Unlike the PYHIN proteins AIM2 and IFI16, which induce inflammasome activation, the cGAMP synthase (cGAS)/cyclic guanosine monophosphate-adenosine monophosphate (cGAMP)/STING pathway mediates the activation of IRF3 and type I interferon responses. KSHV vIRF1 blocks the interaction between STING and TBK1, thereby inhibiting the phosphorylation of STING and concomitant activation [206]. Furthermore, an N-terminally truncated cytoplasmic isoform of LANA forms a complex with cGAS via direct interaction, disturbing cGAS/cGAMP/STING signaling. In addition, the KSHV tegument protein ORF52 has been reported to antagonize host cGAS DNA sensing by directly inhibiting cGAS enzymatic activity [207].

In summary, the proteins encoded by the KSHV genome inhibit the host innate immune system by blocking PRRs and the related pathways or take advantage of the host immunity to establish latency, which promotes KSHV-related disease progression (Fig. 7.4).

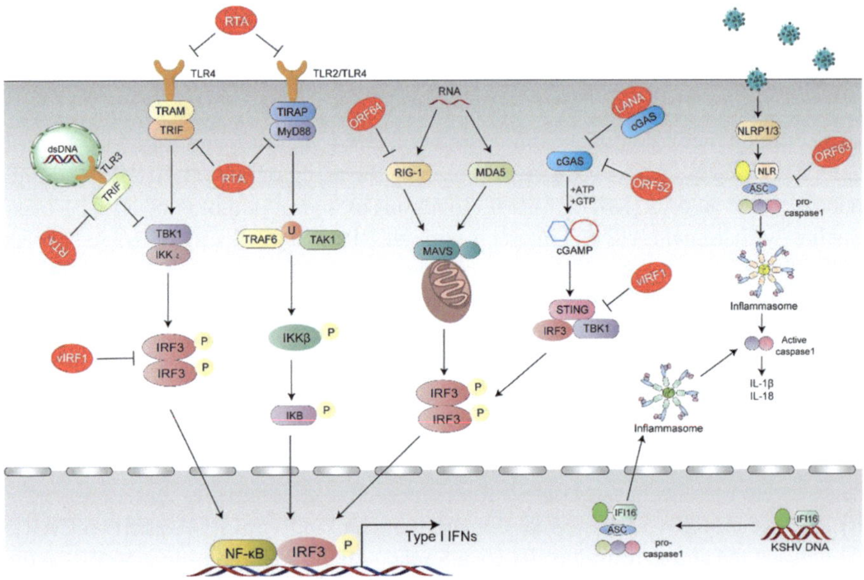

Fig. 7.4 Schematic representation of KSHV-induced immune escape and regulation. TLR signaling uses adaptor proteins such as TRIF and MyD88 to mediate signal transduction. RTA can degrade TRIF and MyD88 via the ubiquitin-proteasome pathway. In addition, RTA downregulates TLR2 and TLR4 protein expression. Upstream of NF-κB and IRF3, the blockade of adaptor and receptors of TLR inhibit pro-inflammatory activity. The activity of the RNA sensor RIG-I can be inhibited by ORF64 through the suppression of K63-linked ubiquitination. The LANA N-terminal truncated isoform can directly bind the DNA sensor cGAS and disturb the cGAS/cGAMP/STING pathway. The NLR pathway leads to IL-1 and IL-18 maturation and secretion. ORF63 serves as a viral homolog of NLRP1 and functions as a dominant negative molecule to block the NLR pathway. vIRF1 also functions as a dominant negative molecule to block IRF3 phosphorylation and translocation; furthermore, vIRF1 blocks STING and TBK1 interaction and subsequent IRF3 activation

7.5.4 KSHV-Induced Cellular Metabolic Alterations

Cancer cells rely on aerobic glycolysis for their proliferation, in contrast to normal cells that generate energy via mitochondrial oxidative phosphorylation. This does not indicate that mitochondria are useless; however, they are indispensable for anabolic activities such as de novo lipid biosynthesis and glutamine-dependent anaplerosis, in what is termed the Warburg effect [208, 209]. Therefore, glucose, fatty acids, and glutamine constitute the cornerstone for tumorigenesis metabolically.

KSHV manipulates these three aspects of metabolism to promote its survival, stress adaptation, and pathogenesis. During latent infection of endothelial cells, KSHV induces a Warburg effect by enhancing aerobic glycolysis and lactic acid production while lowering oxygen consumption. Cells will undergo apoptosis when treated with glycolysis inhibitors, suggesting that this metabolic alteration contributes to cell survival during infection [209]. Mechanistically, the miRNAs encoded by KSHV target and downregulate the HIF prolyl hydroxylase EGLN2 and the heat shock protein HSPA9. The decrease of EGLN2 and HSPA9 stabilizes HIF-1α and reduces mitochondrial numbers [210]. As a downstream effector of HIF-1α, pyruvate kinase2 (PKM2) is upregulated upon KSHV infection [211]. The M2 isoform of pyruvate kinase is characterized by its low affinity for its substrate phosphoenol pyruvate, thereby serving as a rate-limiting checkpoint and facilitating the accumulation of upstream intermediates during glycolysis for cancer anabolic synthesis [209]. Under stress conditions such as nutrient deficiency, however, KSHV miRNAs and vFLIP suppress both aerobic glycolysis and oxidative phosphorylation by downregulating the glucose transporters GLUT1 and GLUT3. The reduced metabolic activities help KSHV adapt to stress and survive in the tumor microenvironment [212]. In a condition opposite to that of excessive glucose provision, the augmented metabolism caused by high glucose induces hydrogen peroxide production, which reduces the expression of silent information regulator 1 (SIRT1). SIRT1 downregulation results in the epigenetic transactivation of KSHV lytic genes and contributes to virus spread and reinfection, promoting KS development [213]. Glutaminolysis is also regulated by KSHV during infection [214]. KSHV induces the Myc/Max and Mondo/Mix heterodimers to upregulate the glutamine transporter SLC1A5, thereby promoting glutamine uptake. Glutamine is required for KSHV-infected cells to supply intermediates for the tri-carboxylic acid (TCA) cycle and macromolecule biosynthesis. SLC1A5 inhibition causes the death of KSHV-infected but not that of mock-infected cells, indicating that the metabolic shift induced by KSHV disturbs the physiological activity of infected cells [214]. With the excessive production of intermediates, KSHV enhances biosynthetic pathways to generate fatty acids and lipid droplet organelles [215], similar to other cancers. Although the KSHV-induced altered metabolism is a relatively new field, growing evidence indicates that KSHV-infected cells display the Warburg effect, which serves as a therapeutic target for treatment (Fig. 7.5).

7.5.5 KSHV-Induced Tumorigenesis

Cancer is characterized by sustained proliferation, resistance against cell death, evasion of growth suppressors, angiogenesis, alterations in metabolism, evasion of immune destruction, and activation of invasion and metastasis [208]. We discussed above how KSHV induces angiogenesis, alters the cell metabolism, and regulates the host immune system. As an established oncogenic virus, KSHV encodes proteins that either activate oncogenes or inhibit tumor suppressors to escape cell death, support proliferation, and help invasion. Here, we discuss further how KSHV transforms cells by targeting multiple host factors.

Two key tumor suppressors, retinoblastoma protein (pRb) and p53 are inactivated by KSHV through multiple ways. pRb prevents excessive cell proliferation by inhibiting cell cycle progression, and pRb inactivation retains host cells in the S

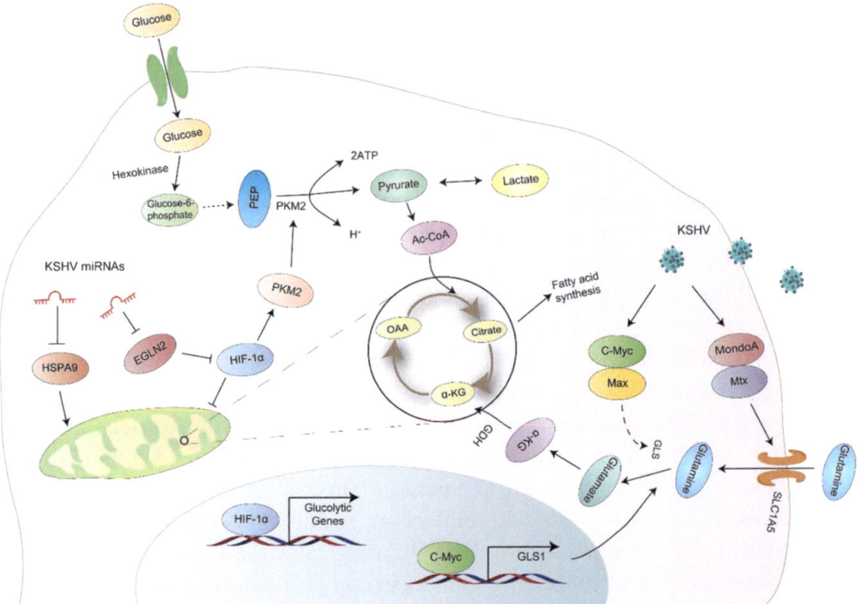

Fig. 7.5 Schematic representation of KSHV-induced cellular metabolic alterations. KSHV-infected cells display the Warburg effect: aerobic glycolysis, reduced mitochondrial oxidative phosphorylation, and glutamine-dependent anaplerosis. KSHV microRNAs target HSPA9, resulting in reduced mitochondrial number and EGLN2, which in turn stabilizes HIF-1α. HIF-1α upregulates pyruvate kinase 2 and delays the generation of Ac-CoA, which enters the TCA cycle. HIF-1α also helps induce glycolytic genes and promotes the glycolysis process. KSHV induces Myc/Max and Mondo/Mix heterodimers to upregulate the glutamine transporter SLC1A5 and facilitate glutamine uptake. c-Myc also induces the expression of glutaminase (GLS1), which converts glutamine to glutamate to be utilized for glutaminolysis

phase. p53 controls genomic stability and prevents genomic mutations, and its inactivation promotes cell immortalization. KSHV LANA interacts with both p53 and pRb, adversely impacting pRb/E2F and p53 transcriptional regulation [216]. LANA also serves as a component of the EC5S ubiquitin complex and targets p53 for degradation [217]. In addition, LANA upregulates Aurora A transcriptional expression, and Aurora A dramatically enhances the binding affinity of LANA for p53, which positively controls LANA-mediated p53 degradation [218]. In addition to LANA, vIRF1 interacts with p53 through its central DNA-binding domain and inhibits the transcriptional activity of p53 [219]. vIRF-3 antagonizes p53 oligomerization and DNA binding by interacting with p53 and inhibiting p53 phosphorylation on serine residues S15 and S20 [220]. K-cyclin, which is uniquely encoded by KSHV, interacts with cyclin-dependent kinase 9 (Cdk9) through its basic domain, which in turn phosphorylates p53 to regulate its function [221]. LANA2 (ORF10.5) inhibits the SUMOylation of p53 by SUMO2, a posttranslational modification of p53 responsible for virus clearance [222]. Furthermore, structural proteins such as ORF22, ORF25, and ORF64 counteract p53-induced apoptosis by suppressing the transactivation of the promoters of p53 target genes [223]. Murine double minute 2 (MDM2), a p53 E3 ubiquitin ligase, targets p53 for ubiquitin-mediated degradation and is bound to KSHV vIRF4 [224]. The interaction leads to reduced MDM2 auto-ubiquitination and stabilizes MDM2. Stabilized MDM2 therefore facilitates proteasome-mediated degradation of p53. In sum, KSHV regulates tumor suppressors such as p53 and pRb at multiple phases.

In parallel with tumor suppressor inactivation, KSHV promotes tumorigenesis by activating oncogenes and related signaling pathways. Two well-known KSHV-encoded oncogenes are LANA and vGPCR. The regulation of p53 by LANA was discussed extensively; however, LANA itself can activate a number of oncogenes. Notch signaling is highly correlated with oncogenesis, and LANA stabilizes intracellular-activated Notch (ICN) by inhibiting the E3 ligase Sel10 [225]. Another critical signaling pathway is mediated by the TGF-β superfamily, which includes TGF-β and BMP. TGF-β inhibits tumorigenesis at an early stage, and KSHV develops multiple mechanisms to repress it. KSHV-encoded miR-K12-11, a homolog of the oncogenic host miR-155, downregulates Smad5 and attenuates TGF-β signaling [129]. On the other hand, LANA silences the TGF-β type II receptor (TbetaRII) by modulating the modification of the TbetaRII promoter [226]. BMP, however, promotes malignancy through its downstream effector inhibitor of DNA-binding (Id) proteins. LANA binds to and sustains BMP-activated p-Smad1 in the nucleus, which leads to aberrant Ids expression and tumorigenesis [227]. vGPCR is involved in KS progression by regulating multiple cellular pathways including MAPK/ERK, PI3K/Src, and the TSC2/mTOR axis [228, 229], which are well-known oncogenic signaling pathways. In addition, vGPCR collaborates with HIV-1 Tat to accelerate KS progression [230], confirming that coinfection with HIV is critical for KSHV-induced pathogenesis.

Collectively, KSHV modulates multiple aspects of cellular activity, which together contribute to KSHV-induced tumorigenesis.

7.5.6 Treatment of KSHV-Related Diseases

Combination antiretroviral therapy (cART) is the most common treatment for HIV-related KS. The incidence of KS has decreased significantly where cART is available. However, cART acts by controlling HIV, as HIV creates a milieu permissive for KSHV to promote KS formation. Regarding KS itself, there is a lack of effective targeted therapeutics. Because it is a highly vascularized tumor, therapeutic approaches that inhibit KS angiogenesis are promising. VEGF-neutralizing antibodies and tyrosine kinase inhibitors show efficacy in KS [230–232]. The cellular signaling pathways hijacked by KSHV to promote tumorigenesis, such as PI3K/Akt/mTOR and Notch, were discussed extensively. Consistent with this, rapamycin (mTOR inhibitor) displays clinical activity against KS [232–234]. Despite the fact that KSHV has been studied for more than two decades, further investigation of its role in disease progression is challenging because of a lack of in vivo animal models. Therefore, everything we learn from the current experimental system may invite misinterpretation. To better develop effective therapeutics, KSHV virology and its related pathogenesis need to be studied in a more physiological system.

7.6 Remarks and Perspectives

In the past two decades, considerable progress has been made in the KSHV field in terms of the mechanisms of viral latency, the functions of viral genes, and the mature experimental systems, such as cell lines for virus production, BAC systems for genetic manipulation, and cell transformation models. However, our understanding of KSHV is still hampered by several difficulties, such as limitations in human cellular transformation models and a lack of animal models for the study of viral infection and oncogenesis. Although the incidence of AIDS-KS has been reduced by the effective control and treatment of AIDS, KSHV infection and its related diseases remain a serious threat to AIDS patients in developing countries and people in KS endemic regions, such as Central Africa and Xinjiang, China. Available treatments are very limited and not curative for these diseases. Meanwhile, there is no available vaccine for the prevention of KSHV infection, and the prevalence rate of KSHV-related diseases remains high in endemic areas. With the wide application of new technologies in the KSHV field, such as gene editing and targeted immune therapy, we believe that the study of KSHV will shed light on the control of viral infection and facilitate the design of targeted therapies for KSHV-related malignancies.

Acknowledgments This work was supported by grants from the Natural Science Foundation for Distinguished Young Scholar (81425017), the Ministry of Science and Technology of China (2016YFA0502100), the Key Project of the National Natural Science Foundation of China (81230037), and the National Institutes of Health (1R01AI116442) to Ke Lan.

References

1. Safai B, Good RA (1981) Kaposi's sarcoma: a review and recent developments. CA Cancer J Clin 31(1):2–12
2. Friedman-Kien AE (1981) Disseminated Kaposi's sarcoma syndrome in young homosexual men. J Am Acad Dermatol 5(4):468–471
3. Wang X et al (2010) Human herpesvirus-8 in northwestern China: epidemiology and characterization among blood donors. Virol J 7:62
4. Dilnur P et al (2001) Classic type of Kaposi's sarcoma and human herpesvirus 8 infection in Xinjiang, China. Pathol Int 51(11):845–852
5. Giraldo G, Beth E, Haguenau F (1972) Herpes-type virus particles in tissue culture of Kaposi's sarcoma from different geographic regions. J Natl Cancer Inst 49(6):1509–1526
6. Walter PR et al (1984) Kaposi's sarcoma: presence of herpes-type virus particles in a tumor specimen. Hum Pathol 15(12):1145–1146

7. Beral V et al (1990) Kaposi's sarcoma among persons with AIDS: a sexually transmitted infection? Lancet 335(8682):123–128
8. Chang Y et al (1994) Identification of herpesvirus-like DNA sequences in AIDS-associated Kaposi's sarcoma. Science 266(5192):1865–1869
9. Wu L et al (2000) Three-dimensional structure of the human herpesvirus 8 capsid. J Virol 74(20):9646–9654
10. Nealon K et al (2001) Lytic replication of Kaposi's sarcoma-associated herpesvirus results in the formation of multiple capsid species: isolation and molecular characterization of A, B, and C capsids from a gammaherpesvirus. J Virol 75(6):2866–2878
11. Lo P et al (2003) Three-dimensional localization of pORF65 in Kaposi's sarcoma-associated herpesvirus capsid. J Virol 77(7):4291–4297
12. Perkins EM et al (2008) Small capsid protein pORF65 is essential for assembly of Kaposi's sarcoma-associated herpesvirus capsids. J Virol 82(14):7201–7211
13. Sathish N, Yuan Y (2010) Functional characterization of Kaposi's sarcoma-associated herpesvirus small capsid protein by bacterial artificial chromosome-based mutagenesis. Virology 407(2):306–318
14. Deng B et al (2008) Cryo-electron tomography of Kaposi's sarcoma-associated herpesvirus capsids reveals dynamic scaffolding structures essential to capsid assembly and maturation. J Struct Biol 161(3):419–427
15. Deng B et al (2007) Direct visualization of the putative portal in the Kaposi's sarcoma-associated herpesvirus capsid by cryoelectron tomography. J Virol 81(7):3640–3644
16. Akula SM et al (2001) Human herpesvirus 8 envelope-associated glycoprotein B interacts with heparan sulfate-like moieties. Virology 284(2):235–249
17. Baghian A et al (2000) Glycoprotein B of human herpesvirus 8 is a component of the virion in a cleaved form composed of amino- and carboxyl-terminal fragments. Virology 269(1):18–25
18. Koyano S et al (2003) Glycoproteins M and N of human herpesvirus 8 form a complex and inhibit cell fusion. J Gen Virol 84(Pt 6):1485–1491
19. Naranatt PP, Akula SM, Chandran B (2002) Characterization of gamma2-human herpesvirus-8 glycoproteins gH and gL. Arch Virol 147(7):1349–1370
20. Wang FZ et al (2001) Human herpesvirus 8 envelope glycoprotein K8.1A interaction with the target cells involves heparan sulfate. J Virol 75(16):7517–7527
21. Zhu FX et al (2005) Virion proteins of Kaposi's sarcoma-associated herpesvirus. J Virol 79(2):800–811
22. Bechtel JT, Winant RC, Ganem D (2005) Host and viral proteins in the virion of Kaposi's sarcoma-associated herpesvirus. J Virol 79(8):4952–4964
23. Sathish N, Wang X, Yuan Y (2012) Tegument proteins of Kaposi's sarcoma-associated herpesvirus and related gamma-herpesviruses. Front Microbiol 3:98
24. Vittone V et al (2005) Determination of interactions between tegument proteins of herpes simplex virus type 1. J Virol 79(15):9566–9571
25. Zhou ZH et al (1999) Visualization of tegument-capsid interactions and DNA in intact herpes simplex virus type 1 virions. J Virol 73(4):3210–3218
26. Chen DH et al (1999) Three-dimensional visualization of tegument/capsid interactions in the intact human cytomegalovirus. Virology 260(1):10–16
27. Dai W et al (2008) Unique structures in a tumor herpesvirus revealed by cryo-electron tomography and microscopy. J Struct Biol 161(3):428–438
28. Dai X et al (2014) Organization of capsid-associated tegument components in Kaposi's sarcoma-associated herpesvirus. J Virol 88(21):12694–12702
29. Russo JJ et al (1996) Nucleotide sequence of the Kaposi sarcoma-associated herpesvirus (HHV8). Proc Natl Acad Sci U S A 93(25):14862–14867
30. Coscoy L (2007) Immune evasion by Kaposi's sarcoma-associated herpesvirus. Nat Rev Immunol 7(5):391–401

31. Lagunoff M, Ganem D (1997) The structure and coding organization of the genomic termini of Kaposi's sarcoma-associated herpesvirus. Virology 236(1):147–154
32. Moore PS, Chang Y (2001) Molecular virology of Kaposi's sarcoma-associated herpesvirus. Philos Trans R Soc Lond Ser B Biol Sci 356(1408):499–516
33. Wen KW, Damania B (2010) Kaposi sarcoma-associated herpesvirus (KSHV): molecular biology and oncogenesis. Cancer Lett 289(2):140–150
34. Chandriani S, Xu Y, Ganem D (2010) The lytic transcriptome of Kaposi's sarcoma-associated herpesvirus reveals extensive transcription of noncoding regions, including regions antisense to important genes. J Virol 84(16):7934–7942
35. Gottwein E (2012) Kaposi's sarcoma-associated herpesvirus microRNAs. Front Microbiol 3:165
36. Sun R et al (1996) Polyadenylated nuclear RNA encoded by Kaposi sarcoma-associated herpesvirus. Proc Natl Acad Sci U S A 93(21):11883–11888
37. Piedade D, Azevedo-Pereira J (2016) The role of microRNAs in the pathogenesis of herpesvirus infection. Virus 8(6):156
38. Fields BN,KD, Howley PM (2013) Fields virology, 6th edn. Wolters Kluwer Health/Lippincott Williams & Wilkins, Philadelphia
39. Perna AM et al (2000) Antibodies to human herpes virus type 8 (HHV8) in general population and in individuals at risk for sexually transmitted diseases in western Sicily. Int J Epidemiol 29(1):175–179
40. Viviano E et al (1997) Human herpesvirus type 8 DNA sequences in biological samples of HIV-positive and negative individuals in Sicily. AIDS 11(5):607–612
41. Whitby D et al (1995) Detection of Kaposi sarcoma associated herpesvirus in peripheral blood of HIV-infected individuals and progression to Kaposi's sarcoma. Lancet 346(8978):799–802
42. Mesri EA, Cesarman E, Boshoff C (2010) Kaposi's sarcoma and its associated herpesvirus. Nat Rev Cancer 10(10):707–719
43. Moore PS, Chang Y (1995) Detection of herpesvirus-like DNA sequences in Kaposi's sarcoma in patients with and without HIV infection. N Engl J Med 332(18):1181–1185
44. Nicholas J et al (1998) Novel organizational features, captured cellular genes, and strain variability within the genome of KSHV/HHV8. J Natl Cancer Inst Monogr 23:79–88
45. Poole LJ et al (1999) Comparison of genetic variability at multiple loci across the genomes of the major subtypes of Kaposi's sarcoma-associated herpesvirus reveals evidence for recombination and for two distinct types of open reading frame K15 alleles at the right-hand end. J Virol 73(8):6646–6660
46. Zong JC et al (1999) High-level variability in the ORF-K1 membrane protein gene at the left end of the Kaposi's sarcoma-associated herpesvirus genome defines four major virus subtypes and multiple variants or clades in different human populations. J Virol 73(5):4156–4170
47. Hughes AL, Hughes MA (2007) Nucleotide substitution at the highly polymorphic K1 locus of human herpesvirus 8 (Kaposi's sarcoma-associated herpesvirus). Infect Genet Evol 7(1):110–115
48. Zong J et al (2002) Genotypic analysis at multiple loci across Kaposi's sarcoma herpesvirus (KSHV) DNA molecules: clustering patterns, novel variants and chimerism. J Clin Virol 23(3):119–148
49. Ramos da Silva S et al (2011) KSHV genotypes A and C are more frequent in Kaposi sarcoma lesions from Brazilian patients with and without HIV infection, respectively. Cancer Lett 301(1):85–94
50. Ramos-da-Silva S et al (2006) Kaposi's sarcoma-associated herpesvirus infection and Kaposi's sarcoma in Brazil. Braz J Med Biol Res 39(5):573–580
51. Ouyang X et al (2014) Genotypic analysis of Kaposi's sarcoma-associated herpesvirus from patients with Kaposi's sarcoma in Xinjiang, China. Viruses 6(12):4800–4810
52. Nsubuga MM et al (2008) Human herpesvirus 8 load and progression of AIDS-related Kaposi sarcoma lesions. Cancer Lett 263(2):182–188
53. Gogineni E et al (2013) Quantitative determinations of anti-Kaposi sarcoma-associated herpesvirus antibody levels in men who have sex with men. Diagn Microbiol Infect Dis 76(1):56–60

54. Hladik W et al (2006) Transmission of human herpesvirus 8 by blood transfusion. N Engl J Med 355(13):1331–1338
55. Aluigi MG et al (1996) KSHV sequences in biopsies and cultured spindle cells of epidemic, iatrogenic and Mediterranean forms of Kaposi's sarcoma. Res Virol 147(5):267–275
56. Kasolo FC, Mpabalwani E, Gompels UA (1997) Infection with AIDS-related herpesviruses in human immunodeficiency virus-negative infants and endemic childhood Kaposi's sarcoma in Africa. J Gen Virol 78(Pt 4):847–855
57. Minhas V, Wood C (2014) Epidemiology and transmission of Kaposi's sarcoma-associated herpesvirus. Virus 6(11):4178–4194
58. Olsen SJ et al (1998) Increasing Kaposi's sarcoma-associated herpesvirus seroprevalence with age in a highly Kaposi's sarcoma endemic region, Zambia in 1985. AIDS 12(14):1921–1925
59. Cao Y et al (2014) High prevalence of early childhood infection by Kaposi's sarcoma-associated herpesvirus in a minority population in China. Clin Microbiol Infect 20(5):475–481
60. Plancoulaine S et al (2000) Human herpesvirus 8 transmission from mother to child and between siblings in an endemic population. Lancet 356(9235):1062–1065
61. Butler LM et al (2011) Human herpesvirus 8 infection in children and adults in a population-based study in rural Uganda. J Infect Dis 203(5):625–634
62. Brayfield BP et al (2004) Distribution of Kaposi sarcoma-associated herpesvirus/human herpesvirus 8 in maternal saliva and breast milk in Zambia: implications for transmission. J Infect Dis 189(12):2260–2270
63. Marcelin AG et al (2004) Quantification of Kaposi's sarcoma-associated herpesvirus in blood, oral mucosa, and saliva in patients with Kaposi's sarcoma. AIDS Res Hum Retrovir 20(7):704–708
64. Ruocco E et al (2013) Kaposi's sarcoma: etiology and pathogenesis, inducing factors, causal associations, and treatments: facts and controversies. Clin Dermatol 31(4):413–422
65. He F et al (2007) Human herpesvirus 8: seroprevalence and correlates in tumor patients from Xinjiang, China. J Med Virol 79(2):161–166
66. Geraminejad P et al (2002) Kaposi's sarcoma and other manifestations of human herpesvirus 8. J Am Acad Dermatol 47(5):641–655. quiz 656-8
67. Xu Y, Ganem D (2007) Induction of chemokine production by latent Kaposi's sarcoma-associated herpesvirus infection of endothelial cells. J Gen Virol 88(Pt 1):46–50
68. Gandhi M, Greenblatt RM (2002) Human herpesvirus 8, Kaposi's sarcoma, and associated conditions. Clin Lab Med 22(4):883–910
69. Said JW et al (1996) Primary effusion lymphoma in women: report of two cases of Kaposi's sarcoma herpes virus-associated effusion-based lymphoma in human immunodeficiency virus-negative women. Blood 88(8):3124–3128
70. Ascoli V et al (1999) Human herpesvirus-8 in lymphomatous and nonlymphomatous body cavity effusions developing in Kaposi's sarcoma and multicentric Castleman's disease. Ann Diagn Pathol 3(6):357–363
71. Soulier J et al (1995) Kaposi's sarcoma-associated herpesvirus-like DNA sequences in multicentric Castleman's disease. Blood 86(4):1276–1280
72. Katano H et al (2000) Expression and localization of human herpesvirus 8-encoded proteins in primary effusion lymphoma, Kaposi's sarcoma, and multicentric Castleman's disease. Virology 269(2):335–344
73. Bechtel JT et al (2003) Host range of Kaposi's sarcoma-associated herpesvirus in cultured cells. J Virol 77(11):6474–6481
74. Jarousse N, Chandran B, Coscoy L (2008) Lack of heparan sulfate expression in B-cell lines: implications for Kaposi's sarcoma-associated herpesvirus and murine gammaherpesvirus 68 infections. J Virol 82(24):12591–12597
75. Chandran B (2010) Early events in Kaposi's sarcoma-associated herpesvirus infection of target cells. J Virol 84(5):2188–2199
76. Inoue N et al (2003) Characterization of entry mechanisms of human herpesvirus 8 by using an Rta-dependent reporter cell line. J Virol 77(14):8147–8152

77. Rappocciolo G et al (2008) Human herpesvirus 8 infects and replicates in primary cultures of activated B lymphocytes through DC-SIGN. J Virol 82(10):4793–4806
78. Raghu H et al (2009) Kaposi's sarcoma-associated herpesvirus utilizes an actin polymerization-dependent macropinocytic pathway to enter human dermal microvascular endothelial and human umbilical vein endothelial cells. J Virol 83(10):4895–4911
79. Akula SM et al (2003) Kaposi's sarcoma-associated herpesvirus (human herpesvirus 8) infection of human fibroblast cells occurs through endocytosis. J Virol 77(14):7978–7990
80. Liao W et al (2003) Kaposi's sarcoma-associated herpesvirus/human herpesvirus 8 transcriptional activator Rta is an oligomeric DNA-binding protein that interacts with tandem arrays of phased A/T-trinucleotide motifs. J Virol 77(17):9399–9411
81. Neipel F, Albrecht JC, Fleckenstein B (1997) Cell-homologous genes in the Kaposi's sarcoma-associated rhadinovirus human herpesvirus 8: determinants of its pathogenicity? J Virol 71(6):4187–4192
82. Cesarman E et al (1996) Kaposi's sarcoma-associated herpesvirus contains G protein-coupled receptor and cyclin D homologs which are expressed in Kaposi's sarcoma and malignant lymphoma. J Virol 70(11):8218–8223
83. Akula SM et al (2001) Human herpesvirus 8 interaction with target cells involves heparan sulfate. Virology 282(2):245–255
84. Wang FZ et al (2003) Human herpesvirus 8 envelope glycoprotein B mediates cell adhesion via its RGD sequence. J Virol 77(5):3131–3147
85. Akula SM et al (2002) Integrin alpha3beta1 (CD 49c/29) is a cellular receptor for Kaposi's sarcoma-associated herpesvirus (KSHV/HHV-8) entry into the target cells. Cell 108(3):407–419
86. Rappocciolo G et al (2006) DC-SIGN is a receptor for human herpesvirus 8 on dendritic cells and macrophages. J Immunol 176(3):1741–1749
87. Kaleeba JA, Berger EA (2006) Kaposi's sarcoma-associated herpesvirus fusion-entry receptor: cystine transporter xCT. Science 311(5769):1921–1924
88. Hahn AS et al (2012) The ephrin receptor tyrosine kinase A2 is a cellular receptor for Kaposi's sarcoma-associated herpesvirus. Nat Med 18(6):961–966
89. Giancotti FG (2000) Complexity and specificity of integrin signalling. Nat Cell Biol 2(1):E13–E14
90. Sharma-Walia N et al (2004) Kaposi's sarcoma-associated herpesvirus/human herpesvirus 8 envelope glycoprotein gB induces the integrin-dependent focal adhesion kinase-Src-phosphatidylinositol 3-kinase-rho GTPase signal pathways and cytoskeletal rearrangements. J Virol 78(8):4207–4223
91. Raghu H et al (2007) Lipid rafts of primary endothelial cells are essential for Kaposi's sarcoma-associated herpesvirus/human herpesvirus 8-induced phosphatidylinositol 3-kinase and RhoA-GTPases critical for microtubule dynamics and nuclear delivery of viral DNA but dispensable for binding and entry. J Virol 81(15):7941–7959
92. Naranatt PP et al (2003) Kaposi's sarcoma-associated herpesvirus induces the phosphatidylinositol 3-kinase-PKC-zeta-MEK-ERK signaling pathway in target cells early during infection: implications for infectivity. J Virol 77(2):1524–1539
93. Veettil MV et al (2006) RhoA-GTPase facilitates entry of Kaposi's sarcoma-associated herpesvirus into adherent target cells in a Src-dependent manner. J Virol 80(23): 11432–11446
94. Naranatt PP et al (2005) Kaposi's sarcoma-associated herpesvirus modulates microtubule dynamics via RhoA-GTP-diaphanous 2 signaling and utilizes the dynein motors to deliver its DNA to the nucleus. J Virol 79(2):1191–1206
95. Hurley JH, Hanson PI (2010) Membrane budding and scission by the ESCRT machinery: it's all in the neck. Nat Rev Mol Cell Biol 11(8):556–566
96. Sharma-Walia N et al (2005) ERK1/2 and MEK1/2 induced by Kaposi's sarcoma-associated herpesvirus (human herpesvirus 8) early during infection of target cells are essential for expression of viral genes and for establishment of infection. J Virol 79(16):10308–10329

97. Cheng F et al (2015) Screening of the human Kinome identifies MSK1/2-CREB1 as an essential pathway mediating Kaposi's sarcoma-associated herpesvirus lytic replication during primary infection. J Virol 89(18):9262–9280
98. Sadagopan S et al (2007) Kaposi's sarcoma-associated herpesvirus induces sustained NF-kappaB activation during de novo infection of primary human dermal microvascular endothelial cells that is essential for viral gene expression. J Virol 81(8):3949–3968
99. Zhao Q et al (2015) Kaposi's sarcoma-associated herpesvirus-encoded replication and transcription activator impairs innate immunity via ubiquitin-mediated degradation of myeloid differentiation factor 88. J Virol 89(1):415–427
100. Cai Q et al (2010) Molecular biology of Kaposi's sarcoma-associated herpesvirus and related oncogenesis. Adv Virus Res 78:87–142
101. Kedes DH et al (1997) Identification of the gene encoding the major latency-associated nuclear antigen of the Kaposi's sarcoma-associated herpesvirus. J Clin Invest 100(10):2606–2610
102. Cai X et al (2005) Kaposi's sarcoma-associated herpesvirus expresses an array of viral microRNAs in latently infected cells. Proc Natl Acad Sci U S A 102(15):5570–5575
103. Kellam P et al (1997) Identification of a major latent nuclear antigen, LNA-1, in the human herpesvirus 8 genome. J Hum Virol 1(1):19–29
104. Rainbow L et al (1997) The 222- to 234-kilodalton latent nuclear protein (LNA) of Kaposi's sarcoma-associated herpesvirus (human herpesvirus 8) is encoded by orf73 and is a component of the latency-associated nuclear antigen. J Virol 71(8):5915–5921
105. Ballestas ME, Chatis PA, Kaye KM (1999) Efficient persistence of extrachromosomal KSHV DNA mediated by latency-associated nuclear antigen. Science 284(5414):641–644
106. Cotter MA 2nd, Robertson ES (1999) The latency-associated nuclear antigen tethers the Kaposi's sarcoma-associated herpesvirus genome to host chromosomes in body cavity-based lymphoma cells. Virology 264(2):254–264
107. Hu J, Garber AC, Renne R (2002) The latency-associated nuclear antigen of Kaposi's sarcoma-associated herpesvirus supports latent DNA replication in dividing cells. J Virol 76(22):11677–11687
108. Lan K et al (2005) Induction of Kaposi's sarcoma-associated herpesvirus latency-associated nuclear antigen by the lytic transactivator RTA: a novel mechanism for establishment of latency. J Virol 79(12):7453–7465
109. Radkov SA, Kellam P, Boshoff C (2000) The latent nuclear antigen of Kaposi sarcoma-associated herpesvirus targets the retinoblastoma-E2F pathway and with the oncogene Hras transforms primary rat cells. Nat Med 6(10):1121–1127
110. Lim C et al (2000) Latency-associated nuclear antigen of Kaposi's sarcoma-associated herpesvirus (human herpesvirus-8) binds ATF4/CREB2 and inhibits its transcriptional activation activity. J Gen Virol 81(Pt 11):2645–2652
111. Chang Y et al (1996) Cyclin encoded by KS herpesvirus. Nature 382(6590):410
112. Li M et al (1997) Kaposi's sarcoma-associated herpesvirus encodes a functional cyclin. J Virol 71(3):1984–1991
113. Direkze S, Laman H (2004) Regulation of growth signalling and cell cycle by Kaposi's sarcoma-associated herpesvirus genes. Int J Exp Pathol 85(6):305–319
114. Sarek G et al (2010) Nucleophosmin phosphorylation by v-cyclin-CDK6 controls KSHV latency. PLoS Pathog 6(3):e1000818
115. Ojala PM et al (1999) Kaposi's sarcoma-associated herpesvirus-encoded v-cyclin triggers apoptosis in cells with high levels of cyclin-dependent kinase 6. Cancer Res 59(19):4984–4989
116. Ojala PM et al (2000) The apoptotic v-cyclin-CDK6 complex phosphorylates and inactivates Bcl-2. Nat Cell Biol 2(11):819–825
117. Liang X et al (2011) Murine gamma-herpesvirus immortalization of fetal liver-derived B cells requires both the viral cyclin D homolog and latency-associated nuclear antigen. PLoS Pathog 7(9):e1002220
118. Ye FC et al (2008) Kaposi's sarcoma-associated herpesvirus latent gene vFLIP inhibits viral lytic replication through NF-kappaB-mediated suppression of the AP-1 pathway: a novel mechanism of virus control of latency. J Virol 82(9):4235–4249

119. Guasparri I, Keller SA, Cesarman E (2004) KSHV vFLIP is essential for the survival of infected lymphoma cells. J Exp Med 199(7):993–1003
120. Israel A (2010) The IKK complex, a central regulator of NF-kappaB activation. Cold Spring Harb Perspect Biol 2(3):a000158
121. Field N et al (2003) KSHV vFLIP binds to IKK-gamma to activate IKK. J Cell Sci 116(Pt 18):3721–3728
122. Brown HJ et al (2003) NF-kappaB inhibits gammaherpesvirus lytic replication. J Virol 77(15):8532–8540
123. Guasparri I, Wu H, Cesarman E (2006) The KSHV oncoprotein vFLIP contains a TRAF-interacting motif and requires TRAF2 and TRAF3 for signalling. EMBO Rep 7(1):114–119
124. McCormick C, Ganem D (2005) The kaposin B protein of KSHV activates the p38/MK2 pathway and stabilizes cytokine mRNAs. Science 307(5710):739–741
125. O'Hara AJ et al (2009) Pre-micro RNA signatures delineate stages of endothelial cell transformation in Kaposi sarcoma. PLoS Pathog 5(4):e1000389
126. O'Hara AJ et al (2009) Tumor suppressor microRNAs are underrepresented in primary effusion lymphoma and Kaposi sarcoma. Blood 113(23):5938–5941
127. Liang D, Lin X, Lan K (2011) Looking at Kaposi's sarcoma-associated herpesvirus-host interactions from a microRNA viewpoint. Front Microbiol 2:271
128. Lin X et al (2012) MicroRNAs and unusual small RNAs discovered in Kaposi's sarcoma-associated herpesvirus virions. J Virol 86(23):12717–12730
129. Liu Y et al (2012) Kaposi's sarcoma-associated herpesvirus-encoded microRNA miR-K12-11 attenuates transforming growth factor beta signaling through suppression of SMAD5. J Virol 86(3):1372–1381
130. Lin X et al (2011) miR-K12-7-5p encoded by Kaposi's sarcoma-associated herpesvirus stabilizes the latent state by targeting viral ORF50/RTA. PLoS One 6(1):e16224
131. Suffert G et al (2011) Kaposi's sarcoma herpesvirus microRNAs target caspase 3 and regulate apoptosis. PLoS Pathog 7(12):e1002405
132. Hansen A et al (2010) KSHV-encoded miRNAs target MAF to induce endothelial cell reprogramming. Genes Dev 24(2):195–205
133. Qin Z et al (2012) KSHV-encoded MicroRNAs: lessons for viral cancer pathogenesis and emerging concepts. Int J Cell Biol 2012:603961
134. Liang D et al (2011) A human herpesvirus miRNA attenuates interferon signaling and contributes to maintenance of viral latency by targeting IKKepsilon. Cell Res 21(5):793–806
135. Krishnan HH et al (2004) Concurrent expression of latent and a limited number of lytic genes with immune modulation and antiapoptotic function by Kaposi's sarcoma-associated herpesvirus early during infection of primary endothelial and fibroblast cells and subsequent decline of lytic gene expression. J Virol 78(7):3601–3620
136. Renne R et al (1996) Lytic growth of Kaposi's sarcoma-associated herpesvirus (human herpesvirus 8) in culture. Nat Med 2(3):342–346
137. Jenner RG et al (2001) Kaposi's sarcoma-associated herpesvirus latent and lytic gene expression as revealed by DNA arrays. J Virol 75(2):891–902
138. Wang SE et al (2003) CCAAT/enhancer-binding protein-alpha is induced during the early stages of Kaposi's sarcoma-associated herpesvirus (KSHV) lytic cycle reactivation and together with the KSHV replication and transcription activator (RTA) cooperatively stimulates the viral RTA, MTA, and PAN promoters. J Virol 77(17):9590–9612
139. Deng H, Young A, Sun R (2000) Auto-activation of the rta gene of human herpesvirus-8/Kaposi's sarcoma-associated herpesvirus. J Gen Virol 81(Pt 12):3043–3048
140. Song MJ, Deng H, Sun R (2003) Comparative study of regulation of RTA-responsive genes in Kaposi's sarcoma-associated herpesvirus/human herpesvirus 8. J Virol 77(17):9451–9462
141. Wang Y et al (2006) Kaposi's sarcoma-associated herpesvirus ori-Lyt-dependent DNA replication: dual role of replication and transcription activator. J Virol 80(24):12171–12186
142. Chen J et al (2009) Genome-wide identification of binding sites for Kaposi's sarcoma-associated herpesvirus lytic switch protein. RTA Virology 386(2):290–302

143. Liang Y, Ganem D (2003) Lytic but not latent infection by Kaposi's sarcoma-associated herpesvirus requires host CSL protein, the mediator of notch signaling. Proc Natl Acad Sci U S A 100(14):8490–8495
144. Li S et al (2016) Fine-tuning of the Kaposi's sarcoma-associated herpesvirus life cycle in neighboring cells through the RTA-JAG1-notch pathway. PLoS Pathog 12(10):e1005900
145. Dittmer DP, Damania B (2013) Kaposi sarcoma associated herpesvirus pathogenesis (KSHV)--an update. Curr Opin Virol 3(3):238–244
146. Li Q et al (2008) Genetic disruption of KSHV major latent nuclear antigen LANA enhances viral lytic transcriptional program. Virology 379(2):234–244
147. Yoo J et al (2012) Opposing regulation of PROX1 by interleukin-3 receptor and NOTCH directs differential host cell fate reprogramming by Kaposi sarcoma herpes virus. PLoS Pathog 8(6):e1002770
148. Cheng F et al (2011) KSHV-initiated notch activation leads to membrane-type-1 matrix metalloproteinase-dependent lymphatic endothelial-to-mesenchymal transition. Cell Host Microbe 10(6):577–590
149. Staskus KA et al (1997) Kaposi's sarcoma-associated herpesvirus gene expression in endothelial (spindle) tumor cells. J Virol 71(1):715–719
150. Boshoff C et al (1995) Kaposi's sarcoma-associated herpesvirus infects endothelial and spindle cells. Nat Med 1(12):1274–1278
151. Jussila L et al (1998) Lymphatic endothelium and Kaposi's sarcoma spindle cells detected by antibodies against the vascular endothelial growth factor receptor-3. Cancer Res 58(8):1599–1604
152. Wang HW et al (2004) Kaposi sarcoma herpesvirus-induced cellular reprogramming contributes to the lymphatic endothelial gene expression in Kaposi sarcoma. Nat Genet 36(7):687–693
153. Kahn HJ, Bailey D, Marks A (2002) Monoclonal antibody D2-40, a new marker of lymphatic endothelium, reacts with Kaposi's sarcoma and a subset of angiosarcomas. Mod Pathol 15(4):434–440
154. Pyakurel P et al (2006) Lymphatic and vascular origin of Kaposi's sarcoma spindle cells during tumor development. Int J Cancer 119(6):1262–1267
155. Hong YK et al (2004) Lymphatic reprogramming of blood vascular endothelium by Kaposi sarcoma-associated herpesvirus. Nat Genet 36(7):683–685
156. Carroll PA, Brazeau E, Lagunoff M (2004) Kaposi's sarcoma-associated herpesvirus infection of blood endothelial cells induces lymphatic differentiation. Virology 328(1):7–18
157. Kaaya EE et al (1995) Heterogeneity of spindle cells in Kaposi's sarcoma: comparison of cells in lesions and in culture. J Acquir Immune Defic Syndr Hum Retrovirol 10(3):295–305
158. Ciufo DM et al (2001) Spindle cell conversion by Kaposi's sarcoma-associated herpesvirus: formation of colonies and plaques with mixed lytic and latent gene expression in infected primary dermal microvascular endothelial cell cultures. J Virol 75(12):5614–5626
159. Abere B, Schulz TF (2016) KSHV non-structural membrane proteins involved in the activation of intracellular signaling pathways and the pathogenesis of Kaposi's sarcoma. Curr Opin Virol 20:11–19
160. Abboud ER et al (2013) Kaposi sarcoma-associated herpesvirus g protein-coupled receptor enhances endothelial cell survival in part by upregulation of bcl-2. Ochsner J 13(1):66–75
161. Sadagopan S et al (2009) Kaposi's sarcoma-associated herpesvirus upregulates angiogenin during infection of human dermal microvascular endothelial cells, which induces 45S rRNA synthesis, antiapoptosis, cell proliferation, migration, and angiogenesis. J Virol 83(7):3342–3364
162. Raggo C et al (2005) Novel cellular genes essential for transformation of endothelial cells by Kaposi's sarcoma-associated herpesvirus. Cancer Res 65(12):5084–5095
163. Flore O et al (1998) Transformation of primary human endothelial cells by Kaposi's sarcoma-associated herpesvirus. Nature 394(6693):588–592
164. Lagunoff M et al (2002) De novo infection and serial transmission of Kaposi's sarcoma-associated herpesvirus in cultured endothelial cells. J Virol 76(5):2440–2448

165. Cancian L, Hansen A, Boshoff C (2013) Cellular origin of Kaposi's sarcoma and Kaposi's sarcoma-associated herpesvirus-induced cell reprogramming. Trends Cell Biol 23(9):421–432
166. Adams RH, Alitalo K (2007) Molecular regulation of angiogenesis and lymphangiogenesis. Nat Rev Mol Cell Biol 8(6):464–478
167. Salahuddin SZ et al (1988) Angiogenic properties of Kaposi's sarcoma-derived cells after long-term culture in vitro. Science 242(4877):430–433
168. Ensoli B et al (1989) AIDS-Kaposi's sarcoma-derived cells express cytokines with autocrine and paracrine growth effects. Science 243(4888):223–226
169. Samaniego F et al (1995) Inflammatory cytokines induce AIDS-Kaposi's sarcoma-derived spindle cells to produce and release basic fibroblast growth factor and enhance Kaposi's sarcoma-like lesion formation in nude mice. J Immunol 154(7):3582–3592
170. Samaniego F et al (1997) Inflammatory cytokines induce endothelial cells to produce and release basic fibroblast growth factor and to promote Kaposi's sarcoma-like lesions in nude mice. J Immunol 158(4):1887–1894
171. Mansouri M et al (2008) Remodeling of endothelial adherens junctions by Kaposi's sarcoma-associated herpesvirus. J Virol 82(19):9615–9628
172. Qian LW et al (2008) Kaposi's sarcoma-associated herpesvirus disrupts adherens junctions and increases endothelial permeability by inducing degradation of VE-cadherin. J Virol 82(23):11902–11912
173. Wang L, Damania B (2008) Kaposi's sarcoma-associated herpesvirus confers a survival advantage to endothelial cells. Cancer Res 68(12):4640–4648
174. DiMaio TA, Gutierrez KD, Lagunoff M (2011) Latent KSHV infection of endothelial cells induces integrin beta3 to activate angiogenic phenotypes. PLoS Pathog 7(12):e1002424
175. Gasperini P et al (2012) Kaposi sarcoma herpesvirus promotes endothelial-to-mesenchymal transition through notch-dependent signaling. Cancer Res 72(5):1157–1169
176. Qian LW et al (2007) Kaposi's sarcoma-associated herpesvirus infection promotes invasion of primary human umbilical vein endothelial cells by inducing matrix metalloproteinases. J Virol 81(13):7001–7010
177. Watanabe T et al (2003) Kaposi's sarcoma-associated herpesvirus latency-associated nuclear antigen prolongs the life span of primary human umbilical vein endothelial cells. J Virol 77(11):6188–6196
178. Murakami Y et al (2006) Ets-1-dependent expression of vascular endothelial growth factor receptors is activated by latency-associated nuclear antigen of Kaposi's sarcoma-associated herpesvirus through interaction with Daxx. J Biol Chem 281(38):28113–28121
179. He M et al (2012) Cancer angiogenesis induced by Kaposi sarcoma-associated herpesvirus is mediated by EZH2. Cancer Res 72(14):3582–3592
180. Wang X et al (2014) Latency-associated nuclear antigen of Kaposi sarcoma-associated herpesvirus promotes angiogenesis through targeting notch signaling effector Hey1. Cancer Res 74(7):2026–2037
181. Bais C et al (1998) G-protein-coupled receptor of Kaposi's sarcoma-associated herpesvirus is a viral oncogene and angiogenesis activator. Nature 391(6662):86–89
182. Sodhi A et al (2000) The Kaposi's sarcoma-associated herpes virus G protein-coupled receptor up-regulates vascular endothelial growth factor expression and secretion through mitogen-activated protein kinase and p38 pathways acting on hypoxia-inducible factor 1alpha. Cancer Res 60(17):4873–4880
183. Aoki Y et al (1999) Angiogenesis and hematopoiesis induced by Kaposi's sarcoma-associated herpesvirus-encoded interleukin-6. Blood 93(12):4034–4043
184. Vart RJ et al (2007) Kaposi's sarcoma-associated herpesvirus-encoded interleukin-6 and G-protein-coupled receptor regulate angiopoietin-2 expression in lymphatic endothelial cells. Cancer Res 67(9):4042–4051
185. Xie J et al (2005) Kaposi's sarcoma-associated herpesvirus induction of AP-1 and interleukin 6 during primary infection mediated by multiple mitogen-activated protein kinase pathways. J Virol 79(24):15027–15037

186. Felcht M et al (2012) Angiopoietin-2 differentially regulates angiogenesis through TIE2 and integrin signaling. J Clin Invest 122(6):1991–2005
187. Wang L et al (2004) The Kaposi's sarcoma-associated herpesvirus (KSHV/HHV-8) K1 protein induces expression of angiogenic and invasion factors. Cancer Res 64(8):2774–2781
188. Bala K et al (2012) Kaposi's sarcoma herpesvirus K15 protein contributes to virus-induced angiogenesis by recruiting PLCgamma1 and activating NFAT1-dependent RCAN1 expression. PLoS Pathog 8(9):e1002927
189. Liu C et al (2001) Human herpesvirus 8 (HHV-8)-encoded cytokines induce expression of and autocrine signaling by vascular endothelial growth factor (VEGF) in HHV-8-infected primary-effusion lymphoma cell lines and mediate VEGF-independent antiapoptotic effects. J Virol 75(22):10933–10940
190. Cai Q et al (2007) A potential alpha-helix motif in the amino terminus of LANA encoded by Kaposi's sarcoma-associated herpesvirus is critical for nuclear accumulation of HIF-1alpha in normoxia. J Virol 81(19):10413–10423
191. Ema M et al (1997) A novel bHLH-PAS factor with close sequence similarity to hypoxia-inducible factor 1alpha regulates the VEGF expression and is potentially involved in lung and vascular development. Proc Natl Acad Sci U S A 94(9):4273–4278
192. Brown LF et al (2000) Expression of Tie1, Tie2, and angiopoietins 1, 2, and 4 in Kaposi's sarcoma and cutaneous angiosarcoma. Am J Pathol 156(6):2179–2183
193. Giffin L et al (2014) Modulation of Kaposi's sarcoma-associated herpesvirus interleukin-6 function by hypoxia-upregulated protein 1. J Virol 88(16):9429–9441
194. Choi YB, Nicholas J (2010) Induction of angiogenic chemokine CCL2 by human herpesvirus 8 chemokine receptor. Virology 397(2):369–378
195. Paul AG, Chandran B, Sharma-Walia N (2013) Cyclooxygenase-2-prostaglandin E2-eicosanoid receptor inflammatory axis: a key player in Kaposi's sarcoma-associated herpes virus associated malignancies. Transl Res 162(2):77–92
196. Cherqui S et al (2007) Lentiviral gene delivery of vMIP-II to transplanted endothelial cells and endothelial progenitors is proangiogenic in vivo. Mol Ther 15(7):1264–1272
197. Stine JT et al (2000) KSHV-encoded CC chemokine vMIP-III is a CCR4 agonist, stimulates angiogenesis, and selectively chemoattracts TH2 cells. Blood 95(4):1151–1157
198. Bussey KA et al (2014) The gammaherpesviruses Kaposi's sarcoma-associated herpesvirus and murine gammaherpesvirus 68 modulate the toll-like receptor-induced proinflammatory cytokine response. J Virol 88(16):9245–9259
199. Ahmad H et al (2011) Kaposi sarcoma-associated herpesvirus degrades cellular toll-interleukin-1 receptor domain-containing adaptor-inducing beta-interferon (TRIF). J Biol Chem 286(10):7865–7872
200. Jacobs SR et al (2013) The viral interferon regulatory factors of kaposi's sarcoma-associated herpesvirus differ in their inhibition of interferon activation mediated by toll-like receptor 3. J Virol 87(2):798–806
201. Gregory SM et al (2011) Discovery of a viral NLR homolog that inhibits the inflammasome. Science 331(6015):330–334
202. Roy A et al (2016) Nuclear innate immune DNA sensor IFI16 is degraded during lytic reactivation of Kaposi's Sarcoma-Associated Herpesvirus (KSHV): role of IFI16 in maintenance of KSHV latency. J Virol 90(19):8822–8841
203. Singh VV et al (2013) Kaposi's sarcoma-associated herpesvirus latency in endothelial and B cells activates gamma interferon-inducible protein 16-mediated inflammasomes. J Virol 87(8):4417–4431
204. West JA et al (2014) An important role for mitochondrial antiviral signaling protein in the Kaposi's sarcoma-associated herpesvirus life cycle. J Virol 88(10):5778–5787
205. Inn KS et al (2011) Inhibition of RIG-I-mediated signaling by Kaposi's sarcoma-associated herpesvirus-encoded deubiquitinase ORF64. J Virol 85(20):10899–10904
206. Ma Z et al (2015) Modulation of the cGAS-STING DNA sensing pathway by gammaherpesviruses. Proc Natl Acad Sci U S A 112(31):E4306–E4315

207. Wu JJ et al (2015) Inhibition of cGAS DNA sensing by a herpesvirus Virion protein. Cell Host Microbe 18(3):333–344
208. Hanahan D, Weinberg RA (2011) Hallmarks of cancer: the next generation. Cell 144(5):646–674
209. Hu H et al (2011) Proteomics revisits the cancer metabolome. Expert Rev Proteomics 8(4):505–533
210. Yogev O et al (2014) Kaposi's sarcoma herpesvirus microRNAs induce metabolic transformation of infected cells. PLoS Pathog 10(9):e1004400
211. Ma T et al (2015) KSHV induces aerobic glycolysis and angiogenesis through HIF-1-dependent upregulation of pyruvate kinase 2 in Kaposi's sarcoma. Angiogenesis 18(4):477–488
212. Zhu Y et al (2016) An oncogenic virus promotes cell survival and cellular transformation by suppressing glycolysis. PLoS Pathog 12(5):e1005648
213. Ye F et al (2016) High glucose induces reactivation of latent Kaposi's sarcoma-associated herpesvirus. J Virol 90:9654–9663
214. Sanchez EL et al (2015) Latent KSHV infected endothelial cells are glutamine addicted and require Glutaminolysis for survival. PLoS Pathog 11(7):e1005052
215. Delgado T et al (2012) Global metabolic profiling of infection by an oncogenic virus: KSHV induces and requires lipogenesis for survival of latent infection. PLoS Pathog 8(8):e1002866
216. Borah S, Verma SC, Robertson ES (2004) ORF73 of herpesvirus saimiri, a viral homolog of Kaposi's sarcoma-associated herpesvirus, modulates the two cellular tumor suppressor proteins p53 and pRb. J Virol 78(19):10336–10347
217. Cai QL et al (2006) EC5S ubiquitin complex is recruited by KSHV latent antigen LANA for degradation of the VHL and p53 tumor suppressors. PLoS Pathog 2(10):e116
218. Cai Q et al (2012) Kaposi's sarcoma herpesvirus upregulates Aurora A expression to promote p53 phosphorylation and ubiquitylation. PLoS Pathog 8(3):e1002566
219. Shin YC et al (2006) Inhibition of the ATM/p53 signal transduction pathway by Kaposi's sarcoma-associated herpesvirus interferon regulatory factor 1. J Virol 80(5):2257–2266
220. Baresova P et al (2014) p53 tumor suppressor protein stability and transcriptional activity are targeted by Kaposi's sarcoma-associated herpesvirus-encoded viral interferon regulatory factor 3. Mol Cell Biol 34(3):386–399
221. Chang PC, Li M (2008) Kaposi's sarcoma-associated herpesvirus K-cyclin interacts with Cdk9 and stimulates Cdk9-mediated phosphorylation of p53 tumor suppressor. J Virol 82(1):278–290
222. Laura MV et al (2015) KSHV latent protein LANA2 inhibits sumo2 modification of p53. Cell Cycle 14(2):277–282
223. Chudasama P et al (2015) Structural proteins of Kaposi's sarcoma-associated herpesvirus antagonize p53-mediated apoptosis. Oncogene 34(5):639–649
224. Lee HR et al (2009) Kaposi's sarcoma-associated herpesvirus viral interferon regulatory factor 4 targets MDM2 to deregulate the p53 tumor suppressor pathway. J Virol 83(13):6739–6747
225. Lan K et al (2007) Kaposi's sarcoma herpesvirus-encoded latency-associated nuclear antigen stabilizes intracellular activated Notch by targeting the Sel10 protein. Proc Natl Acad Sci U S A 104(41):16287–16292
226. Di Bartolo DL et al (2008) KSHV LANA inhibits TGF-beta signaling through epigenetic silencing of the TGF-beta type II receptor. Blood 111(9):4731–4740
227. Liang D et al (2014) Oncogenic herpesvirus KSHV Hijacks BMP-Smad1-Id signaling to promote tumorigenesis. PLoS Pathog 10(7):e1004253
228. Cannon ML, Cesarman E (2004) The KSHV G protein-coupled receptor signals via multiple pathways to induce transcription factor activation in primary effusion lymphoma cells. Oncogene 23(2):514–523
229. Sodhi A et al (2006) The TSC2/mTOR pathway drives endothelial cell transformation induced by the Kaposi's sarcoma-associated herpesvirus G protein-coupled receptor. Cancer Cell 10(2):133–143
230. Guo HG et al (2004) Tumorigenesis by human herpesvirus 8 vGPCR is accelerated by human immunodeficiency virus type 1 Tat. J Virol 78(17):9336–9342

231. Uldrick TS et al (2012) Phase II study of bevacizumab in patients with HIV-associated Kaposi's sarcoma receiving antiretroviral therapy. J Clin Oncol 30(13):1476–1483
232. Dittmer DP, Damania B (2016) Kaposi sarcoma-associated herpesvirus: immunobiology, oncogenesis, and therapy. J Clin Invest 126(9):3165–3175
233. Krown SE et al (2012) Rapamycin with antiretroviral therapy in AIDS-associated Kaposi sarcoma: an AIDS Malignancy Consortium study. J Acquir Immune Defic Syndr 59(5):447–454
234. Stallone G et al (2005) Sirolimus for Kaposi's sarcoma in renal-transplant recipients. N Engl J Med 352(13):1317–1323

Chapter 8
Hepatitis C Virus-Associated Cancers

Zhigang Yi and Zhenghong Yuan

Abstract Most hepatitis C virus (HCV) infection results in persistent infection. Significant portion of chronic HCV-infected patients develop hepatocellular carcinoma (HCC). Chronic hepatitis C is also associated with extrahepatic manifestations, including cryoglobulinemia, lymphoma, insulin resistance, type 2 diabetes, and neurological disorders. The molecular mechanisms of how HCV infection causes liver cancer are largely unknown. HCV replication or viral proteins may perturb cellular hemostasis and induce the generation of reactive oxygen species (ROS); viral components or viral replication products act as agonist to trigger innate immune response and cause chronic inflammation. Within the liver, non-hepatocytes such as hepatic stellate cell (HSC) are activated upon HCV infection to provide the major source of extracellular proteins and play important roles in fibrogenesis. With the great achievements of HCV treatment, especially the direct-acting antivirals (DAAs) against HCV, HCV eradication is possible. However, until now there are only very limited data on the effect of DAA-based anti-HCV treatment on HCC patients.

Keywords Hepatitis C virus • Chronic infection • Direct-acting antivirals • Hepatocellular carcinoma • Fibrogenesis • Steatosis • Inflammation • Hepatic stellate cells • Transforming growth factor • Reactive oxygen species

8.1 Introduction

Hepatitis C virus (HCV), a member of the genus *Hepacivirus* within the *Flaviviridae* family, chronically infects approximately 160 million people worldwide and causes hepatocellular carcinoma (HCC) in a significant proportion of the chronically infected population [1]. HCC represents the leading cause of death from cancer worldwide and is the most common cause of death in patients with cirrhosis [2]. HCV was discovered as the etiologic agent of non-A, non-B hepatitis in 1989 [3].

Z. Yi (✉) • Z. Yuan (✉)
Key Laboratory of Medical Molecular Virology (Ministries of Education and Health), School of Basic Medical Sciences, Fudan University, Shanghai 200032, People's Republic of China
e-mail: zgyi@fudan.edu.cn; zhyuan@shmu.edu.cn

Q. Cai et al. (eds.), *Infectious Agents Associated Cancers: Epidemiology and Molecular Biology*, Advances in Experimental Medicine and Biology 1018,
DOI 10.1007/978-981-10-5765-6_8

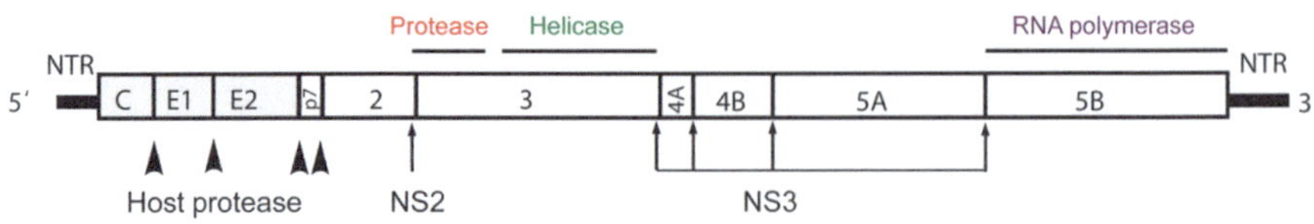

Fig. 8.1 Schematic of HCV genome. The HCV single open reading frame is flanked by the highly conserved 5′ and 3′ untranslated regions (*UTRs*). The 5′ UTR contains an internal ribosome entry site (*IRES*) to start a cap-independent translation. The 3′ UTR is required for RNA replication. The structural proteins core protein (*C*) and envelope glycoproteins (E1 and E2) constitute the virion. The nonstructural proteins NS3 to NS5B assemble viral replicase for viral replication. NS2 is a protease. NS3 contains a protease and an RNA helicase domain. NS5B is the viral RNA-dependent RNA polymerase. The structural proteins are processed by host proteases. HCV NS2 specifically cleave NS2-3 precursor, whereas NS3 processes the other nonstructural proteins

Thereafter, using an assay for detecting circulating antibodies identified, approximately 80% of the infected patients develop chronic infections, and 15% of the infected patients develop acute infections [4]. About 5–20% of the chronically infected patients progress to cirrhosis within 5–20 years, and about 1–2% of the patients with cirrhosis develop HCC per year [5]. HCV together with hepatitis B virus are the predominant cause of HCC all over the world [6].

HCV is a blood-borne virus and transmits mainly by sharing infected needles, receiving infected blood by transfusion, or accidental exposure to infected blood, whereas less efficiently from mother to child or by sexual contact [7]. There are seven main HCV genotypes. The HCV genotypes are epidemiologically distinct with different geographical distributions, which may indicate their recent epidemic spread [8]. In contrast to HBV that patients who get infected during childhood are prone to chronic infection [9], HCV infection at any age can result in chronic infection.

HCV is a positive-sense, single-stranded RNA virus. Its 9.6-kb genome encodes a single open reading frame that is translated into a polypeptide (Fig. 8.1). The polyprotein is cleaved into at least ten individual proteins by the host and viral proteinase [10]. HCV replication takes place in endoplasmic reticulum (ER)-associated viral-specific double-membrane vesicles (DMVs) [11].

Although HCV is likely only to replicate within hepatocytes, HCV infection is commonly associated with extrahepatic manifestations in significant portion of chronically infected patients. Chronic hepatitis C is also associated with cryoglobulinemia, lymphoma, insulin resistance, type 2 diabetes, and neurological disorders [12].

The discovery of new direct-acting antivirals (DAAs) is a breakthrough for anti-HCV treatment. Current treatment can obtain a cure rate that exceeds 90% for certain HCV genotypes [13–15]. Various studies have demonstrated sustained virological response (SVR) in patients with different stages of liver disease including compensated cirrhosis and patients for liver transplantation [16–20]. Anti-HCV therapy with interferon (IFN) may decrease HCC incidence and HCC recurrence and benefit patients [21]. The effect of DAA-based treatment on the HCC incidence and HCC recurrence needs more clinical studies.

8.2 HCV-Associated Hepatocellular Carcinoma (HCC)

Human viral oncogenesis shares common characteristics that oncoviruses are necessary but not sufficient to trigger cancer development, and cancer develops after many years of persistent infection with chronic inflammation or immune suppression [22]. HCV infection induces inflammation and immune-mediated liver damage. During persistent viral infection, chronic inflammation results in prolonged liver damage without virus clearance [23]. Immune-mediated liver damage prompts repeated hepatocellular regeneration and progressive fibrosis, which in turn results in a "cancer field" with genetically altered hepatocyte. Continued hepatocellular turnover may select for cancer stem cells with growth advantages that eventually develop hepatocellular carcinoma [24]. HCV infection or expression of HCV proteins interferes with host cell proliferation and apoptosis. HCV infection triggers oxidative stress, which may contribute to host genetic alteration and chronic inflammation. Within the liver, HCV infection activates hepatic stellate cells (HSCs) to produce excess extracellular matrix (ECM) and modulate fibrosis (Fig. 8.2). In hepatocellular carcinoma progression, the epithelial-to-mesenchymal transition

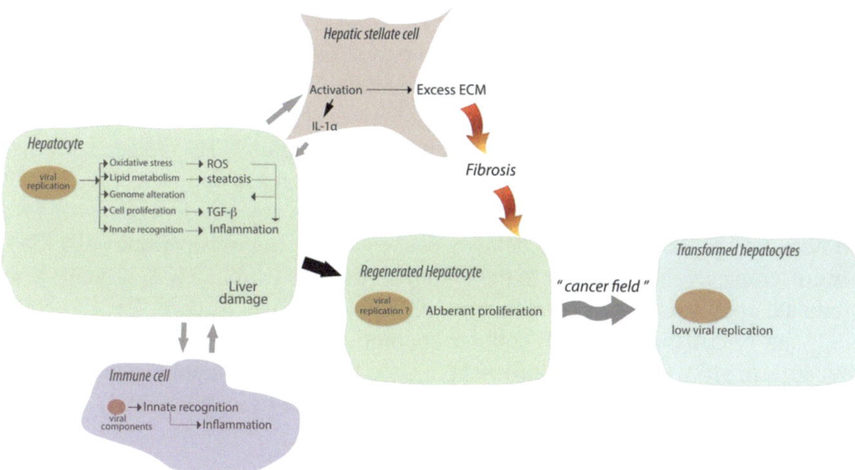

Fig. 8.2 Model for HCV-induced carcinogenesis. HCV infection triggers oxidative stress and production of reactive oxygen species (*ROS*); perturbs lipid metabolism, contributing to steatosis; modulates cellular proliferation signaling pathway, and induces TGF-β production. Viral proteins interfere with components for host DNA repair; viral components are recognized by innate sensors to trigger inflammatory signaling cascades. ROS in turn contributes to host genetic alteration and inflammation. Inflammation triggers adaptive immune response, resulting in immune-mediated liver injury. Liver damage induces hepatocyte regeneration. TGF-β activates hepatic stellate cell to become myoblast to secret excess extracellular matrix (*ECM*), which contributes to fibrosis and influences cancer initiation and progression. The repeated hepatocyte regeneration may select for cancer progenitors with aberrant proliferation to form a "cancer field" which finally develops into hepatocellular carcinoma (*HCC*)

(EMT) plays an important role in early stages of metastasis [25]. HCV infection provokes production of transform growth factor beta (TGF-β) that is a potent inducer of EMT [26, 27]. Thus, HCV infection directly affects infected hepatocyte or indirectly shapes intrahepatic microenvironment to prompt development of hepatocellular carcinoma.

8.2.1 HCV Viral Proteins Interfere with Cell Proliferation and Apoptosis

Oncovirus is commonly not sufficient to drive cancer development, but is necessary for human viral oncogenesis. During many years of persistent infection, viral oncogenes trigger proliferative and antiapoptotic programs to prompt oncogenesis [22]. Numerous studies suggest that HCV proteins trigger proliferative signaling. By means of overexpression in cultured cells, HCV core and NS5A activate pro-survival β-catenin and phosphoinositide 3-kinase (PI3K) signaling pathway [28–31]. HCV NS5B induces tumor suppressor protein (pRb) degradation via ubiquitination of pRb and proteasome-dependent degradation to activate E2F-responsive promoters and cell proliferation [32, 33]. In a transgenic mouse model expressing the entire HCV open reading frame, HCV NS5A activates Akt to stabilize β-catenin, resulting in activation of the c-Myc promoter and aberrant cell cycle arrest [34]. In a HCV core-transgenic mouse model, activated transforming growth factor beta (TGF-β) may drive the cross talk between hepatocytes and stromal environment [35]. However, infection of a cultured HCV strain JFH1 indicates a slowdown in proliferation and delayed cell cycle progression and apoptosis [36, 37]. In agreement with this, HCV NS5B interacts with cyclin-dependent kinase 2-interacting protein (CINP) to delay S-phase progression in hepatoma cells [38]. This discrepancy might be due to the unique of the JFH1 strain. HCV proteins also modulate apoptosis signaling pathway. HCV proteins such as core, E1, E2, NS2, NS3, and NS5A interfere with cellular apoptotic signaling pathway (reviewed in [39, 40]). In HCV chronically infected patients, there are detectable apoptotic cells and caspase activation in the liver that is probably due to immune-mediated liver injury [41, 42].

8.2.2 HCV Infection and Host Genetic Alteration

During hepatocarcinogenesis, accumulative genomic alteration progressively changes the hepatocellular phenotype and eventually renders the hepatocellular carcinoma development [43]. HCV infection of two B-cell lines induces mutations in tumor suppressor and proto-oncogenes, which is evidenced in HCV-associated B-cell lymphomas [44]. HCV-infected PBMCs also show chromosome gaps with a partially overlapped pattern of chromosomal aberrations observed in HCV-associated hepatocellular carcinomas [45]. In these cells, HCV-induced chromosomal aberration is due to the core- and NS3-induced nitric oxide (NO) and reactive oxygen species (ROS) production [45–47]. In cultured cells, HCV NS3/4A interacts with the ATM (ataxia-telangiectasia mutated), a cellular protein that plays an essential role in cellular response to irradiation-induced double-strand DNA breaks, indicating it may impair efficiency of DNA repair [48].

8.2.3 HCV Infection and Chronic Inflammation

In chronic infection, HCV activates immune response that fails to clear viral replication but develops an environment of inflammation and provokes chronic liver damage-induced regeneration of hepatocyte [23]. HCV viral components or replication products act as pathogen-associated molecular pattern (PAMP) to trigger innate immune response and immune modulatory cytokines [49]. HCV infection induces IL-29 and chemokine protein in primary liver cultures [50]. HCV virion may be taken up by intrahepatic macrophages, the Kupffer cells, wherein the viral RNA triggers interleukin-1β (IL-1β) through the NLRP3 inflammasome [51]. When cocultured with HCV-infected cells, hepatic stellate cell (HSC) is activated to prompt HCV-infected hepatocytes to produce pro-inflammatory cytokines, chemokines, and macrophage inflammatory protein 1α (MIP-1α) and MIP-1β [52].

8.2.4 HCV and Transforming Growth Factor Beta (TGF-β)

In hepatocellular carcinoma progression, the epithelial-to-mesenchymal transition (EMT) plays an important role in early stages of metastasis [25]. TGF-β is a potent inducer of EMT and plays an important role in fibrosis [26, 27]. In a HCV core-transgenic mouse model, expression of HCV core activates TGF-β signaling pathway [35]. In cultured cells, HCV infection induces endoplasmic reticulum stress and the unfolded protein response, which in turn triggers the generation of reactive oxygen species (ROS) and activation of the p38 mitogen-activated protein kinase (MAPK) signaling pathway, resulting in production of TGF-β [53, 54]. Jee et al. reported that there are higher expression levels of TGF-β in HCV-infected liver cells from HCV patients, and HCV infection induces TGF-β production through HCV E2-triggered overproduction of glucose-regulated protein 94 (GRP94) [55]. Conversely, overexpression of HCV NS5A protein inhibits TGF beta-mediated signaling pathway in hepatoma cell lines [56].

8.2.5 HCV Induces Oxidative Stress

Altered redox balance is one of the common hallmarks of tumors. Cancer cells exhibit high reactive oxygen species (ROS), which in turn is compensated by an increased antioxidant ability [57]. High ROS can induce DNA damage and genomic instability, which then further active inflammatory response and reprogramming metabolism of cancer cells [58, 59]. HCV infection is frequently associated with oxidative stress [60–62]. In chronically HCV-infected patients, there is frequently elevated iron in serum or in the liver, and the iron overload may play an important role in hepatic oxidative stress [62]. Upon overexpression, HCV core, E1, NS3, and NS5A are potent inducers of oxidative stress [63]. In a HCV core-transgenic mouse, the core protein induces hepatocellular carcinoma probably through altering the oxidant/antioxidant state [64, 65].

Oxidative stress or oxidative stress-generated products activate quiescent stellate cells and cultural fibroblasts to increase production of collagen type I, contributing to hepatic fibrosis. Free radicals and malondialdehyde, a product of lipid peroxidation, activate stellate cells [66]. Ascorbic acid induces lipid peroxidation and reactive aldehydes to stimulate collagen gene expression in cultured fibroblasts [67]. Generation of reactive oxygen species (ROS) may activate inflammation engaging the NLRP3 inflammasome [68] to contribute to fibrosis as discussed above.

8.2.6 Hepatic Stellate Cells (HSCs) and Fibrosis

Cross talk of parenchymal with nonparenchymal cells, along with signaling of inflammation, provides a permissive microenvironment for transition of proliferative hepatocytes into hepatocellular carcinoma [69]. Most HCC patients are associated with fibrosis. Stromal fibroblasts have a profound influence in cancer initiation and progression [70]. Following liver damage, hepatic stellate cells (HSCs) are activated to become matrix-secreting myofibroblasts, providing the major source of extracellular matrix (ECM), and may directly influence HCC through effects on the tumor stroma [71]. Primary human HSCs and HSC cell lines are not permissive for HCV entry or replication [72]. However, hepatoma cells expressing HCV core or HCV-infected cells activate HSC through a TGF-β-dependent signaling pathway [35, 55]. Reciprocally, when cocultured with HCV-infected cells, HSC secrets IL-1α to stimulate HCV-infected hepatocytes to express pro-inflammatory cytokines and chemokines, which may contribute to inflammation-mediated, HCV-related diseases [52].

8.2.7 Chronic HCV Infection and Steatosis

Hepatic steatosis (fatty liver) is characterized by the cytoplasmic accumulation of lipid droplets (LDs). Nonalcoholic fatty liver disease (NAFLD) and nonalcoholic steatohepatitis (NASH) are associated with HCV with approximate prevalence of 55% and 4–10%, respectively. HCV genotype 3 preferentially induces "viral steatosis," with a prevalence of about 86%, whereas genotypes 1 and 2 induce steatosis in about 40 and 50% of infected patients, respectively. HCV-associated NAFLD has a higher prevalence than in the general population and in HBV-infected population [73]. Hepatic steatosis poses an additional risk for HCC in patients with hepatitis C-related cirrhosis [63].

HCV infection perturbs host lipogenesis in a chimpanzee model [74]. Expression of HCV proteins directly alters lipid metabolism in a transgenic mouse model expressing a full length of viral polypeptide [75, 76]. HCV chronically infected patients has a unique composition of triglycerides enriched in carbon monounsaturated (C18:1) fatty acids, suggesting that it may be induced via a virus-specific mechanism [63]. HCV core protein may play an important role in hepatic steatosis. Core protein accumulates on LD [77]. HCV core-transgenic mouse develops HCC and shows hepatic steatosis early in life [64]. In a transgenic mouse model, hepatic overexpression of core protein reduces microsomal triglyceride transfer protein (MTP) activity and the particle size of nascent hepatic triglyceride-rich very low-density lipoproteins (VLDL) [78]. In chronic hepatitis C, hypobetalipoproteinemia that is characterized by a reduced plasma level of apolipoprotein B (apo B)-containing lipoproteins (LDL, VLDL) is observed and associated with steatosis, especially in patients infected with genotype 3 [79].

The mechanism of steatosis that contributes to fibrosis in hepatitis C may be similar to that observed in NAFLD, in which steatosis acts as a "first hit" and together with a "second hit" to progress to inflammation and fibrosis [80]. The oxidative stress may act as a "second hit" and induces lipid peroxidation, generating proinflammation and profibrotic products. A pilot study of antioxidant therapy with d-α-tocopherol significantly decreases the oxidative stress and reduces stellate cell activation and collagen α1 (I) expression, an important step in fibrogenesis [81].

8.2.8 The Cellular Origin of HCC

The origin to the cancer progenitor in HCV-associated HCC is poorly understood. Both infected hepatocytes and uninfected hepatic progenitor cells are capable of regeneration following liver injury to develop HCC founder cells [24]. Immune-mediated liver damage prompts hepatocellular regeneration, which in turn results in a "cancer field" with genetically altered hepatocyte. Whether HCV infects the regenerated hepatocytes is unclear, given that robust interferon production is detected within the HCV-infected hepatocytes from HCV-infected patients [82].

The level of HCV replication within the tumor remains controversial. There are reports showing no differences in the levels of viral RNA between the tumor and nontumorous liver tissues [83, 84], whereas other reports show low levels of viral RNA in the tumor tissue [85, 86], which is probably due to the loss of miR-122, a host factor required for HCV replication [87, 88]. Harouaka et al. reported that HCV infection is dramatically reduced in HCC tissues compared with non-tumor tissue. And the diminished viral replication is not associated with the abundance of miR-122 [89]. Previous studies demonstrate HCV replication is compartmental in the liver and the percentage of infected hepatocytes is low, ranging from 5 to 20% [82, 90–93], which may reflect the cell-to-cell transmission of HCV in vivo [94]. A recent study shows limited intrahepatic compartmentalization in end stage of liver disease [95]. In agreement with this finding, Harouaka et al. found no compartmentalization of HCV replication between non-tumor cells and serum, indicating efficient HCV replication in nontumorous liver tissue [89].

8.2.9 HCV Genotypes and HCC

There are seven main HCV genotypes with distinct geographical distributions [8]. Several studies suggest certain HCV genotypes as an increased risk factor for HCC. HCV genotype 3 may increase the risk of HCC compared to genotype 1 [96], which might be due to its association with steatosis as discussed above. HCV genotype 1b shows a higher risk than genotype 2 to develop HCC [97]. Some mutations in HCV core region of genotype 1a and 1b are associated with high risk of developing HCC [98, 99].

8.3 Association of HCV Infection and Lymphoma

8.3.1 HCV Infection and Mixed Cryoglobulinemia (MC)

Chronic HCV infection is commonly associated with extrahepatic manifestation [12]. Type II mixed cryoglobulinemia (MC) is the most common extrahepatic manifestation in HCV-infected patients, ranging from 10 to 70%. MC is characterized by the presence of cryoglobulins in the circulation. Cryoglobulins are cold-insoluble immune complexes, which precipitate at temperature below 37 °C. Type II MC is characterized by a mixture of polyclonal immunoglobulin (IgG) and monoclonal IgM with rheumatoid factor activity usually against IgG. Cryoglobulins contain rheumatoid factor, polyclonal IgG, and HCV particles and can form intravascular deposition [100].

8.3.2 HCV Infection and Non-Hodgkin Lymphoma (NHL)

Although there is a low incidence of HCV infection (about 2.5%) in non-Hodgkin lymphoma (NHL) patients [101], chronic HCV infection is associated with B-cell non-Hodgkin lymphoma (B-NHL) as HCV eradication by antiviral therapy induces B-NHL regression, suggesting a causal relationship between HCV infection and lymphoma development [102]. HCV MC may represent an antigen-driven B-cell proliferation that occasionally develops NHL [100]. The molecular mechanism of HCV-induced lymphomagenesis may include continuous stimulation of lymphocyte receptors by viral antigens, viral replication in B cells, and permanent B-cell damage [102].

HCV glycoprotein E2 binds the CD81 for virus entry [103]. HCV may bind simultaneously to CD81 and a specific B-cell receptor (BCR) on B cells to trigger B-cell activation and proliferation. Although there are reports to show no binding ability of lymphoma BCRs from B-NHL patients to HCV antigens [104], chronic viral antigen stimulation may play an import role in aberrant B-cell proliferation. Expression of HCV viral proteins in B cells from HCV-infected patients upregulates BCR signaling. HCV nonstructural protein NS3/4A modulates HuR-mediated post-transcriptional regulation of a network of mRNAs that is associated with B-cell lymphoproliferative disorders [105]. The pro-inflammatory cytokines such as IL-6 and some chemokines may also contribute to aberrant B-cell proliferation [106].

HCV infection in B cells is controversial. A B-cell line (SB) from a HCV-infected B-NHL patient supports HCV replication and produces HCV virions. Epstein-Barr virus-immortalized B-cell lines from PBMCs of HCV-positive patients are positive for HCV RNA and protein, suggesting HCV infection in B cells [107]. Using a cell-cultured HCV, Marukian et al. did not observe HCV infection in peripheral blood mononuclear cells (PBMC), and PBMC is unlikely to support HCV viral translation [108].

8.4 Effect of Anti-HCV Treatment on HCC Occurrence and Recurrence

8.4.1 Anti-HCV Treatment with Interferon (IFN)

Before the emergence of direct-acting antivirals before 2011, interferon (IFN)-based therapies are the standard treatment for HCV. The endpoint of treatment is a 6-month-long sustained virological response (SVR) after drug withdrawal. SVR is defined by undetectable HCV RNA below 10–15 international units [IU]/ml. Administration of a combination of pegylated IFN-α with ribavirin for 24 or 48 weeks yielded viral eradication in approximately 80% patients infected with HCV genotypes 2 and 3 or 40–50% of patients infected with HCV genotype 1 [109]. Probably due to the relative low efficacy of IFN therapy, a decreased incidence of HCC or lower recurrence of HCC in HCV-infected patients who receive IFN-based therapy is only observed in more than 5-year follow-up studies. Overall IFN-based therapies significantly benefit HCC patients after curative HCC therapy, which is probably due to the additional inhibitory effects of IFN on liver cancer rather than antiviral effect [21].

8.4.2 Anti-HCV Treatment with Direct-Acting Antivirals (DAAs)

The discovery of new direct-acting antivirals (DAAs) is a breakthrough for HCV treatment. Anti-HCV treatment with DAA can yield sustained virological response (SVR) exceeding 95% in patients with genotype 1 HCV infection and compensated cirrhosis within 12 weeks [18]. Administration of combination of HCV NS5A inhibitor, the nucleotide polymerase inhibitor, and ribavirin to patients with advanced liver disease or post-liver transplantation recurrence produces high rates of SVR12 exceeding 80% [19, 20]. In real-world patients with HCV genotype 1 infection, treatment with NS3/4A protease inhibitor simeprevir and nucleotide polymerase inhibitor sofosbuvir yields overall SVR rate of 84% [15].

Eradication of viral infection by DAA theoretically removes the oncogenic agent and may, like IFN-based therapy, decrease HCC incidence or benefit HCC patients. However a recent study enrolled HCV-related HCC patients who have been successfully treated before for their cancer to receive DAA treatment. After a follow-up of 5.7 months, there is unexpected high rate of tumor recurrence in some patients with HCV eradication [110]. This may be attributed to the sudden elimination of inflammation signals from viral replication and following reduction of immune surveillance [110]. More studies are needed to assess the effect of IFN-free DAA treatment on HCC incidence and recurrence.

8.5 Conclusion and Perspectives

As a cytoplasmic RNA virus, it is possible to eradicate HCV infection in patients by antiviral therapy. The discovery of DAAs makes it possible to successfully cure HCV patients with IFN-free DAA therapies within a short time. Clearance of virus and viral antigens may significantly decrease the incidence of HCC and relieve other extrahepatic manifestations. However, the long-term effect of DAA treatment on HCC incidence and recurrence needs more cohort studies.

Chronic HCV infection induces local inflammations within the liver; HCV infection induces reactive oxygen species and interferes with functions of maintaining host genomic stability; these may generate a microenvironment for aberrant hepatocellular proliferation and help develop hepatocellular carcinoma. The molecular mechanism of how HCV infection induces HCC is still not well understood. If there

Table 8.1 HCV proteins interfere with cellular proliferation signaling pathways

HCV protein	Experiment system	Signaling pathway	Reference
Polyprotein	Transgenic mouse model	c-myc	[34]
Core, E2, NS5A	Transgenic mouse model, HCV replicon, overexpression in cell line	TGF-β	[35, 55, 56]
Core	Transgenic mouse model	Oxidative stress	[65]
Core, NS5A	Overexpression in cell line	wnt/beta-catenin	[28, 29]
NS5A, polyprotein	Overexpression in cell line; HCV subgenomic replicon	PI3K	[30, 31]
NS5B	Overexpression in cell line	Rb, CINP	[32, 38]
NS3/4A	Overexpression in cell line	ATM	[48]

are any driver mutations for HCC development remains to be elucidated in the future studies [24]. The lack of reliable animal model recapitulating the HCV infection and HCC development hampers the studies to address these questions in vivo. The only natural animal model of HCV, chimpanzee, mainly resolves HCV infection. A chimpanzee infected with a cell culture-derived genotype 1a virus with culture-adaptive mutations develops persistent infection with chronic hepatitis, which may provide an opportunity to address the molecular mechanism of HCC [111]. However due to the limitation of accessibility of chimpanzee model, small animal models for HCV infection are needed. HCV transgenic mice expressing HCV core develop HCC without obvious inflammation [64]. Recently an immune-competent mouse model with transgenic human CD81 and occludin genes supports HCV the entire life cycle and develops fibrotic and cirrhotic progression [112]. These animal models may provide tools to dissect the molecular mechanisms of HCC development and identify the key molecular signaling pathways contributing to HCC development.

HCV-associated HCC often develops after many years. It is beneficial to identify the molecular alterations during early time of HCC development for diagnosis and prognosis. Using the high-throughput technologies may help identify the host genetic alterations associated with HCC development [113] (Table 8.1).

Acknowledgments Supported in part by grants from National Basic Research Program of China 2015CB554301 and Sino-German Transregional Collaborative Research and National Natural Science Foundation of China for international collaboration 81461130019.

References

1. Lavanchy D (2009) The global burden of hepatitis C. Liver Int 29(Suppl 1):74–81
2. Forner A, Llovet JM, Bruix J (2012) Hepatocellular carcinoma. Lancet 379:1245–1255
3. Choo QL, Kuo G, Weiner AJ, Overby LR, Bradley DW, Houghton M (1989) Isolation of a cDNA clone derived from a blood-borne non-A, non-B viral hepatitis genome. Science 244:359–362

4. Kuo G, Choo QL, Alter HJ, Gitnick GL, Redeker AG, Purcell RH, Miyamura T, Dienstag JL, Alter MJ, Stevens CE, Tegtmeier GE, Bonino F, Colombo M, Lee WS, Kuo C, Berger K, Shuster JR, Overby LR, Bradley DW, Houghton M (1989) An assay for circulating antibodies to a major etiologic virus of human non-A, non-B-hepatitis. Science 244:362–364
5. Strader DB, Wright T, Thomas DL, Seeff LB, American Association for the Study of Liver D (2004) Diagnosis, management, and treatment of hepatitis C. Hepatology 39:1147–1171
6. de Martel C, Maucort-Boulch D, Plummer M, Franceschi S (2015) World-wide relative contribution of hepatitis B and C viruses in hepatocellular carcinoma. Hepatology 62:1190–1200
7. Wasley A, Alter MJ (2000) Epidemiology of hepatitis C: geographic differences and temporal trends. Semin Liver Dis 20:1–16
8. Simmonds P (2013) The origin of hepatitis C virus. Curr Top Microbiol Immunol 369:1–15
9. McMahon BJ, Alward WL, Hall DB, Heyward WL, Bender TR, Francis DP, Maynard JE (1985) Acute hepatitis B virus infection: relation of age to the clinical expression of disease and subsequent development of the carrier state. J Infect Dis 151:599–603
10. Moradpour D, Penin F, Rice CM (2007) Replication of hepatitis C virus. Nat Rev Microbiol 5:453–463
11. Romero-Brey I, Merz A, Chiramel A, Lee JY, Chlanda P, Haselman U, Santarella-Mellwig R, Habermann A, Hoppe S, Kallis S, Walther P, Antony C, Krijnse-Locker J, Bartenschlager R (2012) Three-dimensional architecture and biogenesis of membrane structures associated with hepatitis C virus replication. PLoS Pathog 8:e1003056
12. Gill K, Ghazinian H, Manch R, Gish R (2016) Hepatitis C virus as a systemic disease: reaching beyond the liver. Hepatol Int 10:415–423
13. Kohli A, Shaffer A, Sherman A, Kottilil S (2014) Treatment of hepatitis C: a systematic review. JAMA 312:631–640
14. Foster GR, Irving WL, Cheung MC, Walker AJ, Hudson BE, Verma S, McLauchlan J, Mutimer DJ, Brown A, Gelson WT, MacDonald DC, Agarwal K, Hcv Research UK (2016) Impact of direct acting antiviral therapy in patients with chronic hepatitis C and decompensated cirrhosis. J Hepatol 64:1224–1231
15. Sulkowski MS, Vargas HE, Di Bisceglie AM, Kuo A, Reddy KR, Lim JK, Morelli G, Darling JM, Feld JJ, Brown RS, Frazier LM, Stewart TG, Fried MW, Nelson DR, Jacobson IM, Group H-TS (2016) Effectiveness of Simeprevir Plus Sofosbuvir, with or without ribavirin, in real-world patients with HCV genotype 1 infection. Gastroenterology 150:419–429
16. Lawitz E, Makara M, Akarca US, Thuluvath PJ, Preotescu LL, Varunok P, Morillas RM, Hall C, Mobashery N, Redman R, Pilot-Matias T, Vilchez RA, Hezode C (2015) Efficacy and safety of ombitasvir, paritaprevir, and ritonavir in an open-label study of patients with genotype 1b chronic hepatitis C virus infection with and without cirrhosis. Gastroenterology 149:971–980. e971
17. Poordad F, Hezode C, Trinh R, Kowdley KV, Zeuzem S, Agarwal K, Shiffman ML, Wedemeyer H, Berg T, Yoshida EM, Forns X, Lovell SS, Da Silva-Tillmann B, Collins CA, Campbell AL, Podsadecki T, Bernstein B (2014) ABT-450/r-ombitasvir and dasabuvir with ribavirin for hepatitis C with cirrhosis. N Engl J Med 370:1973–1982
18. Reddy KR, Bourliere M, Sulkowski M, Omata M, Zeuzem S, Feld JJ, Lawitz E, Marcellin P, Welzel TM, Hyland R, Ding X, Yang J, Knox S, Pang P, Dvory-Sobol H, Subramanian GM, Symonds W, McHutchison JG, Mangia A, Gane E, Mizokami M, Pol S, Afdhal N (2015) Ledipasvir and sofosbuvir in patients with genotype 1 hepatitis C virus infection and compensated cirrhosis: an integrated safety and efficacy analysis. Hepatology 62:79–86
19. Charlton M, Everson GT, Flamm SL, Kumar P, Landis C, Brown RS Jr, Fried MW, Terrault NA, O'Leary JG, Vargas HE, Kuo A, Schiff E, Sulkowski MS, Gilroy R, Watt KD, Brown K, Kwo P, Pungpapong S, Korenblat KM, Muir AJ, Teperman L, Fontana RJ, Denning J, Arterburn S, Dvory-Sobol H, Brandt-Sarif T, Pang PS, McHutchison JG, Reddy KR, Afdhal N, Investigators S (2015) Ledipasvir and Sofosbuvir plus ribavirin for treatment of HCV infection in patients with advanced liver disease. Gastroenterology 149:649–659

20. Poordad F, Schiff ER, Vierling JM, Landis C, Fontana RJ, Yang R, McPhee F, Hughes EA, Noviello S, Swenson ES (2016) Daclatasvir with sofosbuvir and ribavirin for hepatitis C virus infection with advanced cirrhosis or post-liver transplantation recurrence. Hepatology 63:1493–1505
21. Hsu CS, Chao YC, Lin HH, Chen DS, Kao JH (2015) Systematic review: impact of interferon-based therapy on HCV-related hepatocellular carcinoma. Sci Rep 5:9954
22. Mesri EA, Feitelson MA, Munger K (2014) Human viral oncogenesis: a cancer hallmarks analysis. Cell Host Microbe 15:266–282
23. Arzumanyan A, Reis HM, Feitelson MA (2013) Pathogenic mechanisms in HBV- and HCV-associated hepatocellular carcinoma. Nat Rev Cancer 13:123–135
24. Mitchell JK, Lemon SM, McGivern DR (2015) How do persistent infections with hepatitis C virus cause liver cancer? Curr Opin Virol 14:101–108
25. Giannelli G, Koudelkova P, Dituri F, Mikulits W (2016) Role of epithelial to mesenchymal transition in hepatocellular carcinoma. J hepatol 65:798
26. Thiery JP, Acloque H, Huang RY, Nieto MA (2009) Epithelial-mesenchymal transitions in development and disease. Cell 139:871–890
27. Friedman SL (2008) Mechanisms of hepatic fibrogenesis. Gastroenterology 134:1655–1669
28. Liu J, Ding X, Tang J, Cao Y, Hu P, Zhou F, Shan X, Cai X, Chen Q, Ling N, Zhang B, Bi Y, Chen K, Ren H, Huang A, He TC, Tang N (2011) Enhancement of canonical Wnt/beta-catenin signaling activity by HCV core protein promotes cell growth of hepatocellular carcinoma cells. PLoS One 6:e27496
29. Park CY, Choi SH, Kang SM, Kang JI, Ahn BY, Kim H, Jung G, Choi KY, Hwang SB (2009) Nonstructural 5A protein activates beta-catenin signaling cascades: implication of hepatitis C virus-induced liver pathogenesis. J Hepatol 51:853–864
30. Street A, Macdonald A, Crowder K, Harris M (2004) The hepatitis C virus NS5A protein activates a phosphoinositide 3-kinase-dependent survival signaling cascade. J Biol Chem 279:12232–12241
31. Street A, Macdonald A, McCormick C, Harris M (2005) Hepatitis C virus NS5A-mediated activation of phosphoinositide 3-kinase results in stabilization of cellular beta-catenin and stimulation of beta-catenin-responsive transcription. J Virol 79:5006–5016
32. Munakata T, Liang Y, Kim S, McGivern DR, Huibregtse J, Nomoto A, Lemon SM (2007) Hepatitis C virus induces E6AP-dependent degradation of the retinoblastoma protein. PLoS Pathog 3:1335–1347
33. Munakata T, Nakamura M, Liang Y, Li K, Lemon SM (2005) Down-regulation of the retinoblastoma tumor suppressor by the hepatitis C virus NS5B RNA-dependent RNA polymerase. Proc Natl Acad Sci U S A 102:18159–18164
34. Higgs MR, Lerat H, Pawlotsky JM (2013) Hepatitis C virus-induced activation of beta-catenin promotes c-Myc expression and a cascade of pro-carcinogenetic events. Oncogene 32:4683–4693
35. Benzoubir N, Lejamtel C, Battaglia S, Testoni B, Benassi B, Gondeau C, Perrin-Cocon L, Desterke C, Thiers V, Samuel D, Levrero M, Brechot C, Bourgeade MF (2013) HCV core-mediated activation of latent TGF-beta via thrombospondin drives the crosstalk between hepatocytes and stromal environment. J Hepatol 59:1160–1168
36. Walters KA, Syder AJ, Lederer SL, Diamond DL, Paeper B, Rice CM, Katze MG (2009) Genomic analysis reveals a potential role for cell cycle perturbation in HCV-mediated apoptosis of cultured hepatocytes. PLoS Pathog 5:e1000269
37. Kannan RP, Hensley LL, Evers LE, Lemon SM, McGivern DR (2011) Hepatitis C virus infection causes cell cycle arrest at the level of initiation of mitosis. J Virol 85:7989–8001
38. Wang Y, Wang Y, Xu Y, Tong W, Pan T, Li J, Sun S, Shao J, Ding H, Toyoda T, Yuan Z (2011) Hepatitis C virus NS5B protein delays s phase progression in human hepatocyte-derived cells by relocalizing cyclin-dependent kinase 2-interacting protein (CINP). J Biol Chem 286:26603–26615
39. Fischer R, Baumert T, Blum HE (2007) Hepatitis C virus infection and apoptosis. World J Gastroenterol 13:4865–4872

40. McGivern DR, Lemon SM (2011) Virus-specific mechanisms of carcinogenesis in hepatitis C virus associated liver cancer. Oncogene 30:1969–1983
41. Calabrese F, Pontisso P, Pettenazzo E, Benvegnu L, Vario A, Chemello L, Alberti A, Valente M (2000) Liver cell apoptosis in chronic hepatitis C correlates with histological but not biochemical activity or serum HCV-RNA levels. Hepatology 31:1153–1159
42. Bantel H, Lugering A, Poremba C, Lugering N, Held J, Domschke W, Schulze-Osthoff K (2001) Caspase activation correlates with the degree of inflammatory liver injury in chronic hepatitis C virus infection. Hepatology 34:758–767
43. Thorgeirsson SS, Grisham JW (2002) Molecular pathogenesis of human hepatocellular carcinoma. Nat Genet 31:339–346
44. Machida K, Cheng KT, Sung VM, Shimodaira S, Lindsay KL, Levine AM, Lai MY, Lai MM (2004) Hepatitis C virus induces a mutator phenotype: enhanced mutations of immunoglobulin and protooncogenes. Proc Natl Acad Sci U S A 101:4262–4267
45. Machida K, McNamara G, Cheng KT, Huang J, Wang CH, Comai L, Ou JH, Lai MM (2010) Hepatitis C virus inhibits DNA damage repair through reactive oxygen and nitrogen species and by interfering with the ATM-NBS1/Mre11/Rad50 DNA repair pathway in monocytes and hepatocytes. J Immunol 185:6985–6998
46. Machida K, Cheng KT, Sung VM, Lee KJ, Levine AM, Lai MM (2004) Hepatitis C virus infection activates the immunologic (type II) isoform of nitric oxide synthase and thereby enhances DNA damage and mutations of cellular genes. J Virol 78:8835–8843
47. Machida K, Cheng KT, Lai CK, Jeng KS, Sung VM, Lai MM (2006) Hepatitis C virus triggers mitochondrial permeability transition with production of reactive oxygen species, leading to DNA damage and STAT3 activation. J Virol 80:7199–7207
48. Lai CK, Jeng KS, Machida K, Cheng YS, Lai MM (2008) Hepatitis C virus NS3/4A protein interacts with ATM, impairs DNA repair and enhances sensitivity to ionizing radiation. Virology 370:295–309
49. Yi Z, Chen J, Kozlowski M, Yuan Z (2015) Innate detection of hepatitis B and C virus and viral inhibition of the response. Cell Microbiol 17:1295–1303
50. Marukian S, Andrus L, Sheahan TP, Jones CT, Charles ED, Ploss A, Rice CM, Dustin LB (2011) Hepatitis C virus induces interferon-lambda and interferon-stimulated genes in primary liver cultures. Hepatology 54:1913–1923
51. Negash AA, Ramos HJ, Crochet N, Lau DT, Doehle B, Papic N, Delker DA, Jo J, Bertoletti A, Hagedorn CH, Gale M Jr (2013) IL-1beta production through the NLRP3 inflammasome by hepatic macrophages links hepatitis C virus infection with liver inflammation and disease. PLoS Pathog 9:e1003330
52. Nishitsuji H, Funami K, Shimizu Y, Ujino S, Sugiyama K, Seya T, Takaku H, Shimotohno K (2013) Hepatitis C virus infection induces inflammatory cytokines and chemokines mediated by the cross talk between hepatocytes and stellate cells. J Virol 87:8169–8178
53. Lin W, Tsai WL, Shao RX, Wu G, Peng LF, Barlow LL, Chung WJ, Zhang L, Zhao H, Jang JY, Chung RT (2010) Hepatitis C virus regulates transforming growth factor beta1 production through the generation of reactive oxygen species in a nuclear factor kappaB-dependent manner. Gastroenterology 138:2509–2518. 2518 e2501
54. Chusri P, Kumthip K, Hong J, Zhu C, Duan X, Jilg N, Fusco DN, Brisac C, Schaefer EA, Cai D, Peng LF, Maneekarn N, Lin W, Chung RT (2016) HCV induces transforming growth factor beta1 through activation of endoplasmic reticulum stress and the unfolded protein response. Sci Rep 6:22487
55. Jee MH, Hong KY, Park JH, Lee JS, Kim HS, Lee SH, Jang SK (2016) New mechanism of hepatic fibrogenesis: hepatitis C virus infection induces transforming growth factor beta1 production through glucose-regulated protein 94. J Virol 90:3044–3055
56. Choi SH, Hwang SB (2006) Modulation of the transforming growth factor-beta signal transduction pathway by hepatitis C virus nonstructural 5A protein. J Biol Chem 281:7468–7478
57. Panieri E, Santoro MM (2016) ROS homeostasis and metabolism: a dangerous liason in cancer cells. Cell Death Dis 7:e2253

58. Naik E, Dixit VM (2011) Mitochondrial reactive oxygen species drive proinflammatory cytokine production. J Exp Med 208:417–420
59. Gorrini C, Harris IS, Mak TW (2013) Modulation of oxidative stress as an anticancer strategy. Nat Rev Drug Discov 12:931–947
60. Paradis V, Mathurin P, Kollinger M, Imbert-Bismut F, Charlotte F, Piton A, Opolon P, Holstege A, Poynard T, Bedossa P (1997) In situ detection of lipid peroxidation in chronic hepatitis C: correlation with pathological features. J Clin Pathol 50:401–406
61. Barbaro G, Di Lorenzo G, Ribersani M, Soldini M, Giancaspro G, Bellomo G, Belloni G, Grisorio B, Barbarini G (1999) Serum ferritin and hepatic glutathione concentrations in chronic hepatitis C patients related to the hepatitis C virus genotype. J Hepatol 30:774–782
62. Fujita N, Horiike S, Sugimoto R, Tanaka H, Iwasa M, Kobayashi Y, Hasegawa K, Ma N, Kawanishi S, Adachi Y, Kaito M (2007) Hepatic oxidative DNA damage correlates with iron overload in chronic hepatitis C patients. Free Radic Biol Med 42:353–362
63. Yamane D, McGivern DR, Masaki T, Lemon SM (2013) Liver injury and disease pathogenesis in chronic hepatitis C. Curr Top Microbiol Immunol 369:263–288
64. Moriya K, Fujie H, Shintani Y, Yotsuyanagi H, Tsutsumi T, Ishibashi K, Matsuura Y, Kimura S, Miyamura T, Koike K (1998) The core protein of hepatitis C virus induces hepatocellular carcinoma in transgenic mice. Nat Med 4:1065–1067
65. Moriya K, Nakagawa K, Santa T, Shintani Y, Fujie H, Miyoshi H, Tsutsumi T, Miyazawa T, Ishibashi K, Horie T, Imai K, Todoroki T, Kimura S, Koike K (2001) Oxidative stress in the absence of inflammation in a mouse model for hepatitis C virus-associated hepatocarcinogenesis. Cancer Res 61:4365–4370
66. Lee KS, Buck M, Houglum K, Chojkier M (1995) Activation of hepatic stellate cells by TGF alpha and collagen type I is mediated by oxidative stress through c-myb expression. J Clin Invest 96:2461–2468
67. Chojkier M, Houglum K, Solis-Herruzo J, Brenner DA (1989) Stimulation of collagen gene expression by ascorbic acid in cultured human fibroblasts. a role for lipid peroxidation? J Biol Chem 264:16957–16962
68. Tschopp J, Schroder K (2010) NLRP3 inflammasome activation: the convergence of multiple signalling pathways on ROS production? Nature reviews. Immunology 10:210–215
69. Wallace MC, Friedman SL (2014) Hepatic fibrosis and the microenvironment: fertile soil for hepatocellular carcinoma development. Gene Expr 16:77–84
70. Bhowmick NA, Neilson EG, Moses HL (2004) Stromal fibroblasts in cancer initiation and progression. Nature 432:332–337
71. Thompson AI, Conroy KP, Henderson NC (2015) Hepatic stellate cells: central modulators of hepatic carcinogenesis. BMC Gastroenterol 15:63
72. Florimond A, Chouteau P, Bruscella P, Le Seyec J, Merour E, Ahnou N, Mallat A, Lotersztajn S, Pawlotsky JM (2015) Human hepatic stellate cells are not permissive for hepatitis C virus entry and replication. Gut 64:957–965
73. Adinolfi LE, Rinaldi L, Guerrera B, Restivo L, Marrone A, Giordano M, Zampino R (2016) NAFLD and NASH in HCV infection: prevalence and significance in hepatic and extrahepatic manifestations. Int J Mol Sci 17
74. Su AI, Pezacki JP, Wodicka L, Brideau AD, Supekova L, Thimme R, Wieland S, Bukh J, Purcell RH, Schultz PG, Chisari FV (2002) Genomic analysis of the host response to hepatitis C virus infection. Proc Natl Acad Sci U S A 99:15669–15674
75. Alonzi T, Agrati C, Costabile B, Cicchini C, Amicone L, Cavallari C, Rocca CD, Folgori A, Fipaldini C, Poccia F, Monica NL, Tripodi M (2004) Steatosis and intrahepatic lymphocyte recruitment in hepatitis C virus transgenic mice. J Gen Virol 85:1509–1520
76. Lerat H, Honda M, Beard MR, Loesch K, Sun J, Yang Y, Okuda M, Gosert R, Xiao SY, Weinman SA, Lemon SM (2002) Steatosis and liver cancer in transgenic mice expressing the structural and nonstructural proteins of hepatitis C virus. Gastroenterology 122:352–365

77. Barba G, Harper F, Harada T, Kohara M, Goulinet S, Matsuura Y, Eder G, Schaff Z, Chapman MJ, Miyamura T, Brechot C (1997) Hepatitis C virus core protein shows a cytoplasmic localization and associates to cellular lipid storage droplets. Proc Natl Acad Sci U S A 94:1200–1205
78. Perlemuter G, Sabile A, Letteron P, Vona G, Topilco A, Chretien Y, Koike K, Pessayre D, Chapman J, Barba G, Brechot C (2002) Hepatitis C virus core protein inhibits microsomal triglyceride transfer protein activity and very low density lipoprotein secretion: a model of viral-related steatosis. FASEB J: Off Publ Fed Am Soc Exp Biol 16:185–194
79. Serfaty L, Andreani T, Giral P, Carbonell N, Chazouilleres O, Poupon R (2001) Hepatitis C virus induced hypobetalipoproteinemia: a possible mechanism for steatosis in chronic hepatitis C. J Hepatol 34:428–434
80. Lonardo A, Adinolfi LE, Loria P, Carulli N, Ruggiero G, Day CP (2004) Steatosis and hepatitis C virus: mechanisms and significance for hepatic and extrahepatic disease. Gastroenterology 126:586–597
81. Houglum K, Venkataramani A, Lyche K, Chojkier M (1997) A pilot study of the effects of d-alpha-tocopherol on hepatic stellate cell activation in chronic hepatitis C. Gastroenterology 113:1069–1073
82. Wieland S, Makowska Z, Campana B, Calabrese D, Dill MT, Chung J, Chisari FV, Heim MH (2014) Simultaneous detection of hepatitis C virus and interferon stimulated gene expression in infected human liver. Hepatology 59:2121–2130
83. Gerber MA, Shieh YS, Shim KS, Thung SN, Demetris AJ, Schwartz M, Akyol G, Dash S (1992) Detection of replicative hepatitis C virus sequences in hepatocellular carcinoma. Am J Pathol 141:1271–1277
84. Haruna Y, Hayashi N, Kamada T, Hytiroglou P, Thung SN, Gerber MA (1994) Expression of hepatitis C virus in hepatocellular carcinoma. Cancer 73:2253–2258
85. Dash S, Saxena R, Myung J, Rege T, Tsuji H, Gaglio P, Garry RF, Thung SN, Gerber MA (2000) HCV RNA levels in hepatocellular carcinomas and adjacent non-tumorous livers. J Virol Methods 90:15–23
86. Sobesky R, Feray C, Rimlinger F, Derian N, Dos Santos A, Roque-Afonso AM, Samuel D, Brechot C, Thiers V (2007) Distinct hepatitis C virus core and F protein quasispecies in tumoral and nontumoral hepatocytes isolated via microdissection. Hepatology 46:1704–1712
87. Spaniel C, Honda M, Selitsky SR, Yamane D, Shimakami T, Kaneko S, Lanford RE, Lemon SM (2013) microRNA-122 abundance in hepatocellular carcinoma and non-tumor liver tissue from Japanese patients with persistent HCV versus HBV infection. PLoS One 8:e76867
88. Jopling CL, Yi M, Lancaster AM, Lemon SM, Sarnow P (2005) Modulation of hepatitis C virus RNA abundance by a liver-specific MicroRNA. Science 309:1577–1581
89. Harouaka D, Engle RE, Wollenberg K, Diaz G, Tice AB, Zamboni F, Govindarajan S, Alter H, Kleiner DE, Farci P (2016) Diminished viral replication and compartmentalization of hepatitis C virus in hepatocellular carcinoma tissue. Proc Natl Acad Sci U S A 113:1375–1380
90. Gosalvez J, Rodriguez-Inigo E, Ramiro-Diaz JL, Bartolome J, Tomas JF, Oliva H, Carreno V (1998) Relative quantification and mapping of hepatitis C virus by in situ hybridization and digital image analysis. Hepatology 27:1428–1434
91. Lau GK, Davis GL, Wu SP, Gish RG, Balart LA, Lau JY (1996) Hepatic expression of hepatitis C virus RNA in chronic hepatitis C: a study by in situ reverse-transcription polymerase chain reaction. Hepatology 23:1318–1323
92. Liang Y, Shilagard T, Xiao SY, Snyder N, Lau D, Cicalese L, Weiss H, Vargas G, Lemon SM (2009) Visualizing hepatitis C virus infections in human liver by two-photon microscopy. Gastroenterology 137:1448–1458
93. Stiffler JD, Nguyen M, Sohn JA, Liu C, Kaplan D, Seeger C (2009) Focal distribution of hepatitis C virus RNA in infected livers. PLoS One 4:e6661

94. Meredith LW, Harris HJ, Wilson GK, Fletcher NF, Balfe P, McKeating JA (2013) Early infection events highlight the limited transmissibility of hepatitis C virus in vitro. J Hepatol 58:1074–1080
95. Hedegaard DL, Tully DC, Rowe IA, Reynolds GM, Bean DJ, Hu K, Davis C, Wilhelm A, Ogilvie CB, Power KA, Tarr AW, Kelly D, Allen TM, Balfe P, McKeating JA (2016) High resolution sequencing of hepatitis C virus reveals limited intra-hepatic compartmentalization in end stage liver disease. J Hepatol 66:28
96. Kanwal F, Kramer JR, Ilyas J, Duan Z, El-Serag HB (2014) HCV genotype 3 is associated with an increased risk of cirrhosis and hepatocellular cancer in a national sample of U.S. veterans with HCV. Hepatology 60:98–105
97. Raimondi S, Bruno S, Mondelli MU, Maisonneuve P (2009) Hepatitis C virus genotype 1b as a risk factor for hepatocellular carcinoma development: a meta-analysis. J Hepatol 50:1142–1154
98. El-Shamy A, Eng FJ, Doyle EH, Klepper AL, Sun X, Sangiovanni A, Iavarone M, Colombo M, Schwartz RE, Hoshida Y, Branch AD (2015) A cell culture system for distinguishing hepatitis C viruses with and without liver cancer-related mutations in the viral core gene. J Hepatol 63:1323–1333
99. Korba B, Shetty K, Medvedev A, Viswanathan P, Varghese R, Zhou B, Roy R, Makambi K, Ressom H, Loffredo CA (2015) Hepatitis C virus genotype 1a core gene nucleotide patterns associated with hepatocellular carcinoma risk. J Gen Virol 96:2928–2937
100. Charles ED, Dustin LB (2009) Hepatitis C virus-induced cryoglobulinemia. Kidney Int 76:818–824
101. Iqbal T, Mahale P, Turturro F, Kyvernitakis A, Torres HA (2016) Prevalence and association of hepatitis C virus infection with different types of lymphoma. Int J Cancer 138:1035–1037
102. Peveling-Oberhag J, Arcaini L, Hansmann ML, Zeuzem S (2013) Hepatitis C-associated B-cell non-Hodgkin lymphomas. Epidemiology, molecular signature and clinical management. J Hepatol 59:169–177
103. Pileri P, Uematsu Y, Campagnoli S, Galli G, Falugi F, Petracca R, Weiner AJ, Houghton M, Rosa D, Grandi G, Abrignani S (1998) Binding of hepatitis C virus to CD81. Science 282:938–941
104. Ng PP, Kuo CC, Wang S, Einav S, Arcaini L, Paulli M, Portlock CS, Marcotrigiano J, Tarr A, Ball J, Levy R, Levy S (2014) B-cell receptors expressed by lymphomas of hepatitis C virus (HCV)-infected patients rarely react with the viral proteins. Blood 123:1512–1515
105. Dai B, Chen AY, Corkum CP, Peroutka RJ, Landon A, Houng S, Muniandy PA, Zhang Y, Lehrmann E, Mazan-Mamczarz K, Steinhardt J, Shlyak M, Chen QC, Becker KG, Livak F, Michalak TI, Talwani R, Gartenhaus RB (2016) Hepatitis C virus upregulates B-cell receptor signaling: a novel mechanism for HCV-associated B-cell lymphoproliferative disorders. Oncogene 35:2979–2990
106. Vannata B, Zucca E (2014) Hepatitis C virus-associated B-cell non-Hodgkin lymphomas. hematology/the education program of the American Society of Hematology. Am Soc Hematol Educ Program 2014:590–598
107. Sung VM, Shimodaira S, Doughty AL, Picchio GR, Can H, Yen TS, Lindsay KL, Levine AM, Lai MM (2003) Establishment of B-cell lymphoma cell lines persistently infected with hepatitis C virus in vivo and in vitro: the apoptotic effects of virus infection. J Virol 77:2134–2146
108. Marukian S, Jones CT, Andrus L, Evans MJ, Ritola KD, Charles ED, Rice CM, Dustin LB (2008) Cell culture-produced hepatitis C virus does not infect peripheral blood mononuclear cells. Hepatology 48:1843–1850
109. Pawlotsky JM (2013) Treatment of chronic hepatitis C: current and future. Curr Top Microbiol Immunol 369:321–342
110. Reig M, Marino Z, Perello C, Inarrairaegui M, Ribeiro A, Lens S, Diaz A, Vilana R, Darnell A, Varela M, Sangro B, Calleja JL, Forns X, Bruix J (2016) Unexpected high rate of early tumor recurrence in patients with HCV-related HCC undergoing interferon-free therapy. J Hepatol 65:719

111. Yi M, Hu F, Joyce M, Saxena V, Welsch C, Chavez D, Guerra B, Yamane D, Veselenak R, Pyles R, Walker CM, Tyrrell L, Bourne N, Lanford RE, Lemon SM (2014) Evolution of a cell culture-derived genotype 1a hepatitis C virus (H77S.2) during persistent infection with chronic hepatitis in a chimpanzee. J Virol 88:3678–3694
112. Chen J, Zhao Y, Zhang C, Chen H, Feng J, Chi X, Pan Y, Du J, Guo M, Cao H, Chen H, Wang Z, Pei R, Wang Q, Pan L, Niu J, Chen X, Tang H (2014) Persistent hepatitis C virus infections and hepatopathological manifestations in immune-competent humanized mice. Cell Res 24:1050–1066
113. Tornesello ML, Buonaguro L, Izzo F, Buonaguro FM (2016) Molecular alterations in hepatocellular carcinoma associated with hepatitis B and hepatitis C infections. Oncotarget 7:25087–25102

Chapter 9
Human T-Cell Leukemia Virus Type 1 Infection and Adult T-Cell Leukemia

Chi-Ping Chan, Kin-Hang Kok, and Dong-Yan Jin

Abstract Human T-cell leukemia virus type 1 (HTLV-1) is the first retrovirus discovered to cause adult T-cell leukemia (ATL), a highly aggressive blood cancer. HTLV-1 research in the past 35 years has been most revealing in the mechanisms of viral oncogenesis. HTLV-1 establishes a lifelong persistent infection in $CD4^+$ T lymphocytes. The infection outcome is governed by host immunity. ATL develops in 2–5% of infected individuals 30–50 years after initial exposure. HTLV-1 encodes two oncoproteins Tax and HBZ, which are required for initiation of cellular transformation and maintenance of cell proliferation, respectively. HTLV-1 oncogenesis is driven by a clonal selection and expansion process during which both host and viral factors cooperate to impair genome stability, immune surveillance, and other mechanisms of tumor suppression. A better understanding of HTLV-1 biology and leukemogenesis will reveal new strategies and modalities for ATL prevention and treatment.

Keywords Human T-cell leukemia virus type 1 • Adult T-cell leukemia • Tax • HBZ • Humanized mouse model

9.1 Introduction

Human T-cell leukemia virus type 1 (HTLV-1) was discovered in 1980 as the first human retrovirus and the etiological agent of adult T-cell leukemia (ATL) [1, 2]. Since then HTLV-1 research has laid the foundation of viral oncology and human retrovirology [3]. Animal oncogenic retroviruses such as Rous sarcoma virus are

C.-P. Chan • D.-Y. Jin (✉)
School of Biomedical Sciences, The University of Hong Kong,
21 Sassoon Road, Pokfulam, Hong Kong
e-mail: dyjin@hku.hk

K.-H. Kok
Department of Microbiology, The University of Hong Kong,
145 Pokfulam Road, Pokfulam, Hong Kong

Q. Cai et al. (eds.), *Infectious Agents Associated Cancers: Epidemiology and Molecular Biology*, Advances in Experimental Medicine and Biology 1018,
DOI 10.1007/978-981-10-5765-6_9

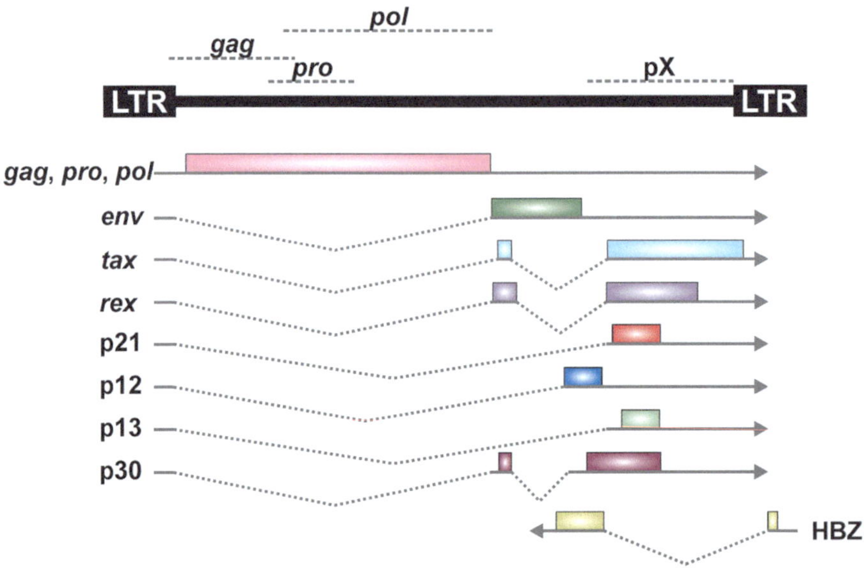

Fig. 9.1 HTLV-1 genome organization. The structural genes *gag*, *pro*, *pol*, and *env* as well as the regulatory genes *tax*, *rex*, p21, p12, p13, p30, and HBZ are shown

replication defective, and they carry viral oncogenes that are originally derived from cellular proto-oncogenes of their host. Unlike these animal retroviruses, HTLV-1 is replication-competent and does not carry any cellular oncogenes. The viral oncoproteins encoded by HTLV-1 are unique, have no cellular counterparts, and are not homologous to any cellular proteins. HTLV-1 has been under intense investigations since HTLV-1 infection causes significant morbidity and mortality in endemic areas. In addition, it also serves as an excellent model for the study of viral oncogenesis. HTLV-1 research has contributed substantially to our understanding of oncogenic viruses and oncogenesis in general.

HTLV-1 is a complex deltaretrovirus that harbors additional regulatory genes in addition to *gag*, *pol*, *pro*, and *env* genes flanked by long terminal repeats (LTR) as found in simple retroviruses (Fig. 9.1). The *gag* gene encodes the major component of the viral capsid. The *pol* and *pro* region provides the reverse transcriptase, protease, and integrase. Interestingly, *gag* and *pol* are produced by ribosome frameshift from a single transcript. The *env* gene codes for a glycoprotein that mediates viral entry. The pX region between *env* and the 3′-LTR encodes Tax, Rex, as well as other accessory proteins p12, p13, p21, and p30 derived from alternatively spliced transcripts. Tax is a viral transactivator that potently activates transcription from the LTR. Rex mediates nuclear export of viral RNA. The additional accessory proteins are dispensable for viral replication and transformation in vitro but are required for viral propagation and persistence in vivo. In particular, p30 counteracts Toll-like receptor signaling and cooperates with c-Myc to promote cellular transformation [4, 5]. p12 and its cleavage product p8 mediate T-cell activation, immune evasion, and cell-to-cell transmission [6, 7]. Distinct to other retroviruses, HTLV-1 also expresses an antisense transcript encoding the helix-basic loop zipper protein HBZ [8, 9]. Tax and HBZ are two viral oncoproteins that cooperate to drive HTLV-1 leukemogenesis. A full discussion of this will be provided below in part 4 of this review.

HTLV-1 has a relative known as HTLV-2. Although they are similar in genome organization and tissue tropism, there is one important difference in pathogenesis: human infection with HTLV-2 is not associated with any malignancy. HTLV-2 and its proteins are therefore commonly used as controls in the study of HTLV-1 oncogenesis. More recently, two new HTLV viruses named HTLV-3 and HTLV-4 have been isolated from Cameroonian hunters of nonhuman primates [10, 11]. Primate counterparts of all four HTLVs have also been identified, and these four pairs of viruses, together with bovine leukemia virus (BLV) and another orphan primate retrovirus, constitute the genus of deltaretroviruses [11, 12]. Whereas infection with HTLV-1 and BLV is associated with leukemia, but HTLV-2 infection is not, it remains to be seen whether HTLV-3 and HTLV-4 are also leukemogenic.

ATL is a heterogeneous disease with four clinical subtypes: acute, lymphoma, chronic, and smoldering. Acute, lymphoma, and unfavorable chronic subtypes are known as aggressive ATL with large tumor burden, blood and lymph node involvement, and hypercalcemia. Favorable chronic and smoldering subtypes are indolent ATL characterized by rash and minimal blood involvement [13]. Smoldering ATL is considered to be an early phase of the disease, which progresses subsequently to acute ATL. Prognosis of aggressive ATL is very poor, with an average survival rate of only a few months.

The mechanisms of HTLV-1 oncogenesis have recently been reviewed [3, 13–16]. In this chapter, we will revisit the topic with an emphasis on new thoughts and findings. We will start with an overview of epidemiology, followed by a brief summary of experimental models in HTLV-1 research. The interplay of the two HTLV-1 oncoproteins and the mechanisms of HTLV-1 oncogenesis will then be discussed in detail. Finally, we will highlight the treatment options and the new approaches to anti-HTLV-1 therapy.

9.2 Epidemiology

Based on sequence and epidemiological analysis, the primate counterparts of HTLVs are well established in their natural hosts. Several lines of evidence including phylogenetic clustering and geographical coincidence support zoonotic transmission of these primate viruses to humans, plausibly through petting and butchering. The process by which HTLVs establish as a human pathogen through adaptive mutations is similar to that demonstrated for human immunodeficiency viruses. The detection of HTLV-3 and HTLV-4 in African hunters of primates has lent further support to this notion [10, 11]. HTLV-1 and HTLV-2 have obviously acquired the ability to transmit from human to human readily. The identification of HTLV-3 and HTLV-4 has provided a golden opportunity to study their human-to-human transmissibility, pathogenicity, and degree of adaptation. Interestingly, each HTLV has its own primate counterpart. Phylogenetic analysis supports at least four independent introductions of virus into human population. Each of this involves a different species of primate virus.

It is estimated that about 20 million people are infected with HTLV-1 worldwide. About one million of them are in Japan, and the adjusted overall prevalence nationwide is approximately 1% [17]. However, the distribution of HTLV-1 carriers within the country is uneven and highly focal. As such, carrier rate in Japanese women in the age of >50 in endemic areas can be as high as 40%, but these areas are surrounded by areas of low to middle prevalence. HTLV-1 is also highly prevalent in African people resided in the Caribbean islands and tropical Africa; Mongoloid people in South America, Central America, and the Middle East; as well as Melanesian people in northern Oceania [18]. Particularly, some Aboriginal Australians have been found to have the highest prevalence (>50%) of HTLV-1. Interestingly, the study of HTLV-1 prevalence in different populations might even provide useful anthropological information concerning their origin, migration routes, and genetic history.

The prevalence of HTLV-1 in healthy blood donors in China is very low, ranging from 0.01 to 0.08% [19, 20]. The rate could be 0.5% or higher in some professional blood donors and drug addicts. Only some of the Chinese individuals who are seropositive for HTLV-1 have been found to have close contact with Japanese people [21]. Interestingly, prevalence rates of HTLV-1 in healthy blood donors in two cities named Ningde and Putian in the coastal Fujian Province are 0.40% and 0.14%, respectively. These significantly higher rates are indicative of some foci of HTLV-1 carriers [19]. The seropositive rate of HTLV-1 of 0.74% in the ethnic group of Xinjiang Uyghurs is also high. A very small number of sporadic tropical spastic paraparesis (TSP) and ATL cases associated with HTLV-1 infection have also been diagnosed in China in recent years. Several sporadic cases of ATL have been found in Hong Kong, where a more advanced disease surveillance system is in place [22]. Some of these victims in Hong Kong had unsafe sex in Southern Japan. The carrier rate in Hong Kong is estimated to be 0.0041%. In Taiwan, the seropositive rate for HTLV-1 was found to range from to 0.058 to 0.48% [23, 24]. Interestingly, the rates in Aborigines and Hakka Taiwanese are higher than in other ethnic groups. Blood donor screening for HTLV-1 has been implemented in Taiwan since 1996. It will be of some interest to determine whether the Taiwanese HTLV-1 strains are closer to those found in Japan or Fujian, which is geographically closer to Taiwan.

Although HTLV-1 can be divided into six genotypes A to F, sequence variations among genotypes are minor and not as significant as in HIV-1. The sequence divergence among HTLV-1 genotypes is much less than that among different HTLV viruses. Genotype A is predominant. The genotype of HTLV-1 in ATL patients and healthy carriers is not found to be different. Neither is there evidence in support of the influence of genotype on pathogenicity or infection outcome.

HTLV-1 is an infection vertically transmitted from mother to child through breastfeeding. The risk of infection acquired through this route can be as high as 30%. Blood transfusion and unsafe sex are two other routes by which HTLV-1 is transmitted, but sexual transmission is not as efficient as in the case of HIV-1. More than 90% of HTLV-1-infected people remain healthy throughout their lifetime. ATL develops in 2–5% of infected individuals after a prolonged latent period of 30–50 years, during which they remain asymptomatic. Once developed, ATL is highly aggressive and fatal, with very limited and unsatisfactory treatment options [25]. A smaller subset of infected people suffers from TSP or HTLV-1-associated myelopathy, a chronic debilitating neurological disease of the spinal cord. It is not common that TSP and ATL develop sequentially in the same individual. During the long process of ATL development, multiple viral, host, and environmental factors are involved. Notably, high HTLV-1 proviral load is the single major risk factor for ATL development in HTLV-1 carriers. Other reported risk factors include advanced age, family history of ATL, male sex, and HTLV-1 infection early in life [26]. Most ATL cases are associated with breastfeeding. The cumulative risks of developing ATL among HTLV-1 carriers are approximately 6% for males and 2% for females. The high predictive value of proviral load suggests that anti-HTLV-1 therapy might be beneficial in the prevention of ATL.

$CD4^+$ T lymphocytes are the primary and preferential target cells of HTLV-1 in vivo, although other cells such as $CD8^+$ T lymphocytes, monocytes, and dendritic cells (DCs) can also be infected. It remains to be clarified whether HTLV-1 might first infect DCs, which pass on the virus to T cells [27]. One recent report has implicated HTLV-1-transformed $CD45RA^+$ T memory stem cells with stemlike properties as the ATL-initiating cells [28]. These cells could serve as the viral reservoir and a barrier for viral eradication by antivirals. Cell-free transmission of HTLV-1 is highly inefficient except for DCs. All major routes of HTLV-1 transmission including breastfeeding, blood transfusion, and sexual intercourse involve the transfer of infected cells residing in the breast milk, blood, and semen. Cell-to-cell transmission of HTLV-1 is achieved through the virological synapse, which involves the interaction between ICAM-1 on infected cells and LFA-1 on target cells and polarization of the microtubule-organizing center induced by Tax protein [29, 30]. Interestingly, virions at the virological synapse are stored as biofilm-like extracellular assemblies [31]. In addition to cell-to-cell contact, HTLV-1 can also be passed on to daughter cells via mitosis. Thus, the HTLV-1 proviral load in vivo might be determined by both mitotic spread and cell-to-cell transmission. In this connection, it will be of great interest to see how the interaction between Tax and mitotic regulators such as MAD1 [32] might influence both processes. In addition, the HTLV-1 proviral load is also affected by host immune response and particularly by cytotoxic T lymphocyte (CTL) response against viral proteins such as Tax and HBZ [33]. Thus, a better understanding of anti-HTLV-1 CTL response might reveal new strategies for prevention of ATL.

9.3 Experimental Models in HTLV-1 Research

Mechanistic studies of HTLV-1 pathogenesis have largely been conducted in transfected cells in which the HTLV-1 protein of interest is overexpressed. Although these cells provide a good model for the study of HTLV-1 proteins, there are several concerns about their relevance to HTLV-1 infection and biology. First, the expression level of the HTLV-1 protein of interest in transfected cells might be much higher than in infected cells. Second, constitutive expression of the HTLV-1 protein of interest could exhaust or sequester its partners and effectors leading to a squelching effect. Regulated or inducible expression is desired. For example, JPX9 cells, in which Tax expression can be induced by Cd^{2+} [34], have proved useful in HTLV-1 research. Third, target cells of HTLV-1 are difficult to transfect. Surrogate models such as HeLa and HEK293 cells with high transfection efficiency are helpful, but they are significantly different from $CD4^+$ T cells in many ways. The concern might be addressed by the use of new transfection reagents tailor-made for T cells. Finally, different HTLV-1 proteins interact with each other to fulfill some functions cooperatively or antagonistically. This might not be reconstituted in transfected cells. Tens if not hundreds of cellular binding partners of HTLV-1 proteins such as Tax and HBZ have been identified. Not all of them have been validated in infected cells. Interpretation of these interaction results should be cautious, bearing in mind the limitations of the transfection system. Whenever possible, T-cell lines such as Jurkat and CEMT4 as well as peripheral blood mononuclear cells (PBMCs) infected with HTLV-1 should be used to verify findings obtained from transfected cells.

Various types of cultured cells including T cells, B cells, DCs, monocytes, endothelial cells, and fibroblasts can be infected in vitro with HTLV-1 through coculture with HTLV-1-infected cells such as MT2 and C8166. These freshly infected cells serve as a good model for acute infection. In contrast, other ATL or derivative cell lines are chronically infected with and transformed by HTLV-1. For example, MT4 and HUT102 cells were derived from ATL patients. MT2 cells were established through coculture of normal cord leukocytes with ATL cells. C8166 cells were obtained by fusion of normal cord leukocytes with ATL cells. Whereas these several lines constitutively express Tax, other lines in which Tax expression has faded but HBZ expression remains robust include ED and TL-Om1 [35, 36]. These cells representative of different phases of infection are widely used in HTLV-1 research. In addition to infection through coculture, infectious molecular clones of HTLV-1 are also available [37]. These clones can be transfected into any cells and spawn HTLV-1 infection in susceptible cells. They greatly facilitate genetic analysis of HTLV-1.

BLV and HTLV-1 share many features in common. Both are transmitted through body fluids requiring cell-to-cell contact. Both are leukemogenic in only a fraction of infected hosts after a prolonged latent period. Thus, BLV serves as a good and relevant model for HTLV-1 research [38]. Particularly, promising results on the prevention of BLV-associated diseases through competitive infection with an attenuated BLV provirus provide useful information as to how proviral load can be reduced with this strategy [39]. BLV can infect several ruminant species with highest prevalence in dairy cattle. Infection of small ruminants such as sheep with BLV has therefore been developed as a productive animal model for both BLV and HTLV-1 research. Sheep can be easily infected and the disease outcome can be observed sooner [40]. By the same reasoning, infection of monkeys such as Japanese macaques with STLV-1 provides useful information about HTLV-1 pathogenesis [41].

Various types of tumor develop in Tax- or HBZ-transgenic mice, but in most cases these are neither leukemia nor lymphoma [42, 43]. Directing the expression of Tax or HBZ more specifically to particular tissues and cells such as thymus and leukocytes is a technical challenge that has only been met partially. Tissue-specific promoters such as those of CD4, CD3ε, Ig, Lck, and granzyme B have been used with some success in the generation of more relevant disease outcomes. Although these transgenic mice are not perfect models for HTLV-1 infection or ATL development, they provide convincing evidence for the oncogenicity of Tax and HBZ proteins, reveal different facets of HTLV-1 oncogenesis, and also serve as platforms for the development of new therapy for ATL. Complementary to transgenic mice, infection of immunocompetent rabbits with HTLV-1 provides another model [38]. However, no disease or symptom related to ATL or TSP can be recapitulated in rabbits.

Many features of ATL can be reproduced in immunocompromised mice engrafted with ATL cells or ATL-derived cell lines [44]. These mouse xenograft models have been used to study HTLV-1 leukemogenesis and to develop anti-HTLV-1 therapy. The immunocompromised mice that have been developed include SCID, NOD-SCID, NSG, NOG, and BRG mice [45]. SCID mice contain a nonsense mutation in the protein kinase required for VDJ recombination of T- and B-cell receptors, leading to a severe combined immunodeficiency (SCID). In NOD-SCID mice, the SCID mutation has been introduced into the nonobese diabetic (NOD) genetic background. This further compromises innate immunity by blocking the function of complements, DCs, and macrophages. Similar to X-linked SCID in human, deficiency in the interleukin-2 receptor common subunit γ (IL2R-γ_C) in mice results in a complete loss of T, B, and NK cells. In NSG and NOG mice, this mutation in IL2R-γ_C has been introduced into the NOD/SCID background. Likewise, BRG mice are deficient for IL2R-γ_C and the recombinase-activating gene 2 (Rag2).

Humanization of immunocompromised mice by reconstituting their immune system through engraftment of human hematopoietic stem cells has not only opened the door for detailed analysis of human immunity but also provided a powerful new tool for the study of human pathogens including HTLV-1 [46]. The mice are engrafted with $CD34^+$ hematopoietic stem cells from human peripheral and cord blood. Because all $CD4^+$ T lymphocytes in these mice are derived from the engrafted human cells, they are excellent models for lymphotropic viruses such as HTLV-1 [47, 48]. These models have already been used successfully to study HTLV-1 infection and oncogenesis [49, 50]. For example, $CD4^+$ T-cell lymphoma was shown to develop in NOD-SCID mice engrafted with human $CD34^+$ cells infected with HTLV-1 [50]. Although the original paper reporting this finding was later retracted by the editors and it remains to be determined whether HTLV-1 infects $CD34^+$ cells, which subsequently differentiated into $CD4^+$ cells, humanized mice still hold great promises to advance HTLV-1 and ATL research with a biologically relevant model. One challenge in this area is to develop a good model for persistent HTLV-1 infection in humanized mice.

9.4 HTLV-1 Oncoproteins

HTLV-1 encodes two major oncoproteins Tax and HBZ. In this part we will first describe existing findings on Tax and HBZ essentially in chronological order. Then we will summarize how they exert their impacts on the hallmarks of cancer. Finally we will discuss their differential roles in HTLV-1 leukemogenesis.

Tax is a 40-kDa transactivator protein serving as the master regulator of HTLV-1 proviral expression from the LTR. To activate HTLV-1 transcription, Tax forms a homodimer to engage CREB and DNA of three cAMP-response element-like 21-bp repeats in the LTR [51–54]. Tax has a minimal transactivation domain [55]. Optimal activity of Tax specifically requires the core TATAA promoter of HTLV-1, CREB, and the 21-bp repeats [56]. Transcriptional coactivators including p300/CREB-binding protein (CBP) and CREB-regulating transcriptional coactivators (CRTCs) are then attracted by Tax [52, 57, 58]. Tax also recruits other regulators and protein modification enzymes to modulate this process [59]. For example, p21-activated kinases are recruited to activate LTR-dependent transcription [60], whereas LKB1 and salt-inducible kinases [61], protein deacetylases SIRT1 [62], as well as T cell-specific transcription factors TCF1 and LEF1 [63] are recruited to medicate negative regulation of proviral transcription.

In addition to CREB, NF-κB is another major cellular transcription factor activated by Tax [16, 64]. Tax interacts with NF-κB regulators such as IκB kinase regulatory subunit IKK-γ [65–67], ubiquitin-editing enzyme A20 [68, 69], ubiquitin-binding adaptor protein TAX1BP1 [70, 71], E2 ubiquitin-conjugating enzyme UBC13 [72], and E3 ubiquitin ligase RNF8 [73] to modulate NF-κB activation through the ubiquitin-proteasome pathway. Notably, Tax is a powerful modulator of K48-linked, K63-linked, and linear ubiquitination of key adaptors of NF-κB signaling such as IKK-γ and TAB2 [73, 74]. Furthermore, Tax can also interact with and stimulate SRF [75, 76] and c-Jun [73, 77] transcription factors resulting in the activation of transcription from serum response elements and AP-1-binding sites.

It is generally accepted that Tax is required for the initiation of HTLV-1-mediated malignant transformation. Expression of Tax alone can sufficiently transform murine fibroblasts [78], immortalize T lymphocytes [79], and induce tumor formation in nude mice and transgenic mice [42, 43, 80]. Through CREB and NF-κB, Tax activates a wide variety of cellular genes that contribute to transformation. Activation of both CREB and NF-κB signaling is required for full-blown transformation induced by Tax [81, 82].

Tax is a multifunctional protein that activates transcription and transformation primarily through protein-protein interaction [59]. Tax is known to interact with a subset of PDZ domain-containing proteins. For example, Tax interacts with TIP1 [83], PDLIM2 [84], and MAGI1 [85] that contain PDZ domains. Another group of Tax-binding proteins contains the coiled-coil motifs that mediate their interaction with Tax [86]. Proteins in this group include mitotic checkpoint protein MAD1 [32], transcriptional repressor GPS2 [87], regulatory subunit IKK-γ of IκB kinase [65–67], centrosomal and ciliary protein TAX1BP2 [88], transcriptional coactivators CRTC1/CRTC2/CRTC3 [57, 58], as well as ubiquitin-binding adaptor protein TAX1BP1 [70, 71]. Tax-binding proteins in both groups are the effectors of Tax in transcriptional regulation and transformation.

HTLV-1 expresses both unspliced and spliced forms of HBZ, with the latter form being more abundant in infected cells [9]. The expression of HBZ appears to be required for HTLV-1 infectivity in vivo [89, 90]. Interestingly, HBZ RNA and protein show differential activity on apoptosis, but both promote cell cycle progression into S phase [87]. HBZ protein was initially identified as a heterodimerization partner of ATF4 [8]. In most cases dimerization of HBZ with ATF4, CREB, c-Jun, and other bZIP transcription factors results in repression of their activity. Thus, HBZ is a negative regulator of proviral transcription and it counteracts the activity of Tax. In addition, HBZ suppresses canonical pathway of NF-κB activation. Collectively, HBZ plays an important role in the proliferation of infected T cells as well as the induction and maintenance of latent infection [3, 16, 89–93].

The hallmarks of cancer include self-sufficiency in growth signal, insensitivity to antigrowth signals, resisting cell death, enabling replicative immortality, evading immune surveillance, genome instability and mutation, as well as tumor-promoting inflammation [94]. Although Tax and HBZ are antagonistic in many scenarios, they cooperate with each other to impinge on the different hallmarks of cancer. Some key examples are summarized below. HBZ activates Wnt signaling to sustain T-cell proliferation [95]. Tax perturbs tumor suppressor function of p53 [96] and Rb [97, 98]. HBZ suppresses apoptosis by targeting FoxO3a that activates proapoptotic genes [99]. Tax suppresses innate antiviral response by preventing TBK1-induced type I interferon production [100]. HBZ induces the expression of immune checkpoint molecule TIGIT to evade T-cell response [101]. Tax impairs DNA damage response [100–105], mitotic checkpoint [32], and centrosome duplication [88] leading to genome instability and a mutator phenotype. HBZ activates the transcription of hTERT to elevate telomerase activity [106]. Whereas HBZ enhances transforming growth factor-β signaling leading to overproduction of IFN-γ [43, 107, 108], Tax activates NF-κB to induce various cytokines [64]. Both result in activation of pro-inflammatory response.

Tax and HBZ play different roles in HTLV-1 oncogenesis. Whereas Tax is required for the initiation of oncogenic transformation, HBZ is essential for the induction and maintenance of HTLV-1 persistence and T-cell proliferation. Consistent with this model, Tax is abundantly expressed in the early stage of infection and transformation, but its expression and activity are suppressed through multiple mechanisms. First, promoter hypermethylation occurs in the 5′-LTR, leading to inhibition of Tax gene transcription [109]. Second, deletions and inactivating mutations are commonly found in the 5′-LTR and Tax coding region in the HTLV-1 genome in ATL cells [3]. Third, Tax recruits a group of inhibitors of proviral transcription such as LKB1, SIRT1, TCF1, and LEF1 through a negative feedback loop [61–63]. Last but not least, the strong CTL response directed against Tax essentially selects for T cells with low or no expression of Tax [33, 110]. In contrast, HBZ is constitutively expressed in all stages of infection and transformation. The differential activation of 5′- and 3′-LTR in the HTLV-1 genome is governed by chromatin insulator CTCF and the CTCF-binding site in the pX region [111].

The expression of Tax and HBZ in infected individuals is highly dynamic. There exist a large number of HTLV-1$^+$ clones, each of which is characterized by a unique integration site of the HTLV-1 provirus in the host genome [112]. The pattern and level of Tax and HBZ expression could vary from one to another clone. They might even be passed on to the daughter cells through mitosis. In particular, Tax expression is known to be influenced by the distance and transcriptional direction of the provirus relative to the host gene in the closest vicinity. Importantly, the abundance of Tax- and HBZ-expressing cells is also governed by the CTL response. HBZ is less immunogenic than Tax [110], but CTL response targeting HBZ can potently suppress T-cell proliferation and has protective effect [113]. As mentioned above, the CTL response confers a survival advantage to HTLV-1$^+$ clones that do not express Tax [33, 110]. From another perspective, HTLV-1-infected cells express cyclin-dependent kinase inhibitors p21 and p27 to high levels and enter cellular senescence in an NF-κB-dependent manner [110]. Cells in which Tax is abundantly expressed, NF-κB activity is high, and HTLV-1 replication is robust would be eliminated by apoptosis. Only latently infected cells with high HBZ expression and low NF-κB activation would survive [16, 114].

9.5 Mechanisms of HTLV-1 Oncogenesis

The long latency period of ATL development indicates that HTLV-1 oncogenesis is a slow and multistage process. Above we describe with examples the impact of Tax and HBZ oncoproteins on the different hallmarks of cancer. The subversion of genome instability, the evasion from immune response, and the induction of pro-inflammatory response are particularly attractive mechanisms that warrant further investigations. These mechanisms are critically important and they contribute to different stages of HTLV-1 oncogenesis. However, we should also bear in mind that ATL does not develop overnight, and it is the collective effect of Tax and HBZ over several decades that ultimately gives rise to ATL. Currently there is no consensus model that could fully explain the process of HTLV-1 oncogenesis.

Although insertional mutagenesis is a widely accepted mechanism for retroviral oncogenesis, how this applies in the case of HTLV-1 remains to be clarified. Integration site analysis in asymptomatic carriers and ATL patients indicates non-random insertion of HTLV-1 provirus into the host genome, with a preference for transcriptional start sites and CpG islands. Although no integration hot spots are found in ATL, a strong bias toward certain binding sites for transcription factors such as STAT1 and p53 is seen [115]. Because HTLV-1 contributes a CTCF site with the potential of forming a chromatin loop with another CTCF site in the host genome [111], it might modify host chromatin structure both in the vicinity and over a long distance, leading to aberrant activation of proto-oncogene, which confers a growth and survival advantage in natural selection. This model could explain why ATL development is a rare accident that occurs only in a small subset of HTLV-1-infected subjects. Further investigations are required to elucidate whether and how CTCF-mediated DNA looping might contribute to HTLV-1 oncogenesis.

Analysis of HTLV-1 clonality by deep sequencing reveals the difference in the frequency distribution of HTLV-1-infected T-cell clones in asymptomatic carriers and ATL patients [112]. In an asymptomatic carrier, there are about 10,000 lower-abundance clones. In a TSP patient, the number increases to about 30,000. These clones contribute substantially to the HTLV-1 proviral load, which is the major risk factor for TSP and also ATL. That is to say, the total number of clones but not the degree of oligoclonal expansion is influential in ATL development. Tax expression is more common in the lower-abundance clones than their high-abundance counterparts. Whether ATL arises from these lower-abundance clones is an issue of debate. Evidence in support of this model comes from integration site analysis and comparison with HTLV-2 [112, 116].

ATL cells are aneuploid and exhibit a mutator phenotype. Various genetic mutations have been found to accumulate in ATL cells, many of which are known to affect NF-κB activation [82]. For example, mutations in CARD11, PRKCB, and PLCG1 are thought to be critical in the activation of NF-κB signaling [117]. Plausibly, some of these mutations might serve as the second or third hit to drive full development of ATL. In light of the requirement of CREB signaling in HTLV-1 oncogenesis, it will not be too surprising if some of the genetic mutations might also be found in the future to have an impact on CREB activation. For instance, mutations of E3 ubiquitin ligase FBW7 have been found to affect Notch signaling [118]. FBW7 is a well-characterized tumor suppressor gene. It will be of interest to see whether these mutations might affect other pathways critically involved in tumor suppression. Through its Tax and HBZ oncoproteins, HTLV-1 can also induce alterations in epigenetic regulators, promoter methylation profiles, and microRNA expression patterns [82, 119]. In this regard, it will be of great importance to determine to what extent the epigenetic and genetic alterations in ATL cells would contribute to leukemogenesis. A working model for HTLV-1 oncogenesis that has incorporated some of the points mentioned above is presented in Fig. 9.2.

9.6 Treatment of ATL

Treatment options for ATL are very limited and unsatisfactory [25, 120, 121]. Indolent ATL can be managed by watchful waiting until disease progression. However, this strategy has been found to result in an even poorer long-term outcome. An alternative treatment for indolent ATL uses a combination of zidovudine

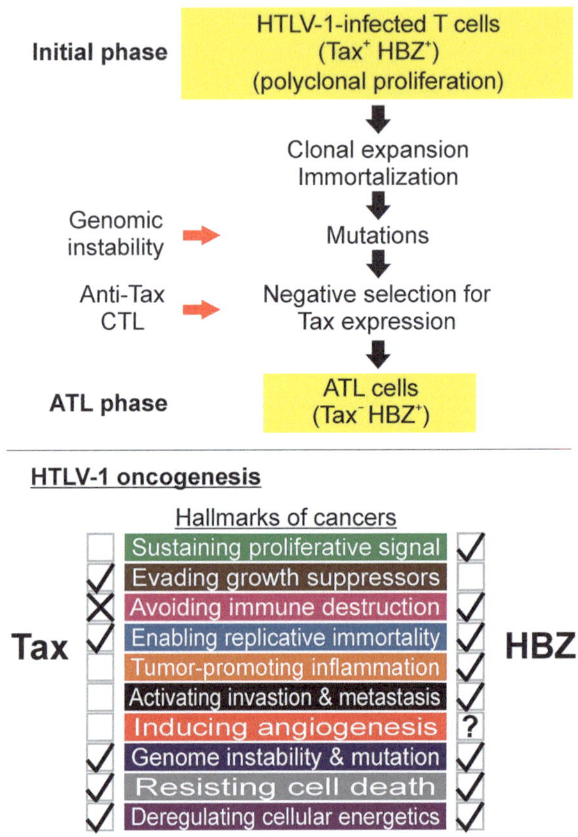

Fig. 9.2 Working model for ATL development. The roles of clonal expansion, genomic instability, and anti-Tax CTL are highlighted. Shown below are the roles of Tax and HBZ on the hallmarks of cancer. At the initial phase, there exists a wide variation in Tax and HBZ expression in HTLV-1-infected T cells. Although some ATL cells might express Tax, abrogation of Tax expression is more common in ATL cells

and interferon-α (AZT/IFN-α). This treatment has been shown to be effective for ATL [122, 123] and has been successfully used in the USA and Europe, but the mechanism of action remains unclear. Whether antiviral or cytotoxic effect is more important needs to be clarified. It is possible that the antiviral effect of AZT relieves viral suppression of IFN signaling, which has immunomodulatory and proapoptotic effect. The AZT/IFN-α therapy is not recommended in Japan pending the final result from an ongoing clinical trial. For patients with aggressive ATL, AZT/IFN-α or intensive chemotherapy is the first-line treatment. Whereas the outcome of AZT/IFN-α treatment is good for leukemia-type ATL [123], chemotherapy is reserved for lymphoma-type ATL. In addition to AZT/IFN-α and chemotherapy, allogeneic hematopoietic stem cell transplantation (allo-HSCT) is a potentially curative option for aggressive ATL. Using unrelated bone marrow and umbilical cord blood as alternative donor source in allo-HSCT has been successful in Japan. Notably, the antitumor effect of allo-HSCT provides the proof of principle for novel immunotherapy of ATL, including immune checkpoint therapy.

An anti-CCR4 monoclonal antibody is a novel targeted therapy for ATL [124]. CCR4 is selectively expressed in regulatory T cells and T helper type 2 cells. It is found in most ATL cells and its expression is induced by HBZ [125]. CCR4 expression is an indicator of poor prognosis [126]. Thus, anti-CCR4 can selectively eliminate ATL cells primarily through antibody-dependent cell-mediated cytotoxicity. Identification of new biomarkers that can be used to select patients who will benefit most from anti-CCR4 antibody might be the next challenge. In addition to anti-CCR4, an antibody against CD25, the α-subunit of IL2R, has also been tested for targeted therapy of ATL [127].

9.7 Concluding Remarks

More than 35 years have passed since the discovery of HTLV-1. Research findings in the field have not only advanced our understanding of HTLV-1 biology and oncogenesis but also provided new strategies and modalities in the management of ATL. A group of international experts in the field has formed a task force under the Global Virus Network and suggested priorities and open questions in HTLV-1 research [128]. Below I would echo their five suggestions as the concluding remarks of this chapter. First, global prevalence of HTLV-1 infection should be reviewed to identify opportunities and means to expand epidemiological studies [17]. A method to reduce mother-to-child transmission by breastfeeding in low-income countries should be developed. Second, biomarkers to predict disease progression should be identified. Searching for driver mutations through deep sequencing [117] should be continued and their clinicopathological significance determined. Third, preventive and therapeutic vaccines should be developed. Fourth, existing drugs should be screened and novel drugs should be developed to improve therapy [25, 120, 121]. Last but not least, basic research should be strengthened. Unraveling mechanisms of viral replication, persistence, and pathogenesis will open insights into novel treatments. This includes studies on HTLV-1 oncoproteins Tax and HBZ that promote viral replication and persistence [14, 16], viral entry, infectious and mitotic cycles of replication, genetic and epigenetic mechanisms that underlie ATL [82], the role of host immunity in the control of HTLV-1 infection [33], as well as HTLV-2, HTLV-3, and HTLV-4 pathogenesis [12]. We are optimistic that better answers to many of these open questions will be obtained in the near future.

Acknowledgments Our research was supported by SK Yee Medical Research Fund (2001) and Hong Kong Research Grants Council (C7011-15R).

References

1. Poiesz BJ, Ruscetti FW, Gazdar AF, Bunn PA, Minna JD, Gallo RC (1980) Detection and isolation of type C retrovirus particles from fresh and cultured lymphocytes of a patient with cutaneous T-cell lymphoma. Proc Natl Acad Sci U S A 77:7415–7419

2. Yoshida M, Miyoshi I, Hinuma Y (1982) Isolation and characterization of retrovirus from cell lines of human adult T-cell leukemia and its implication in the disease. Proc Natl Acad Sci U S A 79:2031–2035
3. Matsuoka M, Jeang KT (2011) Human T-cell leukemia virus type 1 (HTLV-1) and leukemic transformation: viral infectivity, Tax, HBZ and therapy. Oncogene 30:1379–1389
4. Fenizia C, Fiocchi M, Jones K, Parks RW, Ceribelli M, Chevalier SA, Edwards D, Ruscetti F, Pise-Masison CA, Franchini G (2014) Human T-cell leukemia/lymphoma virus type 1 p30, but not p12/p8, counteracts toll-like receptor 3 (TLR3) and TLR4 signaling in human monocytes and dendritic cells. J Virol 88:393–402
5. Romeo MM, Ko B, Kim J, Brady R, Heatley HC, He J, Harrod CK, Barnett B, Ratner L, Lairmore MD, Martinez E, Lüscher B, Robson CN, Henriksson M, Harrod R (2015) Acetylation of the c-MYC oncoprotein is required for cooperation with the HTLV-1 p30(II) accessory protein and the induction of oncogenic cellular transformation by p30(II)/c-MYC. Virology 476:271–288
6. Pise-Masison CA, de Castro-Amarante MF, Enose-Akahata Y, Buchmann RC, Fenizia C, Washington Parks R, Edwards D, Fiocchi M, Alcantara LC Jr, Bialuk I, Graham J, Walser JC, McKinnon K, Galvão-Castro B, Gessain A, Venzon D, Jacobson S, Franchini G (2014) Co-dependence of HTLV-1 p12 and p8 functions in virus persistence. PLoS Pathog 10:e1004454
7. Edwards D, Fukumoto R, de Castro-Amarante MF, Alcantara LC, Galvão-Castro B, Washington Parks R, Pise-Masison C, Franchini G (2014) Palmitoylation and p8-mediated human T-cell leukemia virus type 1 transmission. J Virol 88:2319–2322
8. Gaudray G, Gachon F, Basbous J, Biard-Piechaczyk M, Devaux C, Mesnard JM (2002) The complementary strand of the human T-cell leukemia virus type 1 RNA genome encodes a bZIP transcription factor that down-regulates viral transcription. J Virol 76:12813–12822
9. Satou Y, Yasunaga J, Yoshida M, Matsuoka M (2006) HTLV-I basic leucine zipper factor gene mRNA supports proliferation of adult T cell leukemia cells. Proc Natl Acad Sci U S A 103:720–725
10. Wolfe ND, Heneine W, Carr JK, Garcia AD, Shanmugam V, Tamoufe U, Torimiro JN, Prosser AT, Lebreton M, Mpoudi-Ngole E, McCutchan FE, Birx DL, Folks TM, Burke DS, Switzer WM (2005) Emergence of unique primate T-lymphotropic viruses among central African bushmeat hunters. Proc Natl Acad Sci U S A 102:7994–7999
11. Richard L, Mouinga-Ondémé A, Betsem E, Filippone C, Nerrienet E, Kazanji M, Gessain A (2016) Zoonotic transmission of two new strains of human T-lymphotropic virus type 4 in hunters bitten by a gorilla in central Africa. Clin Infect Dis 63:800–803
12. Gessain A, Rua R, Betsem E, Turpin J, Mahieux R (2013) HTLV-3/4 and simian foamy retroviruses in humans: discovery, epidemiology, cross-species transmission and molecular virology. Virology 435:187–199
13. Panfil AR, Martinez MP, Ratner L, Green PL (2016) Human T-cell leukemia virus-associated malignancy. Curr Opin Virol 20:40–46
14. Matsuoka M, Yasunaga J (2013) Human T-cell leukemia virus type 1: replication, proliferation and propagation by Tax and HTLV-1 bZIP factor. Curr Opin Virol 3:684–691
15. Bangham CR, Ratner L (2015) How does HTLV-1 cause adult T-cell leukaemia/lymphoma (ATL)? Curr Opin Virol 14:93–100
16. Giam CZ, Semmes OJ (2016) HTLV-1 infection and adult T-cell leukemia/lymphoma-a tale of two proteins: Tax and HBZ. Virus 8:E161
17. Satake M, Yamaguchi K, Tadokoro K (2012) Current prevalence of HTLV-1 in Japan as determined by screening of blood donors. J Med Virol 84:327–335
18. Murphy EL (2016) Infection with human T-lymphotropic virus types-1 and -2 (HTLV-1 and -2): implications for blood transfusion safety. Transfus Clin Biol 23:13–19
19. Xie J, Ge S, Zhang Y, Lin Y, Ni H, Zhang J, Chen C (2015) The prevalence of human T-lymphotropic virus infection among blood donors in southeast China, 2004–2013. PLoS Negl Trop Dis 9:e0003685

20. Du J, Chen C, Gao J, Xie J, Rong X, Xu X, Wang Y, Wang F, Li J, Lu Z, Guo W, Li G, Wang Z, Xu D, Weng J, Zhao Z, Weng W, Li H, Du Y, Li S, Zhen C, Liu B, Guo T (2014) History and update of HTLV infection in China. Virus Res 191:134–137
21. Zeng Y, Lan X, Wang B, Fan J, Chen W, Yang T, Liang J, Xu X, Wang Y, Sui Y, Hu R, Hinuma Y (1985) Seroepidemiological studies on human T-cell leukemia antibody in China. Chin J Virol 1:344–348
22. Au WY, Lo JY (2005) HTLV-1-related lymphoma in Hong Kong Chinese. Am J Hematol 78:80–81
23. Wang CH, Chen CJ, Hu CY, You SL, Chu CT, Chou MJ, Essex M, Blattner WA, Liu CH, Yang CS (1988) Seroepidemiology of human T-cell lymphotropic virus type I infection in Taiwan. Cancer Res 48:5042–5044
24. Lu SC, Kao CL, Chin LT, Chen JW, Yang CM, Chang JH, Hsu SC, Chang AC, Chen BH (2001) Seroprevalence and demographic characteristics of HTLV-I among blood donors in Taiwan: 1996–1999. Int J Hematol 74:333–337
25. Cook LB, Taylor GP (2015) Treatment of adult T-cell leukaemia/lymphoma: is the virus a target? Curr Opin Infect Dis 28:583–588
26. Iwanaga M, Watanabe T, Utsunomiya A, Okayama A, Uchimaru K, Koh KR, Ogata M, Kikuchi H, Sagara Y, Uozumi K, Mochizuki M, Tsukasaki K, Saburi Y, Yamamura M, Tanaka J, Moriuchi Y, Hino S, Kamihira S, Yamaguchi K (2010) Human T-cell leukemia virus type I (HTLV-1) proviral load and disease progression in asymptomatic HTLV-1 carriers: a nationwide prospective study in Japan. Blood 116:1211–1219
27. Dutartre H, Clavière M, Journo C, Mahieux R (2016) Cell-free versus cell-to-cell infection by human immunodeficiency virus type 1 and human T-lymphotropic virus type 1: exploring the link among viral source, viral trafficking, and viral replication. J Virol 90:7607–7617
28. Nagai Y, Kawahara M, Hishizawa M, Shimazu Y, Sugino N, Fujii S, Kadowaki N, Takaori-Kondo A (2015) T memory stem cells are the hierarchical apex of adult T-cell leukemia. Blood 125:3527–3535
29. Igakura T, Stinchcombe JC, Goon PK, Taylor GP, Weber JN, Griffiths GM, Tanaka Y, Osame M, Bangham CR (2003) Spread of HTLV-I between lymphocytes by virus-induced polarization of the cytoskeleton. Science 299:1713–1716
30. Nejmeddine M, Bangham CR (2010) The HTLV-1 virological synapse. Virus 2:1427–1447
31. Pais-Correia AM, Sachse M, Guadagnini S, Robbiati V, Lasserre R, Gessain A, Gout O, Alcover A, Thoulouze MI (2010) Biofilm-like extracellular viral assemblies mediate HTLV-1 cell-to-cell transmission at virological synapses. Nat Med 16:83–89
32. Jin DY, Spencer F, Jeang KT (1998) Human T cell leukemia virus type 1 oncoprotein Tax targets the human mitotic checkpoint protein MAD1. Cell 93:81–91
33. Bangham CR (2009) CTL quality and the control of human retroviral infections. Eur J Immunol 39:1700–1712
34. Nagata K, Ohtani K, Nakamura M, Sugamura K (1989) Activation of endogenous c-fos proto-oncogene expression by human T-cell leukemia virus type I-encoded p40lax protein in the human T cell line, Jurkat. J Virol 63:3220–3226
35. Tagaya Y, Taniguchi Y, Naramura M, Okada M, Suzuki N, Kanamori H, Nikaido T, Honjo T, Yodoi J (1987) Transcription of IL-2 receptor gene is stimulated by ATL-derived factor produced by HTLV-I^{+} T cell lines. Immunol Lett 15:221–228
36. Sugamura K, Fujii M, Kannagi M, Sakitani M, Takeuchi M, Hinuma Y (1984) Cell surface phenotypes and expression of viral antigens of various human cell lines carrying human T-cell leukemia virus. Int J Cancer 34:221–228
37. Derse D, Mikovits J, Ruscetti F (1997) X-I and X-II open reading frames of HTLV-I are not required for virus replication or for immortalization of primary T-cells in vitro. Virology 237:123–128
38. Lairmore MD (2014) Animal models of bovine leukemia virus and human T-lymphotrophic virus type-1: insights in transmission and pathogenesis. Ann Rev Anim Biosci 2:189–208

39. Rodríguez SM, Florins A, Gillet N, de Brogniez A, Sánchez-Alcaraz MT, Boxus M, Boulanger F, Gutiérrez G, Trono K, Alvarez I, Vagnoni L, Willems L (2011) Preventive and therapeutic strategies for bovine leukemia virus: lessons for HTLV. Virus 3:1210–1248
40. Barez PY, de Brogniez A, Carpentier A, Gazon H, Gillet N, Gutiérrez G, Hamaidia M, Jacques JR, Perike S, Neelature Sriramareddy S, Renotte N, Staumont B, Reichert M, Trono K, Willems L (2015) Recent advances in BLV research. Virus 7:6080–6088
41. Miura M, Yasunaga J, Tanabe J, Sugata K, Zhao T, Ma G, Miyazato P, Ohshima K, Kaneko A, Watanabe A, Saito A, Akari H, Matsuoka M (2013) Characterization of simian T-cell leukemia virus type 1 in naturally infected Japanese macaques as a model of HTLV-1 infection. Retrovirology 10:118
42. Hasegawa H, Sawa H, Lewis MJ, Orba Y, Sheehy N, Yamamoto Y, Ichinohe T, Katano H, Tsunetsugu-Yokota Y, Takahashi H, Matsuda J, Sata T, Kurata T, Nagashima K, Hall WW (2006) Thymus-derived leukemia-lymphoma in mice transgenic for the Tax gene of human T-lymphotropic virus type I. Nat Med 12:466–472
43. Satou Y, Yasunaga J, Zhao T, Yoshida M, Miyazato P, Takai K, Shimizu K, Ohshima K, Green PL, Ohkura N, Yamaguchi T, Ono M, Sakaguchi S, Matsuoka M (2011) HTLV-1 bZIP factor induces T-cell lymphoma and systemic inflammation in vivo. PLoS Pathog 7:e1001274
44. Zimmerman B, Niewiesk S, Lairmore MD (2010) Mouse models of human T lymphotropic virus type-1-associated adult T-cell leukemia/lymphoma. Vet Pathol 47:677–689
45. Shultz LD, Brehm MA, Garcia-Martinez JV, Greiner DL (2012) Humanized mice for immune system investigation: progress, promise and challenges. Nat Rev Immunol 12:786–798
46. Legrand N, Ploss A, Balling R, Becker PD, Borsotti C, Brezillon N, Debarry J, de Jong Y, Deng H, Di Santo JP, Eisenbarth S, Eynon E, Flavell RA, Guzman CA, Huntington ND, Kremsdorf D, Manns MP, Manz MG, Mention JJ, Ott M, Rathinam C, Rice CM, Rongvaux A, Stevens S, Spits H, Strick-Marchand H, Takizawa H, van Lent AU, Wang C, Weijer K, Willinger T, Ziegler P (2009) Humanized mice for modeling human infectious disease: challenges, progress, and outlook. Cell Host Microbe 6:5–9
47. Marsden MD, Zack JA (2015) Studies of retroviral infection in humanized mice. Virology 479–480:297–309
48. Pérès E, Bagdassarian E, This S, Villaudy J, Rigal D, Gazzolo L, Duc Dodon M (2015) From immunodeficiency to humanization: the contribution of mouse models to explore HTLV-1 leukemogenesis. Virus 7:6371–6386
49. Miyazato P, Yasunaga J, Taniguchi Y, Koyanagi Y, Mitsuya H, Matsuoka M (2006) De novo human T-cell leukemia virus type 1 infection of human lymphocytes in NOD-SCID, common γ-chain knockout mice. J Virol 80:10683–10691
50. Banerjee P, Crawford L, Samuelson E, Feuer G (2010) Hematopoietic stem cells and retroviral infection. Retrovirology 7:8
51. Tie F, Adya N, Greene WC, Giam CZ (1996) Interaction of the human T-lymphotropic virus type 1 Tax dimer with CREB and the viral 21-base-pair repeat. J Virol 70:8368–8374
52. Kwok RP, Laurance ME, Lundblad JR, Goldman PS, Shih HM, Connor LM, Marriott SJ, Goodman RH (1996) Control of cAMP-regulated enhancers by the viral transactivator Tax through CREB and the co-activator CBP. Nature 380:642–646
53. Jin DY, Jeang KT (1997) HTLV-I Tax self-association in optimal trans-activation function. Nucleic Acids Res 25:79–87
54. Baranger AM, Palmer CR, Hamm MK, Giebler HA, Brauweiler A, Nyborg JK, Schepartz A (1995) Mechanism of DNA-binding enhancement by the human T-cell leukaemia virus transactivator Tax. Nature 376:606–608
55. Semmes OJ, Jeang KT (1995) Definition of a minimal activation domain in human T-cell leukemia virus type I Tax. J Virol 69:1827–1833
56. Ching YP, Chun ACS, Chin KT, Zhang ZQ, Jeang KT, Jin DY (2004) Specific TATAA and bZIP requirements suggest that HTLV-1 Tax has transcriptional activity subsequent to the assembly of an initiation complex. Retrovirology 1:18
57. Koga H, Ohshima T, Shimotohno K (2004) Enhanced activation of Tax-dependent transcription of human T-cell leukemia virus type I (HTLV-I) long terminal repeat by TORC3. J Biol Chem 279:52978–52983

58. Siu YT, Chin KT, Siu KL, Choy EYW, Jeang KT, Jin DY (2006) TORC1 and TORC2 coactivators are required for Tax activation of the human T-cell leukemia virus type 1 long terminal repeats. J Virol 80:7052–7059
59. Simonis N, Rual JF, Lemmens I, Boxus M, Hirozane-Kishikawa T, Gatot JS, Dricot A, Hao T, Vertommen D, Legros S, Daakour S, Klitgord N, Martin M, Willaert JF, Dequiedt F, Navratil V, Cusick ME, Burny A, Van Lint C, Hill DE, Tavernier J, Kettmann R, Vidal M, Twizere JC (2012) Host-pathogen interactome mapping for HTLV-1 and -2 retroviruses. Retrovirology 9:26
60. Chan CP, Siu YT, Kok KH, Ching YP, Tang HMV, Jin DY (2013) Group I p21-activated kinases facilitate Tax-mediated transcriptional activation of the human T-cell leukemia virus type 1 long terminal repeats. Retrovirology 10:47
61. Tang HMV, Gao WW, Chan CP, Siu YT, Wong CM, Kok KH, Ching YP, Takemori H, Jin DY (2013) LKB1 tumor suppressor and salt-inducible kinases negatively regulate human T-cell leukemia virus type 1 transcription. Retrovirology 10:40
62. Tang HMV, Gao WW, Chan CP, Cheng Y, Deng JJ, Yuen KS, Iha H, Jin DY (2015) SIRT1 suppresses human T-cell leukemia virus type 1 transcription. J Virol 89:8623–8631
63. Ma G, Yasunaga J, Akari H, Matsuoka M (2015) TCF1 and LEF1 act as T-cell intrinsic HTLV-1 antagonists by targeting Tax. Proc Natl Acad Sci U S A 112:2216–2221
64. Chan JK, Greene WC (2012) Dynamic roles for NF-κB in HTLV-I and HIV-1 retroviral pathogenesis. Immunol Rev 246:286–310
65. Jin DY, Giordano V, Kibler KV, Nakano H, Jeang KT (1999) Role of adapter function in oncoprotein-mediated activation of NF-κB. Human T-cell leukemia virus type I Tax interacts directly with IκB kinase γ. J Biol Chem 274 17402–17405
66. Harhaj EW, Sun SC (1999) IKKγ serves as a docking subunit of the IκB kinase (IKK) and mediates interaction of IKK with the human T-cell leukemia virus Tax protein. J Biol Chem 274:22911–22914
67. Chu ZL, Shin YA, Yang JM, DiDonato JA, Ballard DW (1999) IKKγ mediates the interaction of cellular IκB kinases with the Tax transforming protein of human T cell leukemia virus type 1. J Biol Chem 274:15297–15300
68. De Valck D, Jin DY, Heyninck K, Van de Craen M, Contreras R, Fiers W, Jeang KT, Beyaert R (1999) The zinc finger protein A20 interacts with a novel anti-apoptotic protein which is cleaved by specific caspases. Oncogene 18:4182–4190
69. Pujari R, Hunte R, Thomas R, van der Weyden L, Rauch D, Ratner L, Nyborg JK, Ramos JC, Takai Y, Shembade N (2015) Human T-cell leukemia virus type 1 (HTLV-1) Tax requires CADM1/TSLC1 for inactivation of the NF-κB inhibitor A20 and constitutive NF-κB signaling. PLoS Pathog 11:e1004721
70. Chin KT, Chun AC, Ching YP, Jeang KT, Jin DY (2007) Human T-cell leukemia virus oncoprotein Tax represses nuclear receptor-dependent transcription by targeting coactivator TAX1BP1. Cancer Res 67:1072–1081
71. Journo C, Filipe J, About F, Chevalier SA, Afonso PV, Brady JN, Flynn D, Tangy F, Israël A, Vidalain PO, Mahieux R, Weil R (2009) NRP/Optineurin cooperates with TAX1BP1 to potentiate the activation of NF-κB by human T-lymphotropic virus type 1 Tax protein. PLoS Pathog 5:e1000521
72. Shembade N, Harhaj NS, Yamamoto M, Akira S, Harhaj EW (2007) The human T-cell leukemia virus type 1 Tax oncoprotein requires the ubiquitin-conjugating enzyme Ubc13 for NF-κB activation. J Virol 81:13735–13742
73. Ho YK, Zhi H, Bowlin T, Dorjbal B, Philip S, Zahoor MA, Shih HM, Semmes OJ, Schaefer B, Glover JN, Giam CZ (2015) HTLV-1 Tax stimulates ubiquitin E3 ligase, ring finger protein 8, to assemble lysine 63-linked polyubiquitin chains for TAK1 and IKK activation. PLoS Pathog 11:e1005102
74. Lavorgna A, Harhaj EW (2014) Regulation of HTLV-1 Tax stability, cellular trafficking and NF-κB activation by the ubiquitin-proteasome pathway. Virus 6:3925–3943
75. Fujii M, Tsuchiya H, Chuhjo T, Akizawa T, Seiki M (1992) Interaction of HTLV-1 Tax1 with p67SRF causes the aberrant induction of cellular immediate early genes through CArG boxes. Genes Dev 6:2066–2076

76. Winter HY, Marriott SJ (2007) Human T-cell leukemia virus type 1 Tax enhances serum response factor DNA binding and alters site selection. J Virol 81:6089–6098
77. Jeang KT, Chiu R, Santos E, Kim SJ (1991) Induction of the HTLV-I LTR by Jun occurs through the Tax-responsive 21-bp elements. Virology 181:218–227
78. Smith MR, Greene WC (1991) Type I human T cell leukemia virus Tax protein transforms rat fibroblasts through the cyclic adenosine monophosphate response element binding protein/activating transcription factor pathway. J Clin Invest 88:1038–1042
79. Grassmann R, Dengler C, Müller-Fleckenstein I, Fleckenstein B, McGuire K, Dokhelar MC, Sodroski JG, Haseltine WA (1989) Transformation to continuous growth of primary human T lymphocytes by human T-cell leukemia virus type I X-region genes transduced by a Herpesvirus saimiri vector. Proc Natl Acad Sci U S A 86:3351–3355
80. Soda Y, Jinno A, Tanaka Y, Akagi T, Shimotohno K, Hoshino H (2000) Rapid tumor formation and development of neutrophilia and splenomegaly in nude mice transplanted with human cells expressing human T cell leukemia virus type I or Tax1. Leukemia 14:1467–1476
81. Siu YT, Jin DY (2007) CREB-a real culprit in oncogenesis. FEBS J 274:3224–3232
82. Watanabe T (2017) Adult T-cell leukemia: molecular basis for clonal expansion and transformation of HTLV-1-infected T cells. Blood 129:1071–1081
83. Reynaud C, Fabre S, Jalinot P (2000) The PDZ protein TIP-1 interacts with the Rho effector rhotekin and is involved in Rho signaling to the serum response element. J Biol Chem 275:33962–33968
84. Yan P, Fu J, Qu Z, Li S, Tanaka T, Grusby MJ, Xiao G (2009) PDLIM2 suppresses human T-cell leukemia virus type I Tax-mediated tumorigenesis by targeting Tax into the nuclear matrix for proteasomal degradation. Blood 113:4370–4380
85. Makokha GN, Takahashi M, Higuchi M, Saito S, Tanaka Y, Fujii M (2013) Human T-cell leukemia virus type 1 Tax protein interacts with and mislocalizes the PDZ domain protein MAGI-1. Cancer Sci 104:313–320
86. Chun ACS, Zhou Y, Wong CM, Kung HF, Jeang KT, Jin DY (2000) Coiled-coil motif as a structural basis for the interaction of HTLV type 1 Tax with cellular cofactors. AIDS Res Hum Retrovir 16:1689–1694
87. Jin DY, Teramoto H, Giam CZ, Chun RF, Gutkind JS, Jeang KT (1997) A human suppressor of c-Jun N-terminal kinase 1 activation by tumor necrosis factor α. J Biol Chem 272:25816–25823
88. Ching YP, Chan SF, Jeang KT, Jin DY (2006) The retroviral oncoprotein Tax targets the coiled-coil centrosomal protein TAX1BP2 to induce centrosome overduplication. Nat Cell Biol 8:717–724
89. Arnold J, Yamamoto B, Li M, Phipps AJ, Younis I, Lairmore MD, Green PL (2006) Enhancement of infectivity and persistence in vivo by HBZ, a natural antisense coded protein of HTLV-1. Blood 107:3976–3982
90. Valeri VW, Hryniewicz A, Andresen V, Jones K, Fenizia C, Bialuk I, Chung HK, Fukumoto R, Parks RW, Ferrari MG, Nicot C, Cecchinato V, Ruscetti F, Franchini G (2010) Requirement of the human T-cell leukemia virus p12 and p30 products for infectivity of human dendritic cells and macaques but not rabbits. Blood 116:3809–3817
91. Mitobe Y, Yasunaga J, Furuta R, Matsuoka M (2015) HTLV-1 bZIP factor RNA and protein impart distinct functions on T-cell proliferation and survival. Cancer Res 75:4143–4152
92. Zhao T (2016) The role of HBZ in HTLV-1-induced oncogenesis. Virus 8:E34
93. Ma G, Yasunaga J, Matsuoka M (2016) Multifaceted functions and roles of HBZ in HTLV-1 pathogenesis. Retrovirology 13:16
94. Hanahan D, Weinberg RA (2011) Hallmarks of cancer: the next generation. Cell 144:646–674
95. Ma G, Yasunaga J, Fan J, Yanagawa S, Matsuoka M (2013) HTLV-1 bZIP factor dysregulates the Wnt pathways to support proliferation and migration of adult T-cell leukemia cells. Oncogene 32:4222–4230
96. Zane L, Yasunaga J, Mitagami Y, Yedavalli V, Tang SW, Chen CY, Ratner L, Lu X, Jeang KT (2012) Wip1 and p53 contribute to HTLV-1 Tax-induced tumorigenesis. Retrovirology 9:114

97. Neuveut C, Low KG, Maldarelli F, Schmitt I, Majone F, Grassmann R, Jeang KT (1998) Human T-cell leukemia virus type 1 Tax and cell cycle progression: role of cyclin D-cdk and p110Rb. Mol Cell Biol 18:3620–3632
98. Kehn K, Fuente Cde L, Strouss K, Berro R, Jiang H, Brady J, Mahieux R, Pumfery A, Bottazzi ME, Kashanchi F (2005) The HTLV-I Tax oncoprotein targets the retinoblastoma protein for proteasomal degradation. Oncogene 24:525–540
99. Tanaka-Nakanishi A, Yasunaga J, Takai K, Matsuoka M (2014) HTLV-1 bZIP factor suppresses apoptosis by attenuating the function of FoxO3a and altering its localization. Cancer Res 74:188–200
100. Yuen CK, Chan CP, Fung SY, Wang PH, Wong WM, Tang HMV, Yuen KS, Chan CP, Jin DY, Kok KH (2016) Suppression of type I interferon production by human T-cell leukemia virus type 1 oncoprotein Tax through inhibition of IRF3 phosphorylation. J Virol 90:3902–3912
101. Yasuma K, Yasunaga J, Takemoto K, Sugata K, Mitobe Y, Takenouchi N, Nakagawa M, Suzuki Y, Matsuoka M (2016) HTLV-1 bZIP factor impairs anti-viral immunity by inducing co-inhibitory molecule, T cell immunoglobulin and ITIM domain (TIGIT). PLoS Pathog 12:e1005372
102. Belgnaoui SM, Fryrear KA, Nyalwidhe JO, Guo X, Semmes OJ (2010) The viral oncoprotein Tax sequesters DNA damage response factors by tethering MDC1 to chromatin. J Biol Chem 285:32897–32905
103. Dayaram T, Lemoine FJ, Donehower LA, Marriott SJ (2013) Activation of WIP1 phosphatase by HTLV-1 Tax mitigates the cellular response to DNA damage. PLoS One 8:e55989
104. Chaib-Mezrag H, Lemaçon D, Fontaine H, Bellon M, Bai XT, Drac M, Coquelle A, Nicot C (2014) Tax impairs DNA replication forks and increases DNA breaks in specific oncogenic genome regions. Mol Cancer 13:205
105. Baydoun HH, Cherian MA, Green P, Ratner L (2015) Inducible nitric oxide synthase mediates DNA double strand breaks in human T-cell leukemia virus type 1-induced leukemia/lymphoma. Retrovirology 12:71
106. Kuhlmann AS, Villaudy J, Gazzolo L, Castellazzi M, Mesnard JM, Duc Dodon M (2007) HTLV-1 HBZ cooperates with JunD to enhance transcription of the human telomerase reverse transcriptase gene (hTERT). Retrovirology 4:92
107. Zhao T, Satou Y, Sugata K, Miyazato P, Green PL, Imamura T, Matsuoka M (2011) HTLV-1 bZIP factor enhances TGF-β signaling through p300 coactivator. Blood 118:1865–1876
108. Mitagami Y, Yasunaga J, Kinosada H, Ohshima K, Matsuoka M (2015) Interferon-γ promotes inflammation and development of T-cell lymphoma in HTLV-1 bZIP factor transgenic mice. PLoS Pathog 11:e1005120
109. Miyazato P, Matsuo M, Katsuya H, Satou Y (2016) Transcriptional and epigenetic regulatory mechanisms affecting HTLV-1 provirus. Viruses 8:E171
110. Rowan AG, Suemori K, Fujiwara H, Yasukawa M, Tanaka Y, Taylor GP, Bangham CR (2014) Cytotoxic T lymphocyte lysis of HTLV-1 infected cells is limited by weak HBZ protein expression, but non-specifically enhanced on induction of Tax expression. Retrovirology 11:116
111. Satou Y, Miyazato P, Ishihara K, Yaguchi H, Melamed A, Miura M, Fukuda A, Nosaka K, Watanabe T, Rowan AG, Nakao M, Bangham CR (2016) The retrovirus HTLV-1 inserts an ectopic CTCF-binding site into the human genome. Proc Natl Acad Sci U S A 113:3054305–3054309
112. Cook LB, Melamed A, Niederer H, Valganon M, Laydon D, Foroni L, Taylor GP, Matsuoka M, Bangham CR (2014) The role of HTLV-1 clonality, proviral structure, and genomic integration site in adult T-cell leukemia/lymphoma. Blood 123:3925–3931
113. Sugata K, Yasunaga J, Mitobe Y, Miura M, Miyazato P, Kohara M, Matsuoka M (2015) Protective effect of cytotoxic T lymphocytes targeting HTLV-1 bZIP factor. Blood 126:1095–1105
114. Kuo YL, Giam CZ (2006) Activation of the anaphase promoting complex by HTLV-1 Tax leads to senescence. EMBO J 25:1741–1752

115. Melamed A, Laydon DJ, Gillet NA, Tanaka Y, Taylor GP, Bangham CR (2013) Genome-wide determinants of proviral targeting, clonal abundance and expression in natural HTLV-1 infection. PLoS Pathog 9:e1003271
116. Melamed A, Witkover AD, Laydon DJ, Brown R, Ladell K, Miners K, Rowan AG, Gormley N, Price DA, Taylor GP, Murphy EL, Bangham CR (2014) Clonality of HTLV-2 in natural infection. PLoS Pathog 10:e1004006
117. Kataoka K, Nagata Y, Kitanaka A, Shiraishi Y, Shimamura T, Yasunaga J, Totoki Y, Chiba K, Sato-Otsubo A, Nagae G, Ishii R, Muto S, Kotani S, Watatani Y, Takeda J, Sanada M, Tanaka H, Suzuki H, Sato Y, Shiozawa Y, Yoshizato T, Yoshida K, Makishima H, Iwanaga M, Ma G, Nosaka K, Hishizawa M, Itonaga H, Imaizumi Y, Munakata W, Ogasawara H, Sato T, Sasai K, Muramoto K, Penova M, Kawaguchi T, Nakamura H, Hama N, Shide K, Kubuki Y, Hidaka T, Kameda T, Nakamaki T, Ishiyama K, Miyawaki S, Yoon SS, Tobinai K, Miyazaki Y, Takaori-Kondo A, Matsuda F, Takeuchi K, Nureki O, Aburatani H, Watanabe T, Shibata T, Matsuoka M, Miyano S, Shimoda K, Ogawa S (2015) Integrated molecular analysis of adult T cell leukemia/lymphoma. Nat Genet 47:1304–1315
118. Yeh CH, Bellon M, Pancewicz-Wojtkiewicz J, Nicot C (2016) Oncogenic mutations in the FBXW7 gene of adult T-cell leukemia patients. Proc Natl Acad Sci U S A 113:6731–6736
119. Moles R, Nicot C (2015) The emerging role of miRNAs in HTLV-1 infection and ATLL pathogenesis. Viruses 7:4047–4074
120. Kato K, Akashi K (2015) Recent advances in therapeutic approaches for adult T-cell leukemia/lymphoma. Viruses 7:6604–6612
121. Nasr R, Marçais A, Hermine O, Bazarbachi A (2017) Overview of targeted therapies for adult T-cell leukemia/lymphoma. Methods Mol Biol 1582:197–216
122. Hermine O, Bouscary D, Gessain A, Turlure P, Leblond V, Franck N, Buzyn-Veil A, Rio B, Macintyre E, Dreyfus F, Bazarbachi A (1995) Treatment of adult T-cell leukemia-lymphoma with zidovudine and interferon α. N Engl J Med 332:1749–1751
123. Bazarbachi A, Plumelle Y, Carlos Ramos J, Tortevoye P, Otrock Z, Taylor G, Gessain A, Harrington W, Panelatti G, Hermine O (2010) Meta-analysis on the use of zidovudine and interferon-alfa in adult T-cell leukemia/lymphoma showing improved survival in the leukemic subtypes. J Clin Oncol 28:4177–4183
124. Ishida T, Joh T, Uike N, Yamamoto K, Utsunomiya A, Yoshida S, Saburi Y, Miyamoto T, Takemoto S, Suzushima H, Tsukasaki K, Nosaka K, Fujiwara H, Ishitsuka K, Inagaki H, Ogura M, Akinaga S, Tomonaga M, Tobinai K, Ueda R (2012) Defucosylated anti-CCR4 monoclonal antibody (KW-0761) for relapsed adult T-cell leukemia-lymphoma: a multicenter phase II study. J Clin Oncol 30:837–842
125. Sugata K, Yasunaga J, Kinosada H, Mitobe Y, Furuta R, Mahgoub M, Onishi C, Nakashima K, Ohshima K, Matsuoka M (2016) HTLV-1 viral factor HBZ induces CCR4 to promote T-cell migration and proliferation. Cancer Res 76:5068–5079
126. Ishida T, Utsunomiya A, Iida S, Inagaki H, Takatsuka Y, Kusumoto S, Takeuchi G, Shimizu S, Ito M, Komatsu H, Wakita A, Eimoto T, Matsushima K, Ueda R (2003) Clinical significance of CCR4 expression in adult T-cell leukemia/lymphoma: its close association with skin involvement and unfavorable outcome. Clin Cancer Res 9:3625–3634
127. Kreitman RJ, Stetler-Stevenson M, Jaffe ES, Conlon KC, Steinberg SM, Wilson W, Waldmann TA, Pastan I (2016) Complete remissions of adult T-cell leukemia with anti-CD25 recombinant immunotoxin LMB-2 and chemotherapy to block immunogenicity. Clin Cancer Res 22:310–318
128. Willems L, Hasegawa H, Accolla R, Bangham C, Bazarbachi A, Bertazzoni U, Carneiro-Proietti AB, Cheng H, Chieco-Bianchi L, Ciminale V, Coelho-Dos-Reis J, Esparza J, Gallo RC, Gessain A, Gotuzzo E, Hall W, Harford J, Hermine O, Jacobson S, Macchi B, Macpherson C, Mahieux R, Matsuoka M, Murphy E, Peloponese JM, Simon V, Tagaya Y, Taylor GP, Watanabe T, Yamano Y (2017) Reducing the global burden of HTLV-1 infection: an agenda for research and action. Antivir Res 137:41–48

Chapter 10
Malignancies in HIV-Infected and AIDS Patients

Yongjia Ji and Hongzhou Lu

Abstract Currently, HIV infection and AIDS are still one of the most important epidemic diseases around the world. As early in the initial stage of HIV epidemic, the high incidence of ADCs including Kaposi sarcoma and non-Hodgkin's lymphoma was the substantial amount of disease burden of HIV infection and AIDS. With the increasing accessibility of HAART and improving medical care for HIV infection and AIDS, AIDS-related illness including ADCs has dramatically decreased. Meanwhile, the incidence of NADCs rises in PLWH. Compared with the general population, most of cancers are more likely to attack PLWH, and NADCs in PLWH were characterized as earlier onset and more aggressive. However, the understanding for cancer development in PLWH is still dimness. Herein, we reviewed the current knowledge of epidemiology and pathogenesis for malignancies in PLWH summarized from recent studies. On the basis of that, we discussed the special considerations for cancer treatment in PLWH. As those malignancies could be the major issue for HIV infection or AIDS in the future, we expect enhanced investigations, surveillances, and clinical trial for improving the understanding and management for cancers developed in PLWH.

Keywords HIV • AIDS • Cancer • Malignancy

Y. Ji
Department of Infectious Diseases, Shanghai Public Health Clinical Center, Fudan University, Shanghai 201508, People's Republic of China

H. Lu (✉)
Department of Infectious Diseases, Shanghai Public Health Clinical Center, Fudan University, Shanghai 201508, People's Republic of China

Department of Infectious Diseases, Huashan Hospital Affiliated to Fudan University, Shanghai 200040, People's Republic of China

Department of Internal Medicine, Shanghai Medical College, Fudan University, Shanghai 200032, People's Republic of China
e-mail: luhongzhou@fudan.edu.cn

Q. Cai et al. (eds.), *Infectious Agents Associated Cancers: Epidemiology and Molecular Biology*, Advances in Experimental Medicine and Biology 1018,
DOI 10.1007/978-981-10-5765-6_10

10.1 Introduction

Globally, more than 35 million people are living with human immunodeficiency virus (HIV) infection or acquired immunodeficiency syndrome (AIDS) [1]. Compared with the general population, the population of people living with HIV infection (PLWH) was at higher risk for cancer incidence [2–8]. As early at the opening phase of AIDS epidemic, it was found that non-Hodgkin lymphoma (NHL), cervical premalignant lesions, and Kaposi sarcoma (KS) were strongly associated with immune suppression induced by HIV infection [9]. As a consequence, the US Centers for Disease Control (CDC) has defined KS, certain non-Hodgkin lymphomas, and cervical cancer as AIDS-defining cancers (ADCs) since the 1990s [10].

Over the years with the increasing accessibility to highly active antiretroviral therapy (HAART) and improving medical care for HIV infection and AIDS, the outcome for PLWH has substantially improved, which is largely benefited from the decreasing incidence and mortality rate of AIDS-related illness including opportunistic infections (OI) and ADCs [11]. While with the extension of life span during the HAART era, the spectrum of malignancies occurred in PLWH has significantly transformed [12]. Compared with the general population or people without HIV infection, several non-AIDS-defining cancers (NADCs) such as lung cancer, hepatocellular carcinoma (HCC), and classical Hodgkin lymphoma were found attacking PLWH more frequently [1, 13]. And especially in developed country, malignancy has gradually become the leading cause of deaths in PLWH, and NADCs have replaced ADCs as the major malignancies burden in HIV-infected population [1, 12, 14, 15].

In recent years, the focal interest for malignancies in PLWH was growing, and a large amount of studies in this field were published. In this review, we will discuss the latest advances of epidemiology, pathogenesis, and special consideration for treatment in this field.

10.2 Epidemiology

10.2.1 AIDS-Defining Cancers

ADCs were identified by comparing the risk (standard incidence ratio, SIR) of cancer incidence in PLWH with that in the general population [16]. According to the definition of US Centers for Disease Control (CDC), Kaposi sarcoma, cervical cancer, and specific non-Hodgkin lymphoma (NHL) including primary central nervous system lymphoma (PCNSL), Burkitt's lymphoma (BL), diffuse large B-cell lymphoma (DLBCL), plasmablastic lymphoma (PL), and primary effusion lymphoma (PEL) were categorized as ADCs [10].

For ADCs, several large-scale epidemiology studies have revealed that the SIR in PLWH population is significantly higher than that in general population in both the pre-HAART and HAART eras [1, 7, 17–19]. Kaposi sarcoma and NHL contributed to most cases of ADCs in the pre-HAART era [20]. As reported in the United States, in early phase of HIV epidemic, the risk for Kaposi sarcoma and cervical cancer was about 50,000-fold and eightfold higher in PLWH compared with the general population [20]. And for NHL in pre-HAART era, the most common pathological type of NHL occurred in PLWH was DLBCL and PCNSL took the second place [21]. Compared with the general population, the risk of DLBCL and PCNSL for AIDS patients significantly increased 5000 and 98-fold in the pre-HAART era [20].

After entering into the HAART era, the incidences of ADCs in PLWH have decreased and the outcome got substantial improvement [19, 22]. As studies reported in western country, with the introduction of HAART, Kaposi sarcoma and NHL cases declined by more than 80% and 50%, respectively [1, 20]. However even under such circumstances, ADCs still possess as an important issue in the HAART era, several recent studies indicated that the risks of ADCs are still higher in PLWH than that in the general population [5, 7, 18–20].

Several studies found that HIV infection inducing immune suppression indicated by CD4+ T-cell count is the most important risk factor for ADCs development [7, 23, 24]. As retrospectively analyzed, compared with CD4+ T-cell count less than 100 cells/ml, the SIR for NHL and KS in PLWHA decreased from 145 to 35.8 and 571 to 76 per 100,000 person-years separately when CD4+ T-cell count was more than 500 cells/ml [20]. On the contrary, with the CD4+ T-cell count increased from less than 50 to greater than 250 cells/ml, the incidence of Burkitt's lymphoma increased from 9.6 to 30.7 per 100,000 person-years, which could be correlated with immune reconstitution [25]. As a consequence, Burkitt's lymphoma has replaced PCNSL as the second most common NHL in PLWH during the HAART era [18].

10.2.2 Non-AIDS-Defining Cancers

During the HAART era, NADCs incidences among PLWH increase rapidly. As reported in developed country, NADCs have contributed to more than half of all HIV malignancies, while the number in pre-HAART era was less than 40% [1–3]. Similar with other age-associated diseases, the risk for NADCs is higher in PLWH [26]. And also, the prognosis of HIV-infected patients with NADCs is independently worse than those without HIV infection [27].

There have been numerous studies for exploring risk factors of NADCs in PLWH. Previous studies have identified older age and the longer duration time living with HIV infection as the most important risk factors related with NADC incidence in PLWH [7, 28]. While comparing with people without HIV infection, NADCs occurred at similar ages in PLWH [26]. Previously, several studies have found that HAART administration could be a possible contributor for malignancy development [28–30]. However, there is the view that the lifetime extension benefited by HAART other than its direct effective could be the factor related with

Table 10.1 Oncogenic virus infection associated with malignancy development in PLWH

Virus	Malignancies
HHV-8	KS, NHL (PEL, PL)
EBV	NHL (PCNSL, BL, DLBCL PEL), HL, head and neck carcinoma
HPV	Cervical and anal cancer, head and neck carcinoma
Chronic hepatitis virus (HBV/HCV)	Liver cancer

HHV human herpes virus, *HPV* human papillomavirus, *EBV* Epstein–Barr virus, *HBV* hepatitis B virus, *HCV* hepatitis C virus, *KS* Kaposi sarcoma, *NHL* non-Hodgkin lymphoma, *PEL* primary effusion lymphoma, *PL* plasmablastic lymphoma, *BL* Burkitt's lymphoma, *DLBCL* diffuse large B-cell lymphoma, *PCNSL* primary central nervous system lymphoma

increasing incidence of NADCs [31]. Moreover, the latest epidemiology studies revealed that the early initiation of HAART could diminish the cancer risk and slow the progression of cancer development [32, 33]. The correlation between CD4+ T-cell count and NADCs incidence is controversial depending on different kind of malignancies. For anal cancer, in the setting of patients with CD4+ T-cell count less than 200 cells/ml for more than 5 years, the SIR would rise up [34]. On the contrary, the SIR for Hodgkin lymphoma increases when CD4+ T-cell count rises from 50 to 200 cells/ml [1, 35, 36]. While the incidence of lung cancer is higher among PLWH compared with that in the general population, the CD4+ T-cell count has not been identified as a risk factor [37, 38]. The other risk factors for NADCs development in PLWH included smoking and oncogenic virus infection [39].

10.3 Pathogenesis

10.3.1 Viral Infection

Most cancers including ADCs and NADCs with excess risk among PLWH are associated with oncogenic virus infection (Table 10.1) [39]. Compared with that only less than 5% malignancies in the general population were associated with virus infection, the number in PLWH was reported as up to 40% [40]. This disparity could be attributed to the shared transmission routes of HIV with several oncogenic viruses and immunosuppression induced by HIV infection, which resulted in higher prevalence in PLWH than that in the general population [41–43].

The carcinogenesis of these viruses involves multiple ways, which include regulation of apoptosis and cell life cycle and disturbance of host's tumor-associated genes [44]. And in recent studies, it was suggested that several miRNAs expressed by these oncogenic viruses could be the important promoter for cancer development [45–47]. As for EBV infection, which is highly correlated with various types of lymphoma development in HIV/AIDS setting, the virus expressing MiR-BHRF1-1 and miR-BART1 could involve oncogenesis by inhibiting the tumor suppressor gene p53 and depressing apoptosis by activating antiapoptotic protein BCL-2 [46–48].

Besides, there is growing evidence that HIV could directly participate in the development of malignancies. Tat, the HIV-expressing transactivator protein, is responsible for the activation of viral and cellular genes. In several studies, Tat demonstrated oncogenic effects in vitro and in vivo. After being excreted from cells infected by HIV and accumulated in tissues, tat could induce malignant transformation by interruption of cell proliferation, DNA reparation, and apoptosis [49]. Further, there is evidence suggesting that tat could enhance malignant capacity of other oncogenic viruses such as HPV [42, 50]. Otherwise, it was found that HIV matrix protein (P17) persists in germinal centers even after efficient HAART [51]. And further, several specific variants of HIV P17 were associated with aberrant proliferation of B cell, which may contribute to the development of B-cell lymphoma [52].

10.3.2 Immunosuppression and Inflammation

Host's immunological function impaired by HIV infection has been recognized as an important risk factor for cancer development, which leads to the attenuation of tumor surveillance. An inverse association between CD4+ T-cell count and ADCs risk has been demonstrated by several studies conducted in pre- and early-HAART era [53–55]. As for NADCs, the inverse relationship between CD4 count and cancer risk for NADCs has been reported in recent studies [56, 57]. Nevertheless, the relationship between immunosuppression induced by HIV infection and cancer risk is controversial [58]. Even in patients with virus suppression, low CD4 count is still an important risk factor for cancer incidences [57]. Otherwise, HIV-infected patients with immune reconstitution (CD4+ T-cell counts >500/ml) are still at elevated risks for HL and liver cancer [56]. Similarly, during the HAART era, the incidence of anal cancer was observed as rising in HIV-infected population [34, 59]. These results suggested that early HIV infection inducing immunosuppression-associated carcinogenesis could not be reversed by immunologic function restoration, or the CD4+ T-cell count could not indicate the alterations of host's immune system related with susceptible to malignancy.

Recently, it is found that chronic inflammation could greatly contribute to carcinogenesis in HIV-infected patients. HIV infection could induce a series of immune activation, as B-cell hyperactivation could be induced by HIV replication [60], and CD4+ T-cell exhaustion in mucosal layer of the intestine could evoke host inflammatory response [61]. In clinical observation, even with long-term virological suppression, inflammatory biomarkers remain at high levels in HIV-infected people [59]. And also, the role of inflammation biomarkers for predicting oncogenesis in PLWHA has been confirmed by cohort study [62].

10.4 Treatment

After entering into the HAART era, more HIV-infected patients with malignancies have accepted chemotherapy which is contradicted to immunocompromised patients. Benefiting from efficient HAART and the advancement of anticancer therapy, the outcome for HIV-infected patients with cancers has been improved significantly. Compared with pre-HAART era, the overall survival rate at 5 years for HIV-infected patients with HL and DLBCL has increased to more than 50% [63–66]. And for anal cancer, the prognosis is comparable between PLWH and general population [67]. Considering the drug interaction and special physical condition, the treatment for HIV-infected patients with cancer would be more complicated. We will discuss around the special points of cancer treatment for PLWH, which needs attention of physician and other medical care providers.

10.4.1 Combination of HIV Treatment and Chemotherapy

Currently, it is suggested that all HIV-infected patients with malignancies should continue HAART during chemotherapy [31], and early HAART could diminish the risk for cancer development. However, similar with chemotherapy agents, many drugs for HIV treatment were metabolized by the liver through cytochrome P450 enzyme system, which could interfere the pharmacokinetic of chemotherapy agents and further influence therapeutic efficacy [68, 69]. Ritonavir, the inhibitor of HIV protease, also could exert potential inhibition effective for CYP34A and defer the clearance of specific chemotherapy agents such as vinca alkaloids, taxanes, and alkylating agents [68, 69]. Therefore, the combination of ritonavir and vincristine or vinblastine-based chemotherapy would increase the incidences of chemotherapy-related toxicity such as neuropathy and neutropenia. Similarly, fluconazole is also the inhibitor for CYP3A4 system, which should be avoided in combining with the vinca alkaloid-based chemotherapy [70]. Otherwise, several ART drugs cause overlapping side effect with chemotherapy agents, such as renal and hepatic toxicity, myelosuppression, and peripheral neuropathy [68, 69].

Considering the complex conditions as described above, all these interaction factors should be carefully evaluated before prescribing combination of HAART and chemotherapy. The optimized therapeutic regimen for HIV-infected patients with malignancies should be made by the consensus between the specialists of infectious disease, oncologists, and pharmacists.

10.4.2 HAART Discontinuation During Chemotherapy

As described above, the excessive adverse effects caused by the combination of HAART and chemotherapy could lead to treatment discontinuation. For AIDS patients with lymphoma, there have been investigations evaluating the outcome for discontinuation of HAART during chemotherapy period [65, 71]. During the course of chemotherapy without HAART ranging from 4 to 6 months, the patients' serum HIV viral load rose with the falling of CD4+T-cell count. However, HAART was resumed when chemotherapy is completed; both HIV viral load and CD4+ T-cell count would restore in 6–12 months. And further, the 5-year OS was comparable with that reported by other studies conducted in AIDS patients with lymphoma co-administrated with HAART and chemotherapy [65, 71]. As the lack of studies directly comparing the outcome for chemotherapy with or without HAART, the optimal choice is still in dispute.

10.4.3 Immunological Suppression Associated with Chemotherapy

As for NHL treatment, even co-administrated with HAART, the patients' CD4+ T-cell counts would decline more than half in most cases after taking chemotherapy [72]. Commonly, with the end of chemotherapy, CD4+ T-cell counts will increase to the level before treatment in about 6 months to 1 year [72]. In conditions with radiation therapy, the immunosuppression would be more serious, which might not recover after treatment [73]. This situation might be caused by myelosuppression related with radiation. Pelvic radiation could induce the most severe immunosuppression, and the intestine as the important gathering place of CD4+ T cell could be also affected by abdominal radiation [74]. For this immunosuppression effect of antitumor therapy, it was recommended that the prophylaxis for *Pneumocystis jiroveci* and other opportunistic infections should be administrated in HIV-infected patients with malignancies when initiating chemotherapy or radiotherapy, regardless of CD4+ T-cell count [73]. Meanwhile, for minimizing the risk of opportunistic infections, granulocyte colony-stimulating agents could be administrated on the basis of risk assessments [71, 72, 75].

10.4.4 Immunotherapy

Besides chemotherapy, treatment targeting the modulation of host immune system has emerged as promising options for tumor treatment. So far, FDA has approved CTLA-4 inhibitor and PD-1 inhibitor for specific cancer treatment, respectively, and the IL-2 stimulation has also been in late clinical development [76, 77]. Of note,

PD-1 inhibitor and IL-2 stimulation demonstrate the effectiveness of CD8+ T-cell restoration. Considering the important role of CD8+ T-cell dysfunction in both HIV infection and malignancies, the immunotherapy for AIDS patients with malignancies would be deserved for anticipating [78].

10.5 Conclusion

It has been more than 50 years after the first case of HIV infection was identified, while HIV infection and AIDS are still one of the most important epidemic diseases around the world [79]. In early phase of HIV epidemic, the high incidence of NHL and KS in PLWH was noted, which was recognized relating with specific virus infection. With the improving medical care for AIDS, AIDS-related illness including ADCs has decreased sharply, and NADCs rise in PLWH. Compared with patients without HIV infection or AIDS, NADCs were characterized as earlier onset and more aggressive in PLWH [80].

Until now, the understanding of pathogenesis for cancer especially NADC development in HIV-infected patients is still unclear. Because those patients with HIV/AIDS have not regularly enrolled in clinical trial for cancer treatment, the optimized therapy for PLWH with cancer is controversial [81]. In the future, the enchantment of foundation and clinical research in this field should be expected. Current epidemiology studies indicated that PLWH seems to be more susceptible for most cancers [82]. Considering that, PLWH should be closely monitored for both ADCs and NADCs. And for treatment, multidisciplinary cooperation including specialists of infectious disease, oncologist, and clinical pharmacist is strongly recommended.

References

1. Shiels MS, Pfeiffer RM, Gail MH, Hall HI, Li J, Chaturvedi AK, Bhatia K, Uldrick TS, Yarchoan R, Goedert JJ (2011) Cancer burden in the HIV-infected population in the United States. J Natl Cancer Inst 103:753–762
2. Engels EA, Pfeiffer RM, Goedert JJ, Virgo P, McNeel TS, Scoppa SM, Biggar RJ (2006) Trends in cancer risk among people with AIDS in the United States 1980–2002. AIDS 20:1645–1654
3. Engels EA, Biggar RJ, Hall HI, Cross H, Crutchfield A, Finch JL, Grigg R, Hylton T, Pawlish KS, McNeel TS (2008) Cancer risk in people infected with human immunodeficiency virus in the United States. Int J Cancer 123:187–194
4. Long JL, Engels EA, Moore RD, Gebo KA (2008) Incidence and outcomes of malignancy in the HAART era in an urban cohort of HIV-infected individuals. AIDS. (London, England 22:489
5. Patel P, Hanson DL, Sullivan PS, Novak RM, Moorman AC, Tong TC, Holmberg SD, Brooks JT (2008) Incidence of types of cancer among HIV-infected persons compared with the general population in the United States, 1992–2003. Ann Intern Med 148:728–736

6. Silverberg MJ, Chao C, Leyden WA, Xu L, Tang B, Horberg MA, Klein D, Quesenberry CP Jr, Towner WJ, Abrams DI (2009) HIV infection and the risk of cancers with and without a known infectious cause. AIDS. (London, England 23:2337
7. Clifford GM, Polesel J, Rickenbach M, Dal Maso L, Keiser O, Kofler A, Rapiti E, Levi F, Jundt G, Fisch T (2005) Cancer risk in the Swiss HIV cohort study: associations with immunodeficiency, smoking, and highly active antiretroviral therapy. J Natl Cancer Inst 97:425–432
8. Goedert JJ, Coté TR, Virgo P, Scoppa SM, Kingma DW, Gail MH, Jaffe ES, Biggar RJ, Group A-CMS (1998) Spectrum of AIDS-associated malignant disorders. Lancet 351:1833–1839
9. Grulich AE, van Leeuwen MT, Falster MO, Vajdic CM (2007) Incidence of cancers in people with HIV/AIDS compared with immunosuppressed transplant recipients: a meta-analysis. Lancet 370:59–67
10. Control CfD, Prevention (1993) Impact of the expanded AIDS surveillance case definition on AIDS case reporting – United States, first quarter, 1993. MMWR Morb Mortal Wkly Rep 42:308
11. Consortium WTS (2009) Timing of initiation of antiretroviral therapy in AIDS-free HIV-1-infected patients: a collaborative analysis of 18 HIV cohort studies. Lancet 373:1352–1363
12. Morlat P, Roussillon C, Henard S, Salmon D, Bonnet F, Cacoub P, Georget A, Aouba A, Rosenthal E, May T (2014) Causes of death among HIV-infected patients in France in 2010 (national survey): trends since 2000. AIDS 28:1181–1191
13. Silverberg MJ, Lau B, Achenbach CJ, Jing Y, Althoff KN, D'Souza G, Engels EA, Hessol NA, Brooks JT, Burchell AN, Gill MJ, Goedert JJ, Hogg R, Horberg MA, Kirk GD, Kitahata MM, Korthuis PT, Mathews WC, Mayor A, Modur SP, Napravnik S, Novak RM, Patel P, Rachlis AR, Sterling TR, Willig JH, Justice AC, Moore RD, Dubrow R (2015) Cumulative incidence of cancer among persons with HIV in North America: a cohort study. Ann Intern Med 163:507–518
14. Bonnet F, Burty C, Lewden C, Costagliola D, May T, Bouteloup V, Rosenthal E, Jougla E, Cacoub P, Salmon D (2009) Changes in cancer mortality among HIV-infected patients: the Mortalite 2005 survey. Clin Infect Dis 48:633–639
15. Yarchoan R, Tosato G, Little RF (2005) Therapy insight: AIDS-related malignancies—the influence of antiviral therapy on pathogenesis and management. Nat Clin Pract Oncol 2:406–415
16. Jones ME, Swerdlow AJ (1998) Bias in the standardized mortality ratio when using general population rates to estimate expected number of deaths. Am J Epidemiol 148:1012–1017
17. Powles T, Robinson D, Stebbing J, Shamash J, Nelson M, Gazzard B, Mandelia S, Møller H, Bower M (2009) Highly active antiretroviral therapy and the incidence of non–AIDS-defining cancers in people with HIV infection. J Clin Oncol 27:884–890
18. Shiels MS, Pfeiffer RM, Hall HI, Li J, Goedert JJ, Morton LM, Hartge P, Engels EA (2011) Proportions of Kaposi sarcoma, selected non-Hodgkin lymphomas, and cervical cancer in the United States occurring in persons with AIDS, 1980–2007. JAMA 305:1450–1459
19. Raffetti E, Albini L, Gotti D, Segala D, Maggiolo F, di Filippo E, Saracino A, Ladisa N, Lapadula G, Fornabaio C, Castelnuovo F, Casari S, Fabbiani M, Pierotti P, Donato F, Quiros-Roldan E (2015) Cancer incidence and mortality for all causes in HIV-infected patients over a quarter century: a multicentre cohort study. BMC Public Health 15:235
20. Eltom MA, Jemal A, Mbulaiteye SM, Devesa SS, Biggar RJ (2002) Trends in Kaposi's sarcoma and non-Hodgkin's lymphoma incidence in the United States from 1973 through 1998. J Natl Cancer Inst 94:1204–1210
21. Coté TR, Biggar RJ, Rosenberg PS, Devesa SS, Percy C, Yellin FJ, Lemp G, Hardy C, Geodert JJ, Blattner WA (1997) Non-Hodgkin's lymphoma among people with AIDS: incidence, presentation and public health burden. Int J Cancer 73:645–650
22. Park LS, Tate JP, Sigel K, Rimland D, Crothers K, Gibert C, Rodriguez-Barradas MC, Goetz MB, Bedimo RJ, Brown ST (2016) Time trends in cancer incidence in persons living with HIV/AIDS in the antiretroviral therapy era: 1997–2012. AIDS, London

23. HIV IC, Coutinho R (2000) Highly active antiretroviral therapy and incidence of cancer in human immunodeficiency virus-infected adults. J Natl Cancer Inst 92:1823–1830
24. Frisch M, Biggar RJ, Engels EA, Goedert JJ, Group A-CMRS (2001) Association of cancer with AIDS-related immunosuppression in adults. JAMA 285:1736–1745
25. Ambinder R (2001) Epstein–Barr virus associated lymphoproliferations in the AIDS setting. Eur J Cancer 37:1209–1216
26. Althoff KN, McGinnis KA, Wyatt CM, Freiberg MS, Gilbert C, Oursler KK, Rimland D, Rodriguez-Barradas MC, Dubrow R, Park LS, Skanderson M, Shiels MS, Gange SJ, Gebo KA, Justice AC (2015) Comparison of risk and age at diagnosis of myocardial infarction, end-stage renal disease, and non-AIDS-defining cancer in HIV-infected versus uninfected adults. Clin Infect Dis: Off Publ Infect Dis Soc Am 60:627–638
27. Coghill AE, Shiels MS, Suneja G, Engels EA (2015) Elevated cancer-specific mortality among HIV-infected patients in the United States. J Clin Oncol: Off J Am Soc Clin Oncol 33:2376–2383
28. Burgi A, Brodine S, Wegner S, Milazzo M, Wallace MR, Spooner K, Blazes DL, Agan BK, Armstrong A, Fraser S (2005) Incidence and risk factors for the occurrence of non-AIDS-defining cancers among human immunodeficiency virus-infected individuals. Cancer 104:1505–1511
29. Kowalska JD, Reekie J, Mocroft A, Reiss P, Ledergerber B, Gatell J, Monforte AA, Phillips A, Lundgren JD, Kirk O (2012) Long-term exposure to combination antiretroviral therapy and risk of death from specific causes: no evidence for any previously unidentified increased risk due to antiretroviral therapy. AIDS 26:315–323
30. Rodger AJ, Lodwick R, Schechter M, Deeks S, Amin J, Gilson R, Paredes R, Bakowska E, Engsig FN, Phillips A (2013) Mortality in well controlled HIV in the continuous antiretroviral therapy arms of the SMART and ESPRIT trials compared with the general population. AIDS 27:973–979
31. Rubinstein PG, Aboulafia DM, Zloza A (2014) Malignancies in HIV/AIDS: from epidemiology to therapeutic challenges. AIDS 28:453–465
32. Borges AH, Neuhaus J, Babiker AG, Henry K, Jain MK, Palfreeman A, Mugyenyi P, Domingo P, Hoffmann C, Read TR, Pujari S, Meulbroek M, Johnson M, Wilkin T, Mitsuyasu R (2016) Immediate antiretroviral therapy reduces risk of infection-related cancer during early HIV infection. Clin Infect Dis: Off Publ Infect Dis Soc Am 63:1668–1676
33. Hleyhel M, Belot A, Bouvier AM, Tattevin P, Pacanowski J, Genet P, De Castro N, Berger JL, Dupont C, Lavole A (2015) Trends in survival after cancer diagnosis among HIV-infected individuals between 1992 and 2009. Results from the FHDH-ANRS CO4 cohort. Int J Cancer 137:2443–2453
34. Piketty C, Selinger-Leneman H, Bouvier A-M, Belot A, Mary-Krause M, Duvivier C, Bonmarchand M, Abramowitz L, Costagliola D, Grabar S (2012) Incidence of HIV-related anal cancer remains increased despite long-term combined antiretroviral treatment: results from the French hospital database on HIV. J Clin Oncol 30:4360–4366
35. Biggar RJ, Jaffe ES, Goedert JJ, Chaturvedi A, Pfeiffer R, Engels EA (2006) Hodgkin lymphoma and immunodeficiency in persons with HIV/AIDS. Blood 108:3786–3791
36. Spina M, Carbone A, Gloghini A, Serraino D, Berretta M, Tirelli U (2010) Hodgkin's disease in patients with HIV infection. Adv Hematol 2011
37. Engels EA, Brock MV, Chen J, Hooker CM, Gillison M, Moore RD (2006) Elevated incidence of lung cancer among HIV-infected individuals. J Clin Oncol 24:1383–1388
38. Kirk GD, Merlo C, O'Driscoll P, Mehta SH, Galai N, Vlahov D, Samet J, Engels EA (2007) HIV infection is associated with an increased risk for lung cancer, independent of smoking. Clin Infect Dis 45:103–110
39. Parka LS, Hernandez-Ramirez RU, Silverberg MJ, Crothers KA, Dubrow R (2016) Prevalence of non-HIV cancer risk factors in persons living with HIV/AIDS. AIDS 30:273–291
40. de Martel C, Shiels MS, Franceschi S, Simard EP, Vignat J, Hall HI, Engels EA, Plummer M (2015) Cancers attributable to infections among adults with HIV in the United States. AIDS 29:2173–2181

41. Sulkowski MS (2008) Viral hepatitis and HIV coinfection. J Hepatol 48:353–367
42. Brickman C, Palefsky JM (2015) Human papillomavirus in the HIV-infected host: epidemiology and pathogenesis in the antiretroviral era. Curr HIV/AIDS Rep 12:6–15
43. Martin JN, Ganem DE, Osmond DH, Page-Shafer KA, Macrae D, Kedes DH (1998) Sexual transmission and the natural history of human herpesvirus 8 infection. N Engl J Med 338:948–954
44. McLaughlin-Drubin ME, Munger K (2008) Viruses associated with human cancer. Biochim Biophys Acta (BBA)-Mol Basis Dis 1782:127–150
45. Qi P, Han J, Lu Y, Wang C, Bu F (2006) Virus-encoded microRNAs: future therapeutic targets. Cell Mol Immunol 3:411–419
46. Yeung ML, Bennasser Y, Le SY, Jeang KT (2005) siRNA, miRNA and HIV: promises and challenges. Cell Res 15:935–946
47. Lu J, Getz G, Miska EA, Alvarez-Saavedra E, Lamb J, Peck D, Sweet-Cordero A, Ebert BL, Mak RH, Ferrando AA (2005) MicroRNA expression profiles classify human cancers. Nature 435:834–838
48. Pfeffer S, Voinnet O (2006) Viruses, microRNAs and cancer. Oncogene 25:6211–6219
49. Romani B, Engelbrecht S, Glashoff RH (2010) Functions of tat: the versatile protein of human immunodeficiency virus type 1. J Gen Virol 91:1–12
50. Kim RH, Yochim JM, Kang MK, Shin K-H, Christensen R, Park N-H (2008) HIV-1 tat enhances replicative potential of human oral keratinocytes harboring HPV-16 genome. Int J Oncol 33:777
51. Popovic M, Tenner-Racz K, Pelser C, Stellbrink HJ, van Lunzen J, Lewis G, Kalyanaraman VS, Gallo RC, Racz P (2005) Persistence of HIV-1 structural proteins and glycoproteins in lymph nodes of patients under highly active antiretroviral therapy. Proc Natl Acad Sci U S A 102:14807–14812
52. Dolcetti R, Giagulli C, He W, Selleri M, Caccuri F, Eyzaguirre LM, Mazzuca P, Corbellini S, Campilongo F, Marsico S, Giombini E, Muraro E, Rozera G, De Paoli P, Carbone A, Capobianchi MR, Ippolito G, Fiorentini S, Blattner WA, Lu W, Gallo RC, Caruso A (2015) Role of HIV-1 matrix protein p17 variants in lymphoma pathogenesis. Proc Natl Acad Sci U S A 112:14331–14336
53. Biggar RJ, Chaturvedi AK, Goedert JJ, Engels EA (2007) AIDS-related cancer and severity of immunosuppression in persons with AIDS. J Natl Cancer Inst 99:962–972
54. Franceschi S, Dal Maso L, Rickenbach M, Polesel J, Hirschel B, Cavassini M, Bordoni A, Elzi L, Ess S, Jundt G (2008) Kaposi sarcoma incidence in the Swiss HIV cohort study before and after highly active antiretroviral therapy. Br J Cancer 99:800–804
55. Polesel J, Clifford GM, Rickenbach M, Dal Maso L, Battegay M, Bouchardy C, Furrer H, Hasse B, Levi F, Probst-Hensch NM (2008) Non-Hodgkin lymphoma incidence in the Swiss HIV cohort study before and after highly active antiretroviral therapy. AIDS 22:301–306
56. Hleyhel M, HIV WcotCRGotFHDo (2014) Risk of non-AIDS-defining cancers among HIV-1-infected individuals in France between 1997 and 2009: results from a French cohort. AIDS 28:2109–2118
57. Pacheco Y, Jarrin I, Rosado I, Campins A, Berenguer J, Iribarren J, Rivero M, Muñoz-Medina L, Bernal-Morell E, Gutiérrez F (2015) Increased risk of non-AIDS-related events in HIV subjects with persistent low CD4 counts despite cART in the CoRIS cohort. Antivir Res 117:69–74
58. Dubrow R, Silverberg MJ, Park LS, Crothers K, Justice AC (2012) HIV infection, aging, and immune function: implications for cancer risk and prevention. Curr Opin Oncol 24:506
59. Crum-Cianflone NF, Hullsiek KH, Marconi VC, Ganesan A, Weintrob A, Barthel RV, Agan BK, Group IDCRPHW (2010) Anal cancers among HIV-infected persons: HAART is not slowing rising incidence. AIDS. (London, England 24:535
60. Hunt PW (2012) HIV and inflammation: mechanisms and consequences. Curr HIV/AIDS Rep 9:139–147

61. Marks MA, Rabkin CS, Engels EA, Busch E, Kopp W, Rager H, Goedert JJ, Chaturvedi AK (2013) Markers of microbial translocation and risk of AIDS-related lymphoma. AIDS 27:469–474
62. Borges ÁH, Silverberg MJ, Wentworth D, Grulich AE, Fätkenheuer G, Mitsuyasu R, Tambussi G, Sabin CA, Neaton JD, Lundgren JD (2013) Predicting risk of cancer during HIV infection: the role of inflammatory and coagulation biomarkers. AIDS. (London, England 27:1433
63. Hentrich M, Berger M, Wyen C, Siehl J, Rockstroh JK, Müller M, Fätkenheuer G, Seidel E, Nickelsen M, Wolf T (2012) Stage-adapted treatment of HIV-associated Hodgkin lymphoma: results of a prospective multicenter study. J Clin Oncol:JCO 30:4117–4123. 2012.2041. 8137
64. Xicoy B, Ribera J-M, Miralles P, Berenguer J, Rubio R, Mahillo B, Valencia M-E, Abella E, López-Guillermo A, Sureda A (2007) Results of treatment with doxorubicin, bleomycin, vinblastine and dacarbazine and highly active antiretroviral therapy in advanced stage, human immunodeficiency virus-related Hodgkin's lymphoma. Haematologica 92:191–198
65. Dunleavy K, Little RF, Pittaluga S, Grant N, Wayne AS, Carrasquillo JA, Steinberg SM, Yarchoan R, Jaffe ES, Wilson WH (2010) The role of tumor histogenesis, FDG-PET, and short-course EPOCH with dose-dense rituximab (SC-EPOCH-RR) in HIV-associated diffuse large B-cell lymphoma. Blood 115:3017–3024
66. Sparano JA, Lee JY, Kaplan LD, Levine AM, Ramos JC, Ambinder RF, Wachsman W, Aboulafia D, Noy A, Henry DH (2010) Rituximab plus concurrent infusional EPOCH chemotherapy is highly effective in HIV-associated B-cell non-Hodgkin lymphoma. Blood 115:3008–3016
67. Oehler-Jänne C, Huguet F, Provencher S, Seifert B, Negretti L, Riener M-O, Bonet M, Allal AS, Ciernik IF (2008) HIV-specific differences in outcome of squamous cell carcinoma of the anal canal: a multicentric cohort study of HIV-positive patients receiving highly active antiretroviral therapy. J Clin Oncol 26:2550–2557
68. Rudek MA, Flexner C, Ambinder RF (2011) Use of antineoplastic agents in patients with cancer who have HIV/AIDS. Lancet Oncol 12:905–912
69. Deeken JF, Pantanowitz L, Dezube BJ (2009) Targeted therapies to treat non-AIDS-defining cancers in patients with HIV on HAART therapy: treatment considerations and research outlook. Curr Opin Oncol 21:445–454
70. Ezzat HM, Cheung MC, Hicks LK, Boro J, Montaner JS, Lima VD, Harris M, Leitch HA (2012) Incidence, predictors and significance of severe toxicity in patients with human immunodeficiency virus-associated Hodgkin lymphoma. Leuk Lymphoma 53:2390–2396
71. Little RF, Pittaluga S, Grant N, Steinberg SM, Kavlick MF, Mitsuya H, Franchini G, Gutierrez M, Raffeld M, Jaffe ES, Shearer G, Yarchoan R, Wilson WH (2003) Highly effective treatment of acquired immunodeficiency syndrome-related lymphoma with dose-adjusted EPOCH: impact of antiretroviral therapy suspension and tumor biology. Blood 101:4653–4659
72. Powles T, Imami N, Nelson M, Gazzard BG, Bower M (2002) Effects of combination chemotherapy and highly active antiretroviral therapy on immune parameters in HIV-1 associated lymphoma. AIDS 16:531–536
73. Alfa-Wali M, Allen-Mersh T, Antoniou A, Tait D, Newsom-Davis T, Gazzard B, Nelson M, Bower M (2012) Chemoradiotherapy for anal cancer in HIV patients causes prolonged CD4 cell count suppression. Ann Oncol: Off J Eur Soc Med Oncol/ESMO 23:141–147
74. Ling B, Mohan M, Lackner AA, Green LC, Marx PA, Doyle LA, Veazey RS (2010) The large intestine as a major reservoir for simian immunodeficiency virus in macaques with long-term, nonprogressing infection. J Infect Dis 202:1846–1854
75. Hentrich M, Berger M, Wyen C, Siehl J, Rockstroh JK, Muller M, Fatkenheuer G, Seidel E, Nickelsen M, Wolf T, Rieke A, Schurmann D, Schmidmaier R, Planker M, Alt J, Mosthaf F, Engert A, Arasteh K, Hoffmann C (2012) Stage-adapted treatment of HIV-associated Hodgkin lymphoma: results of a prospective multicenter study. J Clin Oncol: Off J Am Soc Clin Oncol 30:4117–4123
76. Hodi FS, O'Day SJ, McDermott DF, Weber RW, Sosman JA, Haanen JB, Gonzalez R, Robert C, Schadendorf D, Hassel JC, Akerley W, van den Eertwegh AJ, Lutzky J, Lorigan P, Vaubel JM, Linette GP, Hogg D, Ottensmeier CH, Lebbe C, Peschel C, Quirt I, Clark JI, Wolchok JD,

Weber JS, Tian J, Yellin MJ, Nichol GM, Hoos A, Urba WJ (2010) Improved survival with ipilimumab in patients with metastatic melanoma. N Engl J Med 363:711–723
77. Wei XX, Fong L, Small EJ (2015) Prostate cancer immunotherapy with Sipuleucel-T: current standards and future directions. Expert Rev Vaccines 14:1529–1541
78. June CH, Levine BL (2015) T cell engineering as therapy for cancer and HIV: our synthetic future. Philos Trans R Soc Lond Ser B Biol Sci 370:20140374
79. Zhu T, Korber BT, Nahmias AJ, Hooper E, Sharp PM, Ho DD (1998) An African HIV-1 sequence from 1959 and implications for the origin of the epidemic. Nature 391:594–597
80. Engels EA (2009) Non-AIDS-defining malignancies in HIV-infected persons: etiologic puzzles, epidemiologic perils, prevention opportunities. AIDS 23:875–885
81. Little RF (2016) Cancer clinical trials in persons with HIV infection. Curr Opin HIV AIDS
82. Goedert JJ, Hosgood HD, Biggar RJ, Strickler HD, Rabkin CS (2016) Screening for cancer in persons living with HIV infection. Trends Cancer 2:416–428

Chapter 11
Bacterial Infection and Associated Cancers

Caixia Zhu, Yuyan Wang, Cankun Cai, and Qiliang Cai

Abstract Bacterial infections were traditionally not considered as major causes of cancer. However, increasing evidence in the past decades has suggested that several cancers are highly associated with bacterial infection. The bacterial infections have evolved some unique strategies including lateral gene transfer, biofilm and microbiome to induce genome instability and chronic inflammation, as well as escape of immune surveillance for carcinogenesis. Here we summarize and highlight the recent progress on understanding of how bacterial infection plays a role in tumor formation and malignancy.

Keywords Bacterial infection • Cancer

11.1 Introduction

Although viral infection is the main agent of infection-causing cancers in humans, and a number of bacterial pathogens have also been shown to make a significant contribution to cancer [1], research on effects of bacterial infection was left far behind than viral infection. The role of bacterial infection in inducing cancer is still a highly debated subject; in fact, several parameters must be met to be infectious cause of cancer. While the evidence of antibiotics such as aspirin could reduce risks of breast cancer for some time [2, 3], indicating that appropriate bacteria may contribute to the development and progress of particular cancer.

Early observations in 1772, *Mycobacterium tuberculosis* was the first bacterium thought to cause lung cancer, due to active tuberculosis in the lung cancer patient which was more frequently than the general population [4]. However, the *Mycobacterium tuberculosis*-cancer theory failed to stand in many cases test and

Caixia Zhu, Yuyan Wang and Cankun Cai contributed equally to this work.

C. Zhu • Y. Wang • C. Cai • Q. Cai (✉)
Key Laboratory of Medical Molecular Virology (Ministries of Education and Health), School of Basic Medical Sciences, Fudan University, Shanghai 200032, People's Republic of China
e-mail: qiliang@fudan.edu.cn

Q. Cai et al. (eds.), *Infectious Agents Associated Cancers: Epidemiology and Molecular Biology*, Advances in Experimental Medicine and Biology 1018,
DOI 10.1007/978-981-10-5765-6_11

Table 11.1 The association of bacterial infection with various cancers

Bacterium	Related cancers	References
Helicobacter pylori	Gastric adenocarcinoma	[13]
	Mucosa-associated lymphoid tissue (MALT) lymphoma	
Fusobacterium nucleatum	Colorectal cancer	[23, 34, 35]
	Oral squamous cell carcinoma	
	Pancreatic cancer	
Porphyromonas ssp. (*asaccharolytica, gingivalis*)	Colorectal cancer	[23, 34, 35]
	Oral squamous cell carcinoma	
	Pancreatic cancer	
Staphylococci aureus	Mortality of patients with pneumonia?	[39]
Streptococci		
S. VGS	Pediatric acute myeloid leukemia	[41, 42]
S. GBS	Breast cancer	[43]
S. pneumoniae	Leukemia, lymphoma, or myeloma	[44]
S. bovis	Colorectal cancer	[50]
Enterococci faecium	Hematologic malignancies	[52]
Salmonella typhimurium	Gallbladder cancer	[53, 54]

appear to be the results of the malignancy instead of the cause. Despite the early mistake, the possible association of bacterial infection with carcinogenesis continues to be promulgated. A significant breakthrough was made when the chronic infection of *Helicobacter pylori* was identified to cause stomach ulcers followed by onset of gastric carcinomas or MALT (mucosa-associated lymphoid tissue) lymphomas [5, 6]. After the substantial progress in understanding the role of *H. pylori* on carcinogenesis, it has been estimated that bacteria account for at least half of organism infections in patients with malignancy [7]. Here we will address and highlight the relevance of *H. pylori*, oral bacteria, and some gram-positive bacteria with cancers (Table 11.1).

11.2 *Helicobacter pylori* and Cancers

Helicobacter is spiral-shaped, gram-negative bacterium [8, 9] and was firstly isolated and cultured from a human gastric biopsy by Marshall and Warren in 1982 [10]. The seminal discovery of this bacterium and its role in gastritis and peptic ulcer disease led to award of Nobel Prize of Medicine in 2005 for Marshall and Warren. Unlike other viruses and bacteria, *H. pylori* has the ability to colonize in highly acidic environment within the stomach [11]. The majority of *H. pylori* strains have expressed several virulence factors that have evolved to affect host cell signaling pathways, which included CagA (cytotoxin-associated gene A antigen), VacA (vacuolating cytotoxin), BabA (blood group antigen-binding adhesion), OipA (outer inflammatory protein), and IceA.

It has been estimated that nearly half of the world's population is infected with *H. pylori*, and the majority of colonized individuals could develop chronic inflammation. Despite that *H. pylori* colonization does not absolutely cause symptoms [12], long-term carriage of *H. pylori* will significantly increase the risk of developing site-specific diseases. For example, around 10% of the infected individuals will develop peptic ulcer disease, 3% of them will develop gastric adenocarcinoma, and 0.1% of them will develop mucosa-associated lymphoid tissue (MALT) lymphoma [13]. However, due to the fact that individuals infected with *H. pylori* do not necessarily have antibodies against bacteria, or may not be detectable in blood, the number of cancer patients seropositive for *H. pylori* may be underestimated. It has been found that gastric MALT lymphoma can be completely cured by eradication of *H. pylori* at early stage and therefore is considered the first clonal lesion which can be eliminated by treatment with antibiotics [14].

The relationship between *H. pylori* infection and gastric cancer has been widely studied for over four decades. Several studies have now provided clear notion that *H. pylori* infection is significantly associated with gastric cancer and eradication of *H. pylori* could significantly decrease the risk of gastric cancer in infected individuals without premalignant lesions [15–17]. Gastric cancer as the third most common cause of cancer deaths in the world; gastric adenocarcinoma accounts for over 95% of malignant neoplasms of the stomach, followed by gastrointestinal stromal tumors and mucosa-associated lymphoid tissue (MALT) lymphoma [18]. *H. pylori* infection is more prevalent (some up to 80%) in the developing countries where poor hygiene enhances person-to-person transmission through domestic contacts at an early age [8]. A recent prospective and population-based study in China showed that higher education, lifestyle changes, and sanitation habits could influence the rate of infection [9]. Although gastric cancer is relatively rare in United States, the incidence varies in many developed countries, for example, Korea, Mongolia, and Japan are the highest (29.9–41.8 per 100,000 persons), while Canada, Western Europe, and Australia are much lower [8, 19].

11.3 Oral Bacteria and Cancers

Many bacteria including *Bacteroides fragilis*, *Enterococcus faecalis*, *Escherichia coli*, *Fusobacterium nucleatum*, and *Porphyromonas asaccharolytica* have been shown to modulate tumorigenesis in colorectal cancer (CRC) [20, 21]. A recent review has summarized well how oral bacteria potentially induce colorectal cancer [22]; here we will address the key progress on the association between oral bacteria and cancer.

Although *B. fragilis*, *E. faecalis*, and *E. coli* present weak pathogenic features, two oral bacteria *F. nucleatum* and *P. asaccharolytica* were consistently identified in CRC patients and often synergistically promote oral and colon cancer progression [23]. In addition to *F. nucleatum* and *P. asaccharolytica*, other oral strains including *Peptostreptococcus*, *Prevotella*, *Parvimonas*, and *Gemella* were effectively used as biomarkers to detect CRC [24, 25]. Given the fact of the consistent co-occurrence of these oral bacteria in CRC, their potential roles in tumorigenesis were proposed to be synergistic activities of biofilm formation and anaerobic asaccharolytic metabolism. A "driver-passenger" model, namely, a "driver" organism such as *P. gingivalis* and *F. nucleatum* can produce virulence factors to help a "passenger" bacterial load and growth, has been proposed to induce this cancer-associated biofilm formation. The consistence of high abundance of *F. nucleatum* in many CRC and adenoma instead of healthy colon biofilm samples indicates that this oral organism plays an essential role in carcinogenesis [26, 27]. Besides *F. nucleatum*, other oral anaerobic bacteria such as *Leptotrichia* and *Campylobacter* have also been revealed by deep sequencing and pairwise correlation analysis of CRC and normal tissues [28].

Because there are so many anaerobic organisms that exist in colon tissue, it has been proposed that the asaccharolytic metabolism of these oral bacteria may play a role in carcinogenesis. Due to these bacteria usually that digest peptides and amino acids instead of sugar or carbon, they become typically proteolytic. The coordinated metabolism will promote growth of a diverse and cooperative polymicrobial ecosystem and continue the breakdown of host proteins to inhibit immune response [29]. A similar effect was observed when these oral organisms inhabit the colon [30]. Recent studies have shown that *F. nucleatum* can disrupt epithelial junctions through E-cadherin to alter mucosal environment where it facilitates growth of other anaerobic microbes once it localizes in colon tissue [31]. In addition, *F. nucleatum*-mediated degradation of host protein in the mouth and gut will build up a chronic inflammatory microenvironment to promote the development of CRC [32]. Another outcome of these metabolites was found to cause DNA damage in colon tissues by inducing polyamines and genotoxic ROS production, which will facilitate biofilm formation and promote cancer cell proliferation [33].

Given the recent advances in the integrated view of the oral microbiome in colorectal tumorigenesis, it has been accepted that all polymicrobes coordinate in concert rather than just a specific pathogen nor virulence factors to create an inflammatory microenvironment that leads to bacteria-associated cancers. In regard to how the microbe disseminates from the oral cavity to the colon, two hypotheses have been proposed. One possible hypothesis is that the microbe could disseminate from ulcerated gingival tissues into the bloodstream and then is located at colon tissue. The other possible hypothesis is that the oral bacteria are swallowed and colonized at colon tissue. However, these two routes remain to be further studied, and inflamed colon or perturbed community may contribute to colonization of oral microbes.

In addition to colorectal cancer, the oral bacteria including *Porphyromonas* and *Fusobacterium* have also been found to strikingly associate with oral squamous cell carcinoma (OSCC is one of the most common cancers worldwide) and pancreatic cancer [34, 35]. In the epithelial and OSCC cell model, it has been found that *P. gingivalis* infection not only can upregulate the expression of B7-H1 and B7-DC receptors, which contribute to chronic inflammation [36], but also promote cellular invasion of OSCC cells through inducing metalloproteinase MMP-9 expression [37]. Further studies revealed that gingipains, a cysteine proteinase produced by *P. gingivalis*, plays a critical role in this process [37]. Similarly, *F. nucleatum* can also enhance tumor cell proliferation and migration through MMP-9 and MMP-13 [38].

11.4 Gram-Positive Bacteria and Cancer

Several lines of evidence have shown that many gram-positive microbes can cause serious infections and invasive bacterial disease in cancer patients. They include *Staphylococci*, *Streptococci*, and *Enterococci*. To understand the impact of these bacteria in patients with malignancy will help develop cancer therapeutic strategy and improve the survival of the cancer patient. We will describe the most recent progress on these three types of bacteria and their associated cancers below.

Staphylococcus aureus—Whether *S. aureus* is a cause of infection in cancer patients remains to be further demonstrated; it has a high relevance with the mortality of patients with pneumonia [39]. Cancer patients treated with antistaphylococcal antibiotics (i.e., daptomycin or ceftaroline) have shown a generally favorable outcome [40].

Streptococci—Among the *Streptococci*, *viridans group streptococci (VGS)* is the prominent member of the oral microbiome and is often found to correlate with high mortality of patients with pediatric acute myeloid leukemia [41, 42]. β-hemolytic streptococci GBS is found to associate with breast cancer [43], and *S. pneumoniae* affects the malignancy of patients with leukemia, lymphoma, or myeloma [44]. It has been shown that VGS is multidrug resistant including β-lactams, while GBS remains susceptible to β-lactams and could be treated with penicillins or cephalosporins, and *S. pneumonia* is susceptible to levofloxacin and vancomycin [45–47]. In addition, *Streptococcus bovis* in gastrointestinal microflora was found in blood and caused infective endocarditis [48, 49], which has been previously shown to associate with colorectal cancer [50], albeit no *S. bovis* DNA was identified in colorectal neoplastic tissues by using the PCR technique [51].

Enterococci—Although *Enterococci* is generally considered as low-virulence bacteria, *E. faecium* is found to associate with increased risk of mortality of patients with hematologic malignancies [52]. It is known that *E. faecium* is penicillin susceptible, while β-lactam or vancomycin is resistant for therapy of cancer patients.

Another example is *Salmonella typhimurium* (*S. typhi*) and gallbladder cancer (GBC), due to patients who are infected with this bacterium developed GBC more frequently [53, 54]. However, the causative links between this bacterial infection and cancer remain to be further demonstrated.

11.5 Potential Mechanisms of Bacterium-Associated Carcinogenesis

Chromosome instability is a common feature of cancer cells. Despite evidence of epidemiological studies of bacteria-associated cancer is persuasive, the molecular mechanisms of bacterial infection-causing genome instability still remain largely unclear. However, recent studies have shown that bacteria—their eukaryotic endosymbionts lateral gene transfer (LGT)—are a mean of causing genome instability [55]. Although the link between the LGT and tumor-causing ability has not been established, it is speculated that LGT may cause tumorigenesis. For example, in vitro experiments demonstrated that *Bartonella henselae* (a bacterium causing bacillary angiomatosis-peliosis tumor in humans [56]) is able to integrate their plasmid DNA into the human host genome [57]. Bioinformatics analysis revealed high incidence of LGT from *Acinetobacter* and *Pseudomonas* like DNA in the mitochondrial DNA of patients with acute myeloid leukemia [58]. In contrast, very few evidence of integration of *H. pylori* DNA was reported, despite the fact that infection with this bacterium is highly associated with gastric cancer.

Given chronic inflammation which is also a common physiological response of host immunity to microbe infections, and is involved in generation of several mediators such as free radicals, prostaglandins, and cytokines, deregulation of these mediators by bacteria will lead to cell proliferation, angiogenesis, and oncogenic activation. Therefore, the specific modification of the inflammatory response by bacteria will lead to persistent infections and eventually development of cancer cells. For instance, ROS and IL-1β, as products of chronic inflammation, are shown to be upregulated by *H. pylori* in gastric epithelial cells for cell proliferation and angiogenesis [59, 60]. Epithelial cell infection with *Pseudomonas aeruginosa* or *H. pylori* is able to induce VEGF expression and trigger angiogenesis [61, 62]. In addition, *H. pylori* also plays a key role in activation of NF-κB through increasing the expression of IL-8 and TNF-α [63, 64]. Further studies have revealed that Toll-like receptors TLR4-mediated activation of NF-κB signaling facilitate the *H. pylori* colonization [65].

Deregulation of host cell proliferation and apoptosis is another common mechanism targeted by viral infection [66]. Emerging evidence has shown that bacteria could also evolve several strategies to control cell progression. For example, *H. pylori*-encoded CagE could promote the activation of Cyclin D1 in cell cycle [67]. The toxin CNF released by *E. coli* could not only induce G1-S transition and DNA replication but also inhibit cell apoptosis to stimulate cell progression [68, 69]. To block cell apoptosis, *H. pylori*-encoded Cag antigen also induces COX-2 expression to activate Bcl-2 and suppress apoptosis [70].

The evasion of the immune system is another key mechanism utilized by bacteria to promote cell malignancy. It has been demonstrated that bacteria have evolved multiple mechanisms to evade host immune response, including the modulation of bacteria surface, subversion of phagocytes, and blockade of innate immunity. To avoid surface signal recognition by immune system, many bacteria form carbohydrate-rich capsules or incorporate host proteins into capsules to mask their surface antigens from host receptors [71]. Since phagocytosis is one of the main ways used by host to counter bacterial infection, it is not surprising for bacteria to employ different strategies to escape. For example, *Streptococcus pneumonia* and *Staphylococcus aureus* produced immunoglobulin proteases to preclude the capture of antigens [72] or antibody-binding proteins to scavenge opsonizing antibodies [73]. To avoid the acquired immune response, *H. pylori* produce a vacuolating toxin VacA to block T cell proliferation and in turn inhibit the receptor-IL-2 signaling pathway and decrease of activated T cells [74, 75].

11.6 Future Perspective

Although it has been demonstrated that some bacterial infections associate with development of cancer, very few studies demonstrate the genomic instability of cells which was directly caused by bacteria or exposed to bacterial components through blood. More details about the molecular pathways involved in the induction of genomic instability in response to bacterial infection remain to be further explored. It still needs to answer why the causative relationship between the bacterial infection and cancers is only limited for a few cancers and whether the bacterial infection process is required to cause cancer. The association of oral organisms with colorectal cancer indicates that relocation of bacteria in inappropriate tissues and polymicrobial interactions together could be the key for bacteria to cause carcinogenesis. Along with the development of next-generation deep sequencing technology and bioinformatics analysis, it will provide a clear scenario about how bacterial infection contributes to cancer, which will facilitate to develop effective diagnostic and therapeutic strategies against infection-causing cancers.

References

1. Kuper H, Adami HO, Trichopoulos D (2000) Infections as a major preventable cause of human cancer. J Intern Med 248:171–183
2. Ness RB, Cauley JA (2004) Antibiotics and breast cancer – what's the meaning of this? JAMA 291:880–881
3. Harris RE, Beebe-Donk J, Doss H, Burr DD (2005) Aspirin, ibuprofen, and other non-steroidal anti-inflammatory drugs in cancer prevention: a critical review of non-selective COX-2 blockade (review). Oncol Rep 13:559–583
4. Onuigbo WI (1975) Some nineteenth century ideas on links between tuberculous and cancerous diseases of the lung. Br J Dis Chest 69:207–210
5. Parsonnet J, Friedman GD, Vandersteen DP, Chang Y, Vogelman JH, Orentreich N, Sibley RK (1991) Helicobacter pylori infection and the risk of gastric carcinoma. N Engl J Med 325:1127–1131
6. Wotherspoon AC, Ortiz-Hidalgo C, Falzon MR, Isaacson PG (1991) Helicobacter pylori-associated gastritis and primary B-cell gastric lymphoma. Lancet 338:1175–1176
7. Mikulska M, Viscoli C, Orasch C, Livermore DM, Averbuch D, Cordonnier C, Akova M, Fourth European Conference on Infections in Leukemia Group ajvoEEIELN, Esgich/Escmid (2014) Aetiology and resistance in bacteraemias among adult and paediatric haematology and cancer patients. J Infect 68:321–331
8. Mentis A, Lehours P, Megraud F (2015) Epidemiology and diagnosis of helicobacter pylori infection. Helicobacter 20(Suppl 1):1–7
9. Ding Z, Zhao S, Gong S, Li Z, Mao M, Xu X, Zhou L (2015) Prevalence and risk factors of helicobacter pylori infection in asymptomatic Chinese children: a prospective, cross-sectional, population-based study. Aliment Pharmacol Ther 42:1019–1026
10. Marshall BJ, Warren JR (1984) Unidentified curved bacilli in the stomach of patients with gastritis and peptic ulceration. Lancet 1:1311–1315
11. Weeks DL, Eskandari S, Scott DR, Sachs G (2000) A H+−gated urea channel: the link between helicobacter pylori urease and gastric colonization. Science 287:482–485

12. Peek RM Jr, Blaser MJ (2002) Helicobacter pylori and gastrointestinal tract adenocarcinomas. Nature reviews. Cancer 2:28–37
13. Peek RM Jr, Crabtree JE (2006) Helicobacter infection and gastric neoplasia. J Pathol 208:233–248
14. Sena Teixeira Mendes L, DA A, CW A (2014) Helicobacter pylori infection in gastric extra-nodal marginal zone lymphoma of mucosa-associated lymphoid tissue (MALT) lymphoma: a re-evaluation. Gut 63:1526–1527
15. Uemura N, Okamoto S, Yamamoto S, Matsumura N, Yamaguchi S, Yamakido M, Taniyama K, Sasaki N, Schlemper RJ (2001) Helicobacter pylori infection and the development of gastric cancer. N Engl J Med 345:784–789
16. Mera R, Fontham ET, Bravo LE, Bravo JC, Piazuelo MB, Camargo MC, Correa P (2005) Long term follow up of patients treated for helicobacter pylori infection. Gut 54:1536–1540
17. Wong BC, Lam SK, Wong WM, Chen JS, Zheng TT, Feng RE, Lai KC, Hu WH, Yuen ST, Leung SY, Fong DY, Ho J, Ching CK, Chen JS, China Gastric Cancer Study G (2004) Helicobacter pylori eradication to prevent gastric cancer in a high-risk region of China: a randomized controlled trial. JAMA 291:187–194
18. Huret JL, Ahmad M, Arsaban M, Bernheim A, Cigna J, Desangles F, Guignard JC, Jacquemot-Perbal MC, Labarussias M, Leberre V, Malo A, Morel-Pair C, Mossafa H, Potier JC, Texier G, Viguie F, Yau Chun Wan-Senon S, Zasadzinski A, Dessen P (2013) Atlas of genetics and cytogenetics in oncology and haematology in 2013. Nucleic Acids Res 41:D920–D924
19. Song H, Ekheden IG, Zheng Z, Ericsson J, Nyren O, Ye W (2015) Incidence of gastric cancer among patients with gastric precancerous lesions: observational cohort study in a low risk western population. BMJ 351:h3867
20. Zackular JP, Baxter NT, Iverson KD, Sadler WD, Petrosino JF, Chen GY, Schloss PD (2013) The gut microbiome modulates colon tumorigenesis. MBio 4:e00692–e00613
21. Sears CL, Garrett WS (2014) Microbes, microbiota, and colon cancer. Cell Host Microbe 15:317–328
22. Flynn KJ, Baxter NT, Schloss PD (2016) Metabolic and community synergy of oral bacteria in colorectal cancer. mSphere 1
23. Binder Gallimidi A, Fischman S, Revach B, Bulvik R, Maliutina A, Rubinstein AM, Nussbaum G, Elkin M (2015) Periodontal pathogens Porphyromonas gingivalis and Fusobacterium nucleatum promote tumor progression in an oral-specific chemical carcinogenesis model. Oncotarget 6:22613–22623
24. Yu J, Feng Q, Wong SH, Zhang D, Liang QY, Qin Y, Tang L, Zhao H, Stenvang J, Li Y, Wang X, Xu X, Chen N, Wu WK, Al-Aama J, Nielsen HJ, Kiilerich P, Jensen BA, Yau TO, Lan Z, Jia H, Li J, Xiao L, Lam TY, Ng SC, Cheng AS, Wong VW, Chan FK, Xu X, Yang H, Madsen L, Datz C, Tilg H, Wang J, Brunner N, Kristiansen K, Arumugam M, Sung JJ, Wang J (2017) Metagenomic analysis of faecal microbiome as a tool towards targeted non-invasive biomarkers for colorectal cancer. Gut 66:70–78
25. Zeller G, Tap J, Voigt AY, Sunagawa S, Kultima JR, Costea PI, Amiot A, Bohm J, Brunetti F, Habermann N, Hercog R, Koch M, Luciani A, Mende DR, Schneider MA, Schrotz-King P, Tournigand C, Tran Van Nhieu J, Yamada T, Zimmermann J, Benes V, Kloor M, Ulrich CM, von Knebel DM, Sobhani I, Bork P (2014) Potential of fecal microbiota for early-stage detection of colorectal cancer. Mol Syst Biol 10:766
26. Whitmore SE, Lamont RJ (2014) Oral bacteria and cancer. PLoS Pathog 10:e1003933
27. Dejea CM, Wick EC, Hechenbleikner EM, White JR, Mark Welch JL, Rossetti BJ, Peterson SN, Snesrud EC, Borisy GG, Lazarev M, Stein E, Vadivelu J, Roslani AC, Malik AA, Wanyiri JW, Goh KL, Thevambiga I, Fu K, Wan F, Llosa N, Housseau F, Romans K, Wu X, McAllister FM, Wu S, Vogelstein B, Kinzler KW, Pardoll DM, Sears CL (2014) Microbiota organization is a distinct feature of proximal colorectal cancers. Proc Natl Acad Sci U S A 111:18321–18326
28. Warren RL, Freeman DJ, Pleasance S, Watson P, Moore RA, Cochrane K, Allen-Vercoe E, Holt RA (2013) Co-occurrence of anaerobic bacteria in colorectal carcinomas. Microbiome 1:16

29. Hajishengallis G (2015) Periodontitis: from microbial immune subversion to systemic inflammation. Nature reviews. Immunology 15:30–44
30. Donaldson GP, Lee SM, Mazmanian SK (2016) Gut biogeography of the bacterial microbiota. Nature reviews. Microbiology 14:20–32
31. Rubinstein MR, Wang X, Liu W, Hao Y, Cai G, Han YW (2013) Fusobacterium nucleatum promotes colorectal carcinogenesis by modulating E-cadherin/beta-catenin signaling via its FadA adhesin. Cell Host Microbe 14:195–206
32. Kostic AD, Chun E, Robertson L, Glickman JN, Gallini CA, Michaud M, Clancy TE, Chung DC, Lochhead P, Hold GL, El-Omar EM, Brenner D, Fuchs CS, Meyerson M, Garrett WS (2013) Fusobacterium nucleatum potentiates intestinal tumorigenesis and modulates the tumor-immune microenvironment. Cell Host Microbe 14:207–215
33. Johnson CH, Dejea CM, Edler D, Hoang LT, Santidrian AF, Felding BH, Ivanisevic J, Cho K, Wick EC, Hechenbleikner EM, Uritboonthai W, Goetz L, Casero RA Jr, Pardoll DM, White JR, Patti GJ, Sears CL, Siuzdak G (2015) Metabolism links bacterial biofilms and colon carcinogenesis. Cell Metab 21:891–897
34. Nagy KN, Sonkodi I, Szoke I, Nagy E, Newman HN (1998) The microflora associated with human oral carcinomas. Oral Oncol 34:304–308
35. Michaud DS (2013) Role of bacterial infections in pancreatic cancer. Carcinogenesis 34:2193–2197
36. Groeger S, Domann E, Gonzales JR, Chakraborty T, Meyle J (2011) B7-H1 and B7-DC receptors of oral squamous carcinoma cells are upregulated by Porphyromonas gingivalis. Immunobiology 216:1302–1310
37. Inaba H, Sugita H, Kuboniwa M, Iwai S, Hamada M, Noda T, Morisaki I, Lamont RJ, Amano A (2014) Porphyromonas gingivalis promotes invasion of oral squamous cell carcinoma through induction of proMMP9 and its activation. Cell Microbiol 16:131–145
38. Uitto VJ, Baillie D, Wu Q, Gendron R, Grenier D, Putnins EE, Kanervo A, Firth JD (2005) Fusobacterium nucleatum increases collagenase 3 production and migration of epithelial cells. Infect Immun 73:1171–1179
39. Bodro M, Gudiol C, Garcia-Vidal C, Tubau F, Contra A, Boix L, Domingo-Domenech E, Calvo M, Carratala J (2014) Epidemiology, antibiotic therapy and outcomes of bacteremia caused by drug-resistant ESKAPE pathogens in cancer patients. Support Care Cancer: Off J Multinatl Assoc Support Care Cancer 22:603–610
40. Rolston KV, Besece D, Lamp KC, Yoon M, McConnell SA, White P (2014) Daptomycin use in neutropenic patients with documented gram-positive infections. Support Care Cancer: Off J Multinatl Assoc Support Care Cancer 22:7–14
41. Dix D, Cellot S, Price V, Gillmeister B, Ethier MC, Johnston DL, Lewis V, Michon B, Mitchell D, Stobart K, Yanofsky R, Portwine C, Silva M, Bowes L, Zelcer S, Brossard J, Traubici J, Allen U, Beyene J, Sung L (2012) Association between corticosteroids and infection, sepsis, and infectious death in pediatric acute myeloid leukemia (AML): results from the Canadian infections in AML research group. Clin Infect Dis: Off Publ Infect Dis Soc Am 55:1608–1614
42. Lewis V, Yanofsky R, Mitchell D, Dix D, Ethier MC, Gillmeister B, Johnston D, Michon B, Stobart K, Portwine C, Silva M, Cellot S, Price V, Bowes L, Zelcer S, Brossard J, Beyene J, Sung L (2014) Predictors and outcomes of viridans group streptococcal infections in pediatric acute myeloid leukemia: from the Canadian infections in AML research group. Pediatr Infect Dis J 33:126–129
43. Shelburne SA 3rd, Tarrand J, Rolston KV (2013) Review of streptococcal bloodstream infections at a comprehensive cancer care center, 2000–2011. J Infect 66:136–146
44. Domenech A, Ardanuy C, Grau I, Calatayud L, Pallares R, Fenoll A, Brueggemann AB, Linares J (2014) Evolution and genetic diversity of the Spain23F-ST81 clone causing adult invasive pneumococcal disease in Barcelona (1990–2012). J Antimicrob Chemother 69:924–931
45. Han SB, Bae EY, Lee JW, Lee DG, Chung NG, Jeong DC, Cho B, Kang JH, Kim HK (2013) Clinical characteristics and antimicrobial susceptibilities of viridans streptococcal bacteremia

during febrile neutropenia in patients with hematologic malignancies: a comparison between adults and children. BMC Infect Dis 13:273
46. Phares CR, Lynfield R, Farley MM, Mohle-Boetani J, Harrison LH, Petit S, Craig AS, Schaffner W, Zansky SM, Gershman K, Stefonek KR, Albanese BA, Zell ER, Schuchat A, Schrag SJ, Active Bacterial Core surveillance/Emerging Infections Program N (2008) Epidemiology of invasive group B streptococcal disease in the United States, 1999–2005. JAMA 299:2056–2065
47. Richter SS, Heilmann KP, Dohrn CL, Riahi F, Diekema DJ, Doern GV (2013) Pneumococcal serotypes before and after introduction of conjugate vaccines, United States, 1999–2011(1.) Emerg Infect Dis 19:1074–1083
48. Noble CJ (1978) Carriage of group D streptococci in the human bowel. J Clin Pathol 31:1182–1186
49. Kupferwasser I, Darius H, Muller AM, Mohr-Kahaly S, Westermeier T, Oelert H, Erbel R, Meyer J (1998) Clinical and morphological characteristics in Streptococcus bovis endocarditis: a comparison with other causative microorganisms in 177 cases. Heart 80:276–280
50. Mc CW, Mason JM 3rd (1951) Enterococcal endocarditis associated with carcinoma of the sigmoid; report of a case. J Med Assoc State Ala 21:162–166
51. Tjalsma H, Scholler-Guinard M, Lasonder E, Ruers TJ, Willems HL, Swinkels DW (2006) Profiling the humoral immune response in colon cancer patients: diagnostic antigens from Streptococcus bovis. Int J Cancer 119:2127–2135
52. Zhou X, Arends JP, Span LF, Friedrich AW (2013) Algorithm for pre-emptive glycopeptide treatment in patients with haematologic malignancies and an Enterococcus faecium bloodstream infection. Antimicrob Resist Infect Control 2:24
53. Welton JC, Marr JS, Friedman SM (1979) Association between hepatobiliary cancer and typhoid carrier state. Lancet 1:791–794
54. Tewari M, Mishra RR, Shukla HS (2010) Salmonella typhi and gallbladder cancer: report from an endemic region. Hepatobiliary Pancreat Dis Int: HBPD INT 9:524–530
55. Robinson KM, Sieber KB, Dunning Hotopp JC (2013) A review of bacteria-animal lateral gene transfer may inform our understanding of diseases like cancer. PLoS Genet 9:e1003877
56. Koehler JE, Sanchez MA, Garrido CS, Whitfeld MJ, Chen FM, Berger TG, Rodriguez-Barradas MC, LeBoit PE, Tappero JW (1997) Molecular epidemiology of bartonella infections in patients with bacillary angiomatosis-peliosis. N Engl J Med 337:1876–1883
57. Schroder G, Schuelein R, Quebatte M, Dehio C (2011) Conjugative DNA transfer into human cells by the VirB/VirD4 type IV secretion system of the bacterial pathogen Bartonella henselae. Proc Natl Acad Sci U S A 108:14643–14648
58. Riley DR, Sieber KB, Robinson KM, White JR, Ganesan A, Nourbakhsh S, Dunning Hotopp JC (2013) Bacteria-human somatic cell lateral gene transfer is enriched in cancer samples. PLoS Comput Biol 9:e1003107
59. Handa O, Naito Y, Yoshikawa T (2011) Redox biology and gastric carcinogenesis: the role of Helicobacter pylori. Redox Rep: Commun Free Radic Res 16:1–7
60. El-Omar EM, Carrington M, Chow WH, McColl KE, Bream JH, Young HA, Herrera J, Lissowska J, Yuan CC, Rothman N, Lanyon G, Martin M, Fraumeni JF Jr, Rabkin CS (2000) Interleukin-1 polymorphisms associated with increased risk of gastric cancer. Nature 404:398–402
61. Martin C, Thevenot G, Danel S, Chapron J, Tazi A, Macey J, Dusser DJ, Fajac I, Burgel PR (2011) Pseudomonas aeruginosa induces vascular endothelial growth factor synthesis in airway epithelium in vitro and in vivo. Eur Respir J 38:939–946
62. Takayama S, Takahashi H, Matsuo Y, Okada Y, Takeyama H (2010) Effect of Helicobacter bilis infection on human bile duct cancer cells. Dig Dis Sci 55:1905–1910
63. Mori N, Wada A, Hirayama T, Parks TP, Stratowa C, Yamamoto N (2000) Activation of intercellular adhesion molecule 1 expression by Helicobacter pylori is regulated by NF-kappaB in gastric epithelial cancer cells. Infect Immun 68:1806–1814

64. Isomoto H, Mizuta Y, Miyazaki M, Takeshima F, Omagari K, Murase K, Nishiyama T, Inoue K, Murata I, Kohno S (2000) Implication of NF-kappaB in Helicobacter pylori-associated gastritis. Am J Gastroenterol 95:2768–2776
65. Hold GL, Rabkin CS, Chow WH, Smith MG, Gammon MD, Risch HA, Vaughan TL, McColl KE, Lissowska J, Zatonski W, Schoenberg JB, Blot WJ, Mowat NA, Fraumeni JF Jr, El-Omar EM (2007) A functional polymorphism of toll-like receptor 4 gene increases risk of gastric carcinoma and its precursors. Gastroenterology 132:905–912
66. Coschi CH, Dick FA (2012) Chromosome instability and deregulated proliferation: an unavoidable duo. Cell Mol Life Sci: CMLS 69:2009–2024
67. Lax AJ (2005) Opinion: bacterial toxins and cancer – a case to answer? Nature reviews. Microbiology 3:343–349
68. Patyar S, Joshi R, Byrav DS, Prakash A, Medhi B, Das BK (2010) Bacteria in cancer therapy: a novel experimental strategy. J Biomed Sci 17:21
69. Oswald E, Sugai M, Labigne A, Wu HC, Fiorentini C, Boquet P, O'Brien AD (1994) Cytotoxic necrotizing factor type 2 produced by virulent Escherichia coli modifies the small GTP-binding proteins Rho involved in assembly of actin stress fibers. Proc Natl Acad Sci U S A 91:3814–3818
70. Lax AJ, Thomas W (2002) How bacteria could cause cancer: one step at a time. Trends Microbiol 10:293–299
71. Finlay BB, McFadden G (2006) Anti-immunology: evasion of the host immune system by bacterial and viral pathogens. Cell 124:767–782
72. Sarantis H, Grinstein S (2012) Subversion of phagocytosis for pathogen survival. Cell Host Microbe 12:419–431
73. Nitsche-Schmitz DP, Johansson HM, Sastalla I, Reissmann S, Frick IM, Chhatwal GS (2007) Group G streptococcal IgG binding molecules FOG and protein G have different impacts on opsonization by C1q. J Biol Chem 282:17530–17536
74. Torres VJ, VanCompernolle SE, Sundrud MS, Unutmaz D, Cover TL (2007) Helicobacter pylori vacuolating cytotoxin inhibits activation-induced proliferation of human T and B lymphocyte subsets. J Immunol 179:5433–5440
75. Gebert B, Fischer W, Weiss E, Hoffmann R, Haas R (2003) Helicobacter pylori vacuolating cytotoxin inhibits T lymphocyte activation. Science 301:1099–1102

Chapter 12
Parasite-Associated Cancers (Blood Flukes/Liver Flukes)

Meng Feng and Xunjia Cheng

Abstract Parasitic infection remains as a persistent public health problem and can be carcinogenic. Three helminth parasites, namely, *Clonorchis sinensis* (liver fluke) and *Opisthorchis viverrini* as well as *Schistosoma haematobium* (blood fluke), are classified as Group 1 carcinogens by the World Health Organization's International Agency for Research on Cancer (IARC Infection with liver flukes (*Opisthorchis viverrini, Opisthorchis felineus* and *Clonorchis sinensis*), World Health Organization, International Agency for Research on Cancer, 2011). Infection by these parasites is frequently asymptomatic and is thus rarely diagnosed at early exposure. Persistent infection can cause severe cancer complications. Until now, the cellular and molecular mechanisms linking fluke infections to cancer formation have yet to be defined, although many studies have focused on these mechanisms in recent years, and numerous findings were made in various aspects of parasite-associated cancers. Herein, we only introduce the fluke-induced cholangiocarcinoma (CCA) and bladder carcinoma and mainly focus on key findings in the last 5 years.

Keywords Parasitic infection • Helminth parasites • Fluke-induced cholangiocarcinoma • Bladder carcinoma • Cancer

12.1 Cholangiocarcinoma and Liver Flukes

The small liver flukes *O. viverrini* and *C. sinensis* are helminths that can affect humans. These flukes are particularly prevalent in the Southeast and East Asia, particularly in countries such as Thailand, Lao PDR, Vietnam, and Cambodia [1–4]. These countries have a strikingly high incidence of CCA (hepatic cancer of the bile duct epithelium) and approximately 700 million people at risk of infection.

M. Feng • X. Cheng (✉)
Department of Medical Microbiology and Parasitology, Institute of Biomedical Sciences, School of Basic Medical Sciences, Fudan University, Shanghai, People's Republic of China
e-mail: xjcheng@shmu.edu.cn

Q. Cai et al. (eds.), *Infectious Agents Associated Cancers: Epidemiology and Molecular Biology*, Advances in Experimental Medicine and Biology 1018,
DOI 10.1007/978-981-10-5765-6_12

Moreover, 10 and 35 million people are infected by *O. viverrini* and *C. sinensis*, respectively [2]. Thus, liver fluke infections are considered a serious global public health concern.

O. viverrini is represented in literature as a cause of hepatobiliary disease often referred to clinically as "opisthorchiasis," particularly CCA [5, 6]. The disease is acquired by ingesting contaminated raw or undercooked freshwater fish or fish products, such as fish sauces and fermented fishes. The flukes reside in the small intrahepatic bile ducts and cause chronic inflammation and, eventually, CCA. Similarly, *C. sinensis* is considered as an agent of CCA. However, the detailed mechanisms in the pathogenesis of CCA caused by *C. sinensis* are less clear than those caused by *O. viverrini*. Meanwhile, infections caused by *Opisthorchis felineus*, another kind of liver fluke, have no observed association with CCA so far [7]. *O. felineus* holds a highly close phylogenetic relationship with *O. viverrini*. Moreover, the biology of the former and the medical and epidemiological implications of the related infection are suggested to be essentially identical to those of *O. viverrini* and *C. sinensis*. Thus, further studies are warranted to assess the contribution of opisthorchiasis to CCA.

O. viverrini is endemic in the northeastern region of Thailand, which has the highest worldwide incidence of CCA. This geographical coincidence typically supports the liver fluke–CCA causal relationship. Pathophysiological research demonstrated the capability of *O. viverrini* to induce cancer in artificially infected laboratory animals (hamsters) [8]. The main liver fluke–CCA causal mechanisms were observed to be involved in the mechanical damage caused by parasite activities and immunopathology due to infection-related inflammation. Each of these effects has several minor counterparts. In brief, liver flukes generate tumor microenvironments that result in CCA.

12.1.1 Risk Factors for Liver Fluke-Associated Cholangiocarcinoma

To date, studies that were able to attain the valuable risk factors of liver fluke-associated CCA are few, indicating that the disease factors are exceedingly complex. A meta-analysis found that liver fluke infection is significantly associated with cholangitis, cholecystitis, cholelithiasis, hepatocellular carcinoma, and CCA [9]. Moreover, heavy infection was observed to be significantly associated with high incidence of hepatobiliary pathological changes.

Apart from heavy infection, risk factors were reported to affect the liver fluke–CCA relation. Liver fluke infections may exhibit hepatobiliary abnormalities and chronic infection that may lead to CCA development. A study on a population in the northeast of Thailand found that the prevalence of *O. viverrini* infection has decreased because of a liver fluke control program that has been implemented over the decades [10]. However, the prevalence of PDF remained high. In 55,246 subjects, the overall prevalence of PDF was 33.0 %, and males were at higher risk in developing PDF than females. The study also showed that old age (≥70 years) and hepatitis B are associated with increased PDF risk. By contrast, the number of praziquantel treatments and diabetes mellitus cases were significantly associated with decreased PDF risk. Another report suggested that fermented food consumption can exacerbate cholangitis and cholangiofibrosis, which are risk factors for CCA-associated opisthorchiasis [11].

12.1.2 Cholangiocarcinoma-Associated Gene Variation

Genetic changes have been widely reported to be associated with liver fluke-related CCA. Ong et al. reported that whole-exome and targeted sequencing not only can verify frequent mutations in known CCA-related genes, including TP53 (44 %), KRAS (16.7 %), and SMAD4 (16.7 %), but also can identify somatic mutations in 10 newly implicated genes, including MLL3 (14.8 %), ROBO2 (9.3 %), RNF43 (9.3 %), PEG3 (5.6 %), and GNAS oncogene (9.3 %) [12]. Gene functions can be grouped according to the deactivation of histone modifiers, activation of G protein signaling, and genome stability loss. Another report compared CAA associated with *O. viverrini* with those not associated with *O. viverrini*. The study revealed mutations in novel CCA-related genes associated with chromatin remodeling (BAP1, ARID1A, MLL3, and IDH1/2), WNT signaling (RNF43 and PEG3), and KRAS/G protein signaling (GNAS and ROBO2) [13].

Genetic variations are frequently reported in liver fluke-related CCA. In addition, when the epigenetic changes and miRNAs were characterized in liver fluke-related CCA, they were found to hold potential as diagnostic or prognostic biomarkers. Sriraksa et al. reported that aberrant hypermethylation of a certain loci is a common event in liver fluke-related CCA and thus may potentially contribute to cholangiocarcinogenesis. In their results, a number of CpG islands (OPCML, SFRP1, HIC1, PTEN, and DcR1) showed frequent hypermethylation, and 91 % of CCA were methylated in at least one CpG island [14]. Furthermore, patients with methylated DcR1 exhibited significantly longer overall survival than those without. Runglawan et al. revealed that the levels of urinary miR-192 and miR-21 are higher in the risk group of subjects than those in healthy individuals. This result suggests that increased miR-192 and miR-21 levels in host urine may provide better predictive values in areas endemic for *O. viverrini* than they do in nonendemic regions [11].

Liver fluke-induced chronic inflammation plays a crucial role in cholangiocarcinogenesis through distinct signatures of genetic, epigenetic, and transcriptional alterations. These alterations indicate a unique pathogenic process in liver fluke-related CCA and thus may hold potential clinical implications in CAA diagnostics, therapeutics, and prevention.

12.1.3 Cholangiocarcinoma-Associated Parasite Proteins

Liver fluke-secreted proteins were previously shown to accelerate human cholangiocytes. The endocytosis of liver fluke proteins by host epithelial cells affects the pathways and induces the parasites to cause a highly devastating form of cancer, that is, CCA. Although the detailed mechanisms by which cells internalize liver fluke-secreted proteins remains unclear, recent studies implied that liver fluke proteins have a role in pathogenesis and highlighted an approach for vaccine development against this infectious cancer.

O. viverrini excretory/secretory products (OvESs) have been a focus of study. These products are especially internalized by biliary cells and postulated to be responsible for the chronic inflammation and proliferation of cholangiocytes. The physical attachment of *O. viverrini* to the biliary epithelium causes damage by releasing highly immunogenic OvES. When internalized preferentially by liver cell lines, OvESs induce liver cell proliferation and promote IL6 secretion [15].

Proteins secreted by *O. viverrini* accelerate wound resolution in human cholangiocytes. Smout et al. demonstrated that a gene encoding granulin-like growth factor (Ov-GRN-1), which was derived from recombinant *O. viverrini*, induces angiogenesis, an essential mechanism for malignant tumor development. In addition, Monica et al. showed that Ov-GRN-1 induces angiogenesis and accelerates wound healing in mice. In fact, wound healing and cancer progression have remarkable similarities, such as new blood vessel growth in a process called angiogenesis. Thus, determining the effect of Ov-GRN-1 in CCA progression is valuable [16].

Recent reports have highlighted the presence of secreted extracellular vesicles (EVs) in helminths. In particular, *O. viverrini* secretes EVs that induce a proinflammatory or tumorigenic phenotype in human cholangiocytes. Chaiyadet et al. demonstrated that *O. viverrini* EVs are released in the secreted products of carcinogenic liver flukes. These EVs are then internalized by cholangiocytes, subsequently driving cell proliferation and IL-6 secretion and promoting an inflammatory but simultaneously modulatory environment. These processes ultimately facilitate the CCA development [17].

Thioredoxin from *O. viverrini* was reported to inhibit the oxidative stress-induced apoptosis of bile duct epithelial cells and cholangiocytes. Immunolocalization revealed the presence of liver fluke thioredoxin within the cholangiocytes. The cells exposed to thioredoxin were observed to downregulate the apoptotic genes in mitogen-activated protein kinase pathway and upregulate antiapoptosis-related genes, including apoptosis signaling kinase 1, caspase 9, caspase 8, caspase 3, and survivin. These results suggest that *O. viverrini* thioredoxin can inhibit apoptosis and facilitate carcinogenesis. As such, thioredoxin may play an important role for liver fluke oxidoreductase in *O. viverrini*-induced CCA [18].

12.1.4 Immunopathology, Tissue Damage, and Cholangiocarcinoma

Cholangiocarcinoma (CCA) incidence related to chronic *O. viverrini* infection is a multifactorial process encompassing immunopathological mechanism, tissue damage, and signal pathway activation. Immune-mediated pathogenesis in response to liver fluke infection is a major driving force of CCA onset. Liver fluke-associated CCA involves different pro- and anti-inflammatory cytokines that may instigate cancer development. In *O. viverrini* infection, elevated total serum IL-6, and IL-6 production stimulated by *O. viverrini* in PBMC have been reported in infected individuals with advanced periductal fibrosis. Surapaitoon et al. reported 11 cytokine profiles, including those of IFN-γ, IL-1β, IL-2, IL-4, IL-5, IL-6, IL-8, IL-10, IL-12p70, TNF-α, and LT-α, which were increased in CCA incidence associated with *O. viverrini* infection relative to the same profiles in uninfected normal controls. The results suggest the dysregulation of immune response in liver fluke-associated CCA [19].

Oxidative tissue damage caused by the free radicals released from effector cells has been observed to surround infected bile ducts. Thanee et al. suggested that the accumulation of CD44v suppresses p38MAPK expression in transforming bile duct cells and is linked to poor prognosis in CCA patients [20]. CD44 is a single transmembrane protein involved in cancer development and expressed in a wide variety of isoforms. Moreover, the expression of CD44v on a cell surface stabilizes xCT and promotes glutathione synthesis as defense against reactive oxygen species (ROS). This process enhances cancer development and increases chemotherapy resistance. Apinya et al. identified the oxysterols Triol and 3K4 as potential mediators of cholangiocarcinogenesis. Triol and 3K4 induce DNA damage and cell apoptosis via mitochondrial-dependent mechanisms. Chronic liver fluke infection increases the production of ROS/RNS when chronic inflammation occurs in the biliary system. Oxysterols and free radicals can induce biliary epithelial cell apoptosis. Ineffective DNA repair and persistent exposure to DNA damaging agents select for resistant cells that clonally expand to become malignant.

Signaling pathways involved in chronic inflammation regulate the progression of liver fluke-associated CCA. Yothaisong et al. reported a significant activation of the PI3K/AKT signaling pathway with PTEN suppression in human CCA tissues. Watcharin et al. suggested that the PKA signaling pathway and the switching of PKA regulatory subunits from Prkar2b/PKAII to Prkar1a/PKAI are involved in the development of CCA via altered cell transformation and proliferation [13]. Furthermore, the overexpression of Prkar1a can lead to the secretion of extracellular PKA (ECPKA) outside the CCA cells. Thus, the signal pathway activation may be responsible for the CCA development induced by *O. viverrini*. Targeting the components of these particular pathways may prove beneficial for the development of effective treatment, early diagnostics, and prevention strategies for CCA.

Inflammation is a well-known key component of tumor microenvironments. CCA is induced by chronic inflammation caused by the combination of several mechanical damage types. Immunosuppressive prednisolone was found to enhance early CCA in Syrian hamsters with liver fluke infection. This result suggests that host immune responses occur to ameliorate pathology and are also crucially associated with the pathogenesis of *O. viverrini* infection. Thus, imbalanced host immunity may enhance cancer-related inflammation.

12.1.5 Liver Fluke-Associated Microbiome and Cholangiocarcinoma

Microbial interaction with host cells can influence the health of the host considerably and has been implicated in liver fluke-associated CCA. The bacterial families *Dietziaceae*, *Pseudomonadaceae*, and *Oxalobacteraceae* are distinct microbiomes that dominate bile duct tissues. Chng et al. compared groups associated with those not associated with *O. viverrini* and then identified the enrichment for specific enteric bacteria (*Bifidobacteriaceae*, *Enterobacteriaceae*, and *Enterococcaceae*). They found that *Bifidobacteriaceae* was enriched and dominant in the *O. viverrini* microbiome and thus provided a mechanistic link to the parasite. Functional analysis revealed that altered microbiota increases the production of bile acids and ammonia in *O. viverrini* tissues and thus is linked to carcinogenesis [21]. These findings denote that parasitic infections and tissue microenvironments can influence each other and both can contribute to cancer.

12.2 Bladder Carcinoma and Blood Flukes

Schistosomiasis is a neglected tropical disease transmitted to humans from freshwater snails [22]. This disease is caused by a blood fluke of the genus *Schistosoma*. Schistosomiasis is considered as the most important helminthiasis that causes high rates of morbidity and mortality. Schistosomes affect at least 76 countries and 200–400 million people worldwide. *S. haematobium* is the causative agent of urogenital schistosomiasis and is responsible for two-thirds of the 200–400 million cases of human schistosomiasis worldwide [22, 23]. This infection is also associated with a high incidence of squamous cell carcinoma of the bladder, which is prevalent in the developing world [24, 25].

S. haematobium cercariae penetrate the skin and then transform into schistosomula. After infecting the subcutaneous tissue, the schistosomula enter the circulation and travel to the lungs and liver, where they achieve sexual maturity before entering into the vesical venous plexus. The eggs released from paired adults travel to the bladder, become trapped in the bladder wall, and release antigens and other metabolites. The eggs then lodge into the tissues and produce granulomatous inflammation that can lead to fibrosis [26, 27].

Long-term urinary schistosomiasis has been associated with the development of bladder cancer [28]. This disease is the leading cause of bladder cancer and occurs following years of chronic inflammation, fibrosis, and hyperproliferation. As a result, bladder cancer has become a significant health problem in the rural areas of Africa and the Middle East, where *S. haematobium* is prevalent [29, 30]. This information supports the association between malignant transformation and infection caused by this blood fluke. A study on an adult rural population in a Ghana region with endemic urinary schistosomiasis revealed the potential schistosome-associated bladder cancer problem and increasing association among age, severe bladder abnormalities, and occurrence of cancer biomarkers. Data on the epidemiological extent in different geographical areas estimate a schistosome-associated bladder cancer incidence of 3–4 cases per 100,000 [31]. This value suggests that schistosome-associated bladder cancer is an important public health concern in areas where *S. haematobium* is prevalent.

12.2.1 Bladder Carcinoma-Associated Parasite Genes

Bladder cancer is often difficult to diagnose without invasive measures, such as cystoscopy. However, benefitting from the development of molecular diagnostic techniques, some biomarkers were recognized as candidates for diagnosis and prognosis of this neglected tropical disease-linked cancer. In a recent study, liquid chromatography–mass spectrometry analysis was performed on urine from Angolans diagnosed with urogenital schistosomiasis and schistosome-associated bladder cancer. The metabolites were analyzed and expected to provide deep insight into the schistosome-associated bladder cancer. The analysis revealed numerous estrogen-like metabolites, including catechol estrogen quinones, CEQ-DNA-adducts, and novel metabolites derived directly from 8-oxo-7,8-dihydro-2′-deoxyguanosine, which were identified in urine of all patients [32].

A correlation was observed among the frequency of the biomarkers of bladder cancer associated with *S. haematobium*, those with p53, and sialylated glycans. The correlation highlights a missing link between infection and cancer development. The eggs of *S. haematobium* express sLea and sLex antigens in mimicry of human leukocyte glycosylation and thus may play a role in colonization and disease dissemination [33]. These results may facilitate early identification of infected patients at a high risk of developing bladder cancer and guide the future development of noninvasive diagnostic tests.

12.2.2 Bladder Carcinoma-Associated Parasite Proteins

Chronic infection with *S. haematobium* is associated with squamous cell carcinoma of the bladder. However, the molecular mechanisms underlying this association are poorly understood. Previous data revealed that the soluble extracts of mixed-sex adult *S. haematobium* worms (SWAP) are tumorigenic [22]. Moreover, the estrogen-related molecules in SWAP downregulate the estrogen receptors (ERs) in estrogen-responsive cells [22, 34]. Schistosome estrogens present in the sera of schistosomiasis patients repress the transcription of ERs in urothelial cells [34].

The soluble egg antigens of *S. haematobium* (Sh-SEA) exhibit potent tumorigenic capacity, because the *S. haematobium* eggs are in the developmental stage wherein they can directly cause urogenital disease during schistosomiasis haematobia. The findings confirmed that Sh-SEA can stimulate cell proliferation, reduce apoptosis, and increase the oxidative stress of urothelial cells. Furthermore, the presence of catechol estrogens in Sh-SEA might induce bladder cancer. These catechol estrogens are formed by a prospective estrogen–DNA adduct-mediated pathway in the *S. haematobium* eggs [22].

12.2.3 Immunopathology, Tissue Damage, and Bladder Carcinoma

Schistosomes elicit chronic inflammatory responses in both humans and mice. In *S. haematobium* infection, the prolonged inflammatory response is thought to contribute to the development of squamous cell carcinoma. Furthermore, the dependence of schistosomes on host factors for successful infection is evolutionarily conserved in *S. haematobium*. When infecting its host, schistosomes use common host mechanisms. In addition, schistosomes use immune signals for its development. The contributions of the host genes, which are discrete from immune system genes, must be understood, because these contributions are necessary for parasite establishment and development. Previous studies addressing the host–parasite interactions during schistosomiasis focused on a subset of immune response genes used to mount a Th1/Th2 response during infection. These genes include critical immune response genes, such as IL-4, IL-6, and IL-10, which control the Th1/Th2 response. In brief, regulatory pathways accommodate host permissiveness to schistosome establishment and productive schistosome infection and parasitism [35].

Schistosomiasis haematobia is a chronic infection. The adult and egg-producing schistosomes can live for many years. Thus, reinfections frequently occur and thus can lead to bladder cancer. To discuss the basic mechanisms that are potentially common in cancers, many studies focused on the role of both estrogens and ERs on the carcinogenesis associated with urogenital schistosomiasis. Botelho et al. observed a noteworthy elevation in estradiol serum levels [34]. They also observed that the serum levels of the luteinizing and follicle-stimulating hormones remained normal. Thus, they hypothesized that excess estradiol is external to the host. The molecule responsible for the effect may be an estradiol-like molecule derived from *S. haematobium*. This molecule is an antagonist of estradiol and thus repressed the transcriptional activity of the ERs. ER transcriptional activity was suppressed in urothelial cells, and ER expression was also inhibited in the mouse bladders in response to *S. haematobium* infection. These findings revealed the estrogen metabolism and ER signaling pathways associated with cancer induction in the context of *S. haematobium* infection [36, 37].

Another kind of hormonal and estrogenic molecule was identified in the Sh-SEA. The majority of these compounds are catechol estrogens, which are formed through the hydroxylation of steroid aromatic ring A. The hydroxylation of both C-2 and C-3 on a steroid ring is apparent and is subjected to further oxidation into an estradiol-2,3-quinone [38, 39]. The genotoxic effects of estrogen metabolites may be attributed to the oxidation of catechol estrogens to quinones, followed by redox cycling and formation of reactive oxygen species, which react with DNA. The metabolism of estrogens and production of depurinating estrogen–DNA adducts can be implicated in a pathway underlying host cell DNA damage promoted by *S. haematobium* and eventually lead to cell transformation. The carcinogenic effect of this estrogen–DNA adduct-mediated pathway may explain the link between chronic schistosomiasis haematobia and squamous cell carcinoma of the bladder [22].

At present, many mechanisms remain unclear, and further studies are necessary to understand how schistosomiasis haematobia leads to the squamous cell carcinoma of the bladder.

12.3 Other Parasite-Associated Cancers

Although certain cancer-related parasites belong to helminths, recent works have reported the association between unicellular protozoa and tumor. Despite its small size, protozoa can cause a series of diseases and public health problems worldwide. Their ability to induce tumor must be given sufficient attention. Herein, we list two kinds of tumors potentially caused by unicellular protozoa.

12.3.1 Toxoplasma and Brain Tumors

Toxoplasmosis is caused by *Toxoplasma gondii* and is a highly prevalent parasitic disease. This condition is estimated to affect a third of the world's human population [40, 41]. Two recent studies highlighted a positive correlation between the prevalence of brain tumors and *T. gondii* at the national and international scales [42, 43]. Unfortunately, these studies are correlative, and the links between *T. gondii* and cancer are complex. Thus, a causality between *T. gondii* and brain tumors was not attained. Even so, further research could shed light on the possible mechanisms underlying this association.

12.3.2 Trichomonas and Prostate Cancer (PCA)

Trichomonas vaginalis is an extracellular flagellated parasitic protozoan that causes a relatively common parasitic sexually transmitted infection. The disease holds an annual incidence of over 3 million. The majority of infections is asymptomatic in men and thus is often undiagnosed and untreated. These untreated cases are hypothesized to promote chronic persistent prostatic infection and resultant urethritis and prostatitis [44, 45].

T. vaginalis was previously suspected to be associated with the development of PCA [46]. For decades, a couple of clinical studies seemed to support this correlation. However, two recent studies gave negative results. A study that included 146 men with advanced prostate cancer demonstrated that *T. vaginalis* serostatus is not associated with increased risk of metastatic or fatal prostate cancer (odds ratio <1) [47]. This result does not support the increased risk of advanced or fatal prostate cancer in men infected with *T. vaginalis*. In another study, Zhu et al. suggested that the culture supernatant of *T. vaginalis* inhibits prostate cancer growth by disrupting the proliferation and promotion of apoptosis [48].

Epidemiologic evidence for the association of *T. vaginalis* with prostate cancer is inconsistent, and the role of *T. vaginalis* in PCA development remains controversial. Thus, further study may help elucidate the association between PCA and *T. vaginalis* infection.

References

1. IARC. (2011) Infection with liver flukes (*Opisthorchis viverrini, Opisthorchis felineus* and *Clonorchis sinensis*), World Health Organization, International Agency for Research on Cancer
2. Petney TN, Andrews RH, Saijuntha W, Wenz-Mucke A, Sithithaworn P (2013) The zoonotic, fish-borne liver flukes *Clonorchis sinensis, Opisthorchis felineus* and *Opisthorchis viverrini*. Int J Parasitol 43(12–13):1031–1046
3. Sithithaworn P, Andrews RH, Nguyen VD, Wongsaroj T, Sinuon M, Odermatt P et al (2012) The current status of opisthorchiasis and clonorchiasis in the Mekong Basin. Parasitol Int 61(1):10–16
4. Sripa B, Bethony JM, Sithithaworn P, Kaewkes S, Mairiang E, Loukas A et al (2011) Opisthorchiasis and opisthorchis-associated cholangiocarcinoma in Thailand and Laos. Acta Trop 120(Suppl 1):S158–S168
5. Sithithaworn P, Yongvanit P, Duenngai K, Kiatsopit N, Pairojkul C (2014) Roles of liver fluke infection as risk factor for cholangiocarcinoma. J Hepatobiliary Pancreat Sci 21(5):301–308
6. Sripa B, Brindley PJ, Mulvenna J, Laha T, Smout MJ, Mairiang E et al (2012) The tumorigenic liver fluke *Opisthorchis viverrini* – multiple pathways to cancer. Trends Parasitol 28(10):395–407
7. Fedorova OS, Kovshirina YV, Kovshirina AE, Fedotova MM, Deev IA, Petrovskiy FI, et al. Opisthorchis felineus infection and cholangiocarcinoma in the Russian Federation: a review of medical statistics. Parasitol Int. 2016;S1383–5769(16):30236–30237

8. Wilcox BA, Echaubard P. Balancing biomedical and ecological perspectives in research framing of liver fluke and cholangiocarcinoma in NE Thailand. Parasitol Int. 2016;S1383–5769(16):30258–30256
9. Xia J, Jiang SC, Peng HJ (2015) Association between liver fluke infection and hepatobiliary pathological changes: a systematic review and meta-analysis. PLoS One 10(7):e0132673
10. Intajarurnsan S, Khuntikeo N, Chamadol N, Thinkhamrop B, Promthet S (2016) Factors associated with periductal fibrosis diagnosed by ultrasonography screening among a high risk population for cholangiocarcinoma in Northeast Thailand. Asian Pac J Cancer Prev 17(8):4131–4136
11. Silakit R, Loilome W, Yongvanit P, Thongchot S, Sithithaworn P, Boonmars T, et al. Urinary microRNA-192 and microRNA-21 as potential indicators for liver fluke-associated cholangiocarcinoma risk group. Parasitol Int. 2015;S1383–5769(15):00165–00168
12. Ong CK, Subimerb C, Pairojkul C, Wongkham S, Cutcutache I, Yu W et al (2012) Exome sequencing of liver fluke-associated cholangiocarcinoma. Nat Genet 44(6):690–693
13. Yothaisong S, Thanee M, Namwat N, Yongvanit P, Boonmars T, Puapairoj A et al (2014) Opisthorchis Viverrini infection activates the PI3K/ AKT/PTEN and Wnt/β-catenin signaling pathways in a Cholangiocarcinogenesis model. Asian Pac J Cancer Prev 15(23):10463–10468
14. Sriraksa R, Zeller C, El-Bahrawy MA, Dai W, Daduang J, Jearanaikoon P et al (2011) CpG-island methylation study of liver fluke-related cholangiocarcinoma. Br J Cancer 104(8):1313–1318
15. Chaiyadet S, Smout M, Johnson M, Whitchurch C, Turnbull L, Kaewkes S et al (2015) Excretory/secretory products of the carcinogenic liver fluke are endocytosed by human cholangiocytes and drive cell proliferation and IL6 production. Int J Parasitol 45(12):773–781
16. Botelho MC, Alves H, Richter J (2016) Wound healing and cancer progression in *Opisthorchis viverrini* associated cholangiocarcinoma. Parasitol Res 115(7):2913–2914
17. Chaiyadet S, Sotillo J, Smout M, Cantacessi C, Jones MK, Johnson MS et al (2015) Carcinogenic liver fluke secretes extracellular vesicles that promote cholangiocytes to adopt a tumorigenic phenotype. J Infect Dis 212(10):1636–1645
18. Matchimakul P, Rinaldi G, Suttiprapa S, Mann VH, Popratiloff A, Laha T et al (2015) Apoptosis of cholangiocytes modulated by thioredoxin of carcinogenic liver fluke. Int J Biochem Cell Biol 65:72–80
19. Surapaitoon A, Suttiprapa S, Khuntikeo N, Pairojkul C, Sripa B (2017) Cytokine profiles in *Opisthorchis viverrini* stimulated peripheral blood mononuclear cells from cholangiocarcinoma patients. Parasitol Int 66(1):889–892
20. Thanee M, Loilome W, Techasen A, Sugihara E, Okazaki S, Abe S et al (2016) CD44 variant-dependent redox status regulation in liver fluke-associated cholangiocarcinoma: a target for cholangiocarcinoma treatment. Cancer Sci 107(7):991–1000
21. Chng KR, Chan SH, Ng AH, Li C, Jusakul A, Bertrand D et al (2016) Tissue microbiome profiling identifies an enrichment of specific enteric bacteria in Opisthorchis Viverrini associated cholangiocarcinoma. EBioMedicine 8:195–202
22. Botelho MC, Vale N, Gouveia MJ, Rinaldi G, Santos J, Santos LL et al (2013) Tumour-like phenotypes in urothelial cells after exposure to antigens from eggs of *Schistosoma haematobium*: an oestrogen-DNA adducts mediated pathway? Int J Parasitol 43(1):17–26
23. van der Werf MJ, de Vlas SJ, Brooker S, Looman CW, Nagelkerke NJ, Habbema JD et al (2003) Quantification of clinical morbidity associated with schistosome infection in sub-Saharan Africa. Acta Trop 86(2–3):125–139
24. Parkin DM (2006) The global health burden of infection-associated cancers in the year 2002. Int J Cancer 118(12):3030–3044
25. Hodder SL, Mahmoud AA, Sorenson K, Weinert DM, Stein RL, Ouma JH et al (2000) Predisposition to urinary tract epithelial metaplasia in *Schistosoma haematobium* infection. Am J Trop Med Hyg 63(3–4):133–138
26. Stavitsky AB (2004) Regulation of granulomatous inflammation in experimental models of schistosomiasis. Infect Immun 72(1):1–12

27. Pearce EJ, MacDonald AS (2002) The immunobiology of schistosomiasis. Nat Rev Immunol 2(7):499–511
28. Mostafa MH, Sheweita SA, O'Connor PJ (1999) Relationship between schistosomiasis and bladder cancer. Clin Microbiol Rev 12(1):97–111
29. Zheng YL, Amr S, Saleh DA, Dash C, Ezzat S, Mikhail NN et al (2012) Urinary bladder cancer risk factors in Egypt: a multicenter case-control study. Cancer Epidemiol Biomark Prev 21(3):537–546
30. Salem S, Mitchell RE, El-Alim El-Dorey A, Smith JA, Barocas DA (2011) Successful control of schistosomiasis and the changing epidemiology of bladder cancer in Egypt. BJU Int 107(2):206–211
31. Shiff C, Veltri R, Naples J, Quartey J, Otchere J, Anyan W et al (2006) Ultrasound verification of bladder damage is associated with known biomarkers of bladder cancer in adults chronically infected with *Schistosoma haematobium* in Ghana. Trans R Soc Trop Med Hyg 100(9):847–854
32. Gouveia MJ, Santos J, Brindley PJ, Rinaldi G, Lopes C, Santos LL et al (2015) Estrogen-like metabolites and DNA-adducts in urogenital schistosomiasis-associated bladder cancer. Cancer Lett 359(2):226–232
33. Santos J, Fernandes E, Ferreira JA, Lima L, Tavares A, Peixoto A et al (2014) P53 and cancer-associated sialylated glycans are surrogate markers of cancerization of the bladder associated with *Schistosoma haematobium* infection. PLoS Negl Trop Dis 8(12):e3329
34. Botelho MC, Alves H, Barros A, Rinaldi G, Brindley PJ, Sousa M (2015) The role of estrogens and estrogen receptor signaling pathways in cancer and infertility: the case of schistosomes. Trends Parasitol 31(6):246–250
35. Nair SS, Bommana A, Bethony JM, Lyon AJ, Ohshiro K, Pakala SB et al (2011) The metastasis-associated protein-1 gene encodes a host permissive factor for schistosomiasis, a leading global cause of inflammation and cancer. Hepatology 54(1):285–295
36. Botelho MC, Soares R, Vale N, Ribeiro R, Camilo V, Almeida R et al (2010) *Schistosoma haematobium*: identification of new estrogenic molecules with estradiol antagonistic activity and ability to inactivate estrogen receptor in mammalian cells. Exp Parasitol 126(4):526–535
37. Botelho MC, Ribeiro R, Vale N, Oliveira P, Medeiros R, Lopes C et al (2012) Inactivation of estrogen receptor by *Schistosoma haematobium* total antigen in bladder urothelial cells. Oncol Rep 27(2):356–362
38. Cavalieri EL, Stack DE, Devanesan PD, Todorovic R, Dwivedy I, Higginbotham S et al (1997) Molecular origin of cancer: catechol estrogen-3,4-quinones as endogenous tumor initiators. Proc Natl Acad Sci U S A 94(20):10937–10942
39. Cavalieri EL, Rogan EG (2011) Unbalanced metabolism of endogenous estrogens in the etiology and prevention of human cancer. J Steroid Biochem Mol Biol 125(3–5):169–180
40. Ewald PW (2009) An evolutionary perspective on parasitism as a cause of cancer. Adv Parasitol 68:21–43
41. Vittecoq M, Elguero E, Lafferty KD, Roche B, Brodeur J, Gauthier-Clerc M et al (2012) Brain cancer mortality rates increase with *Toxoplasma gondii* seroprevalence in France. Infect Genet Evol 12(2):496–498
42. Thirugnanam S, Rout N, Gnanasekar M (2013) Possible role of *Toxoplasma gondii* in brain cancer through modulation of host microRNAs. Infect Agent Cancer 8(1):8
43. Thomas F, Lafferty KD, Brodeur J, Elguero E, Gauthier-Clerc M, Missé D (2012) Incidence of adult brain cancers is higher in countries where the protozoan parasite *Toxoplasma gondii* is common. Biol Lett 8(1):101–103
44. Dirkx M, Boyer MP, Pradhan P, Brittingham A, Wilson WA (2014) Expression and characterization of a β-fructofuranosidase from the parasitic protist *Trichomonas vaginalis*. BMC Biochem 15:12
45. Satterwhite CL, Torrone E, Meites E, Dunne EF, Mahajan R, Ocfemia MC et al (2013) Sexually transmitted infections among US women and men: prevalence and incidence estimates, 2008. Sex Transm Dis 40(3):187–193

46. Siegel RL, Miller KD, Jemal A (2015) Cancer statistics, 2015. CA Cancer J Clin 65(1):5–29
47. Shui IM, Kolb S, Hanson C, Sutcliffe S, Rider JR, Stanford JL (2016) *Trichomonas vaginalis* infection and risk of advanced prostate cancer. Prostate 76(7):620–623
48. Zhu Z, Davidson KT, Brittingham A, Wakefield MR, Bai Q, Xiao H et al (2016) *Trichomonas vaginalis*: a possible foe to prostate cancer. Med Oncol 33(10):115

Chapter 13
Prion Protein Family Contributes to Tumorigenesis via Multiple Pathways

Xiaowen Yang, Zhijun Cheng, Lihua Zhang, Guiru Wu, Run Shi, Zhenxing Gao, and Chaoyang Li

Abstract A wealth of evidence suggests that proteins from prion protein (PrP) family contribute to tumorigenesis in many types of cancers, including pancreatic ductal adenocarcinoma (PDAC), breast cancer, glioblastoma, colorectal cancer, gastric cancer, melanoma, etc. It is well documented that PrP is a biomarker for PDAC, breast cancer, and gastric cancer. However, the underlying mechanisms remain unclear. The major reasons for cancer cell-caused patient death are metastasis and multiple drug resistance, both of which connect to physiological functions of PrP expressing in cancer cells. PrP enhances tumorigenesis by multiple pathways. For example, PrP existed as pro-PrP in most of the PDAC cell lines, thus increasing cancer cell motility by binding to cytoskeletal protein filamin A (FLNa). Using PDAC cell lines BxPC-3 and AsPC-1 as model system, we identified that dysfunction of glycosylphosphatidylinositol (GPI) anchor synthesis machinery resulted in

X. Yang • Z. Cheng
Department of the First Abdominal Surgery, Jiangxi Tumor Hospital, Nanchang, Jiangxi 330029, People's Republic of China

L. Zhang
Department of Pathology, Zhongda Hospital, Southeast University, Nanjing 210009, People's Republic of China

G. Wu • R. Shi
Wuhan Institute of Virology, Chinese Academy of Sciences, University of Chinese Academy of Sciences, 44 xiao hong shan zhong qu, Wuhan, Hubei Province 430071, People's Republic of China

Z. Gao
State Key Laboratory of Virology, Institute of Virology, Chinese Academy of Sciences, Wuhan, Hubei Province 430071, People's Republic of China

C. Li (✉)
State Key Laboratory of Virology, Wuhan Institute of Virology, Chinese Academy of Sciences, Wuhan, Hubei 430071, People's Republic of China

Hubei Collaborative Innovation Center for Industrial Fermentation, Wuhan, Hubei 430068, People's Republic of China
e-mail: cyli@wh.iov.cn

Q. Cai et al. (eds.), *Infectious Agents Associated Cancers: Epidemiology and Molecular Biology*, Advances in Experimental Medicine and Biology 1018, DOI 10.1007/978-981-10-5765-6_13

the biogenesis of pro-PrP. In addition, in cancer cells without FLNa expression, pro-PrP can modify cytoskeleton structure by affecting cofilin/F-actin axis, thus influencing cancer cell movement. Besides pro-PrP, we showed that GPI-anchored unglycosylated PrP can elevate cell mobility by interacting with VEGFR2, thus stimulating cell migration under serum-free condition. Besides affecting cancer cell motility, overexpressed PrP or doppel (Dpl) in cancer cells has been shown to increase cell proliferation, multiple drug resistance, and angiogenesis, thus, proteins from PrP gene family by affecting important processes via multiple pathways for cancer cell growth exacerbating tumorigenesis.

Keywords Prion protein • Doppel protein • Tumorigenesis • Signal transduction • Cell motility

13.1 Introduction

The prion gene family is comprised of four genes, *PRNP*, *PRND*, *SPRN*, and *PRNT*, among which *PRNP*, *PRND*, and *PRNT* genes are located within a 55 kb region on 20p13 locus (*PRNP*-20 kb-*PRND*-3 kb-*PRNT*) in human. *PRNP*, *PRND*, and *SPRN* encode prion protein (PrP), doppel (Dpl) which is homologous to the C-terminus of PRP and shadoo (Sho) which is homologous to the N-terminus of PrP, respectively, whereas *PRNT* is probably a pseudogene due to multiple large transposon insertions [1]. Positive immunoreaction against antibody specific to Prt, the potential protein product of *PRNT*, was reported in the testis and ejaculated spermatozoa of ovine [2]. However, it remains to be seen that Prt is indeed expressing in those tissues. *PRNP* and *SPRN* are present in all vertebrates, whereas *PRND* is detected in tetrapods but not in avians probably due to its lost in an early ancestor of birds [1, 3, 4]. The presence of *PRNT* in mammals is controversial; some reported that *PRNT* is only identified in primates, whereas others argued that it is highly conserved in mammals [1, 3, 4]. The evolutionary origin of *PRNP* remains ambiguous although multiple lines of evidence implicated that PrP may evolve from ZIP family of metal ion transporters [5]. It is likely that during the emergence of metazoan, a prion-like ectodomain in a subbranch of ZIP genes composing of ZIP5, ZIP6, and ZIP10 was generated by a cysteine-flanked core domain inserting in a preexisting ZIP ancestor gene, which then presumably evolved as the founder gene for prion [6]. The amino acid sequence of PrP can be divided into two domains, the N-terminus flexible domain and the C-terminus globular domain. PrP is expressed in many cells, such as neuron cells in central nervous system (CNS), epithelial cells in the gut and intestine, smooth muscle cells of blood vessel, islet cells of the pancreas, monocytes and lymphocytes, etc. It's generally undetectable in hepatocytes, ductal epithelial cells in the pancreas, and the kidney. The regulation of tissue-specific PrP expression is not clear. It has been shown that the 1.2 kb sequence upstream of *PRNP* contains promoter and repressor elements with major promoter activity adjacent to the 5′ region of exons 1 and 2 whereas suppressor activity adjacent to the 5′ region of intron 1 [7].

Transcriptional factors such as metal transcription factor-1 (MTF-1), p53, Sp1, Sp-3, and cleaved activating transcription factor 6 α (ΔATF6α) and spliced X-box protein-1 (sXBP1) have all been reported to increase PrP expression, whereas Yin Yang-1 (YY1), nuclear factor-erythroid 2-related factor-2 (Nrf-2), and Hes-1 reduced PrP expression [8–13]. In addition to transcriptional regulation, about 70% of rat brain PrP mRNA is associated with polysomes and thus is subject to translational regulation [14]. Whether PrP mRNA in other species has the same level of polysome association remains to be investigated. More importantly, this association implies that the cells may need a quick synthesis of PrP under certain conditions. However, not much is known on this aspect. Contrary to translational regulation, a lot has been learned for posttranslational modification of prion protein. The newly synthesized human PrP protein is a 254 amino acid polypeptide with a N-terminal leader peptide (amino acids 1–22) for the endoplasmic reticulum (ER) and a C-terminal glycosylphosphatidylinositol (GPI) signaling sequence (GPI-PSS) (amino acids 232–254). The N-terminal leader signal is cleaved upon entry into ER. However, due to its low efficacy to guide PrP into ER, cytosolic PrP is detected in many cell types [15]. It remains to be determined whether this cytosolic PrP is pro-PrP or not. Similar to the inefficiency of the N-terminal signal, in vitro experiment proved that the C-terminal GPI-PSS of PrP is of low efficiency in replacing GPI anchor compared to other GPI-anchored proteins [16]. Besides N-terminal and C-terminal signals, human PrP may also be modified with N-linked glycans on N181 and N197 and intramolecular disulfide bond connecting C179 and C214. The significance of N-linked glycans and intramolecular disulfide bonds remains elucidated for the physiological function of PrP, although it is reported that intramolecular disulfide bonds are required for the alpha-helical conformation of recombinant PrP [17]. Like other GPI-anchored proteins, PrP is tethered on the outer leaflet of the cholesterol-enriched cell membrane domain called lipid raft by means of its GPI anchor. Lipid raft is the platform to integrate cellular signaling, and some membrane receptors have been shown recruited into lipid raft to be activated [18, 19]. Although the physiological functions of PrP in vivo remain unidentified, the location of PrP in lipid raft implicates that PrP is involved in signal transduction or adhesion. By antibody crosslinking, it has been shown that PrP can activate tyrosine kinase Fyn in murine 1C11 neuronal differentiation model [20]. Besides antibody crosslinking, PrP can also stimulate signaling pathway by interacting with its physiological partners. For example, when binding to stress-inducible protein 1, PrP may activate mTOR [21], a pathway involved in cancer cell growth. PrP also plays roles in other signaling pathways important for cancer cell growth, such as anti-apoptotic pathway. For example, in breast cancer cells, MCF-7, PrP inhibits pro-apoptotic Bax conformational change occurring initially in Bax activation to prevent cell death [22]. In addition, cancer cells expressing PrP have accelerated cell cycle. PrP has been shown transcriptionally activating cyclin D1 expression to promote G1/S transition of human gastric cancer cells SGC 7901 and AGS [23]. Since PrP can be detected in the nuclei of some cancer cells, it remains to be determined if PrP can directly bind to the promoter region of cyclin D1, thus activating its expression. More importantly, PrP expression level is positively correlated with multidrug

Table 13.1 Reported tumors with PrP family protein expression

Prion or prion-like protein	Cancer type	References
PrP	Gastric cancer	[90–94]
PrP	Breast cancer	[10, 42, 70]
Pro-PrP	Pancreatic cancer	[46]
Pro-PrP	Melanoma	[30]
PrP	Glioblastoma	[95, 96]
PrP and Dpl	Osteosarcoma	[97]
PrP and Dpl	Colon cancer	[89, 98, 99]
Dpl	Lung cancer	[89]
PrP	Head and neck squamous cell carcinoma	[100]
PrP	Oral squamous cells carcinoma	[87]

resistance in cancer cells as shown in SGC7901/ADR gastric cancer cells or MCF-7/ADR breast cancer cells [24, 25]. Besides cancer cell lines, PrP has been reported to be upregulated in many types of cancers (Table 13.1). Together, these results suggest that PrP participates into many important aspects of tumorigenesis.

In this review, we will focus on physiological functions of PrP implied in cancer cell migration and growth and dissect how these functions contribute to tumorigenesis.

13.2 PrP Participates in Cell Adhesion and Migration

Most cancer patients die due to cancer cell metastasis which includes cell adhesion and migration. The role of PrP playing in cell adhesion and migration might be conserved evolutionary. Unlike mammalian PrP which only composed of one PrP gene, zebra fish has two PrP genes named PrP1 and PrP2. PrP-downregulated zebra fish embryo showed arrested gastrulation due to deficient morphogenetic cell movements [26]. In addition, using morpholino knockdown and PrP2-targeted inactivation, [27] found rosette formation defects in PrP2 morphants, which suggests an abnormal primordium organization and a loss of cell adhesion during migration of the primordium. Such a defect is accompanied by mislocation of adherens junction proteins, E-cadherin and ß-catenin. In a systematic integrin adhesome profiling study, PrP is one of the molecules in consensus adhesion [28]. This is also supported by other studies. In a large-scale quantitative and comparative interactome investigation, Watts et al. identified a physical association between PrP and ß1 integrins in mouse neuroblastoma N2a cells [29]. In addition, we showed in melanoma cell line A7 that PrP existed in the functional protein complex with FLNa and ß1 integrins, of which the binding between pro-PrP and FLNa enhanced the association between FLNa and ß1 integrins, thus promoting cell adhesion and migration [30]. The above observation is consistent with the fact that PrP is located at the outer cellular

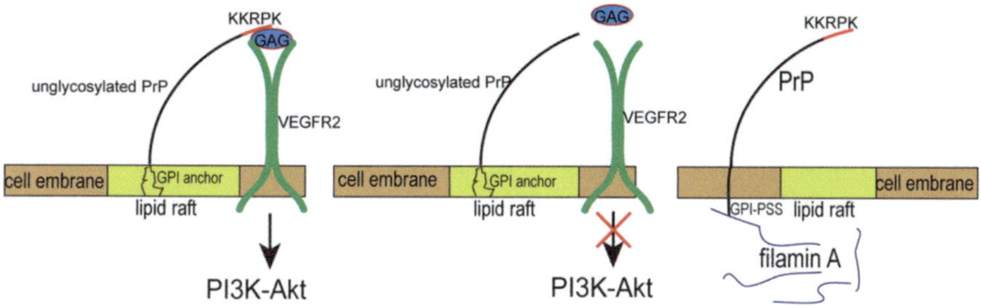

Fig. 13.1 Cartoon of several PrP isoforms expressed on cells and the potential interacting partners for the PrP to initiate cell migration. GPI-anchored unglycosylated PrP binds GAG and VEGFR2 via its KKRPK, activating PI3K-Akt signaling pathway. In this model, PrP is required to localize in lipid raft (*left panel*)

membrane. In fact, several types of PrP isoforms were implicated in cell adhesion and migration (Fig. 13.1). Due to its cell surface location, GPI-anchored PrP is implicated in cell-cell adhesion; in 2002, elevated PrP expression increases cation-independent cell aggregation, which could be reduced after treatment of the cells with phosphatidylinositol-specific phospholipase C to release cell surface PrP [31]. Previously, the PrP fragment which induces cell migration was attributed to amino acids 106–126, which induces migration of monocyte-derived dendritic cell regulated by neuropeptide substance P [32]. In addition, PrP peptide (106–126) between the concentration range 160–640 ng/ml induces murine macrophage cell line Ana-1 chemotaxis in a concentration-dependent manner, and the migration of macrophage can be inhibited by inhibitors for multiple signaling pathways, implying that PrP-stimulated macrophage motility is mediated by multiple manners [33]. A similar phenomenon was observed when BV-2 microglia cells were treated with PrP fragment (106–126) at the concentration range of 25–100 μM. This peptide stimulates microglia cell chemotaxis [34]. However, induction of cell migration by PrP fragment may be simply due to cells that respond to oxidative stress imposed by the peptide. A better approach to assess whether PrP facilitates cell adhesion or migration is to knock down/knock out the expression of PrP and detect alteration of cell behavior. In brain endothelial cells, PrP colocates with platelet endothelial cell adhesion molecule-1 (PECAM-1) in lipid raft domains. Antibody specific to PrP has the same efficacy as anti-PECAM-1 antibodies to block the transmigration of U937 human monocytes as well as freshly isolated monocytic cells when preincubated with U937 or hCMEC/D3 cells [35], suggesting that PrP is facilitating monocyte transmigration. Dpl plays a similar role in astrocytomas. When Dpl expression was reduced in IPDDC-A2 cell line, cellular migration was greatly reduced as assayed by wound healing [36]. When PrP expression was knocked down in a mouse brain microvascular endothelial cell line bEND.3, cell migration into the damaged regions but not cell proliferation was greatly reduced, implicating that PrP may play a role in neurovascular unit recovery from brain injury such as an ischemic insult [37]. Elevation of epidermal growth factor (EGF) or EGF receptor (EGFR) signaling often affects cellular processes, including increasing cell motility and reducing cell adhesion by modifying turnover of E-cadherin/catenin complex [38]. Proper maintenance of E-cadherin/catenin complex is necessary to prevent extensive cell overlap, thus maintaining cell adhesion [39]. PrP by involving in E-cadherin mediated adherens junction (AJ) formation in A431 cells regulating phosphorylation of catenin, thus affecting cell adhesion [40]. More importantly, PrP expression is positively correlated with an increased risk of metastasis in colorectal cancer. Knocking down PrP expression or treatment of the cancer stem cells with PrP-specific monoclonal antibodies significantly reduces cancer cell metastasis via ERK2 pathway [41]. PrP is expressed in the breast cancer cell line MCF7/Adr which showed a reduced migration in vitro when PrP is downregulated by siRNA [42]. Knocking down PrP expression reduces the formation of uropod and monocyte adhesion but increases cell motility on endothelial cell under shear stress most likely through RhoA-mediated pathway [43]. This observation is especially interesting considering that when cancer cells migrate out of the primary site, cancer cells must reduce adhesion but increase migration capability. The functions can be achieved simply by PrP downregulation. Considering that KKRPK the five positively charged amino acids at the far most N-terminus of PrP has been shown to bind glycosaminoglycans (GAGs), such as heparin or sulfated heparan [44], it is not surprising that PrP participates in cell adhesion and migration. We recently proved that CHO cells expressing GPI-anchored unglycosylated PrP adhere and migrate faster than GPI-anchored glycosylated PrP and this migration depends on the interactions between GAG and the N-terminal KKRPK domain of PrP because cells expressing similar level of KKRPK truncated unglycosylated PrP migrate much slower [45]. Unlike previous reports which did not define other binding partners in the interaction, we proved that VEGFR2 is the receptor involved in the protein complex containing PrP and GAG because activation of VEGFR2 signaling is greatly reduced when the cells were treated with EFGR2-specific inhibitor or when VEGFR2 was downregulated [45]. We now provide further evidence that VEGFR2 is directly involved in migration of these CHO cell. Knocking down VEGFR2 or treating the unglycosylated PrP expressing CHO cells with VEGFR2-specific inhibitor significantly reduced cell migration using wound-healing assays under serum-free condition (Fig. 13.2). In addition to regulation of migration by the GPI-anchored PrP^C, we identified an isoform of PrP in pancreatic cancer, and melanoma cells also contribute to cell migration [30, 46]. In this isoform, the GPI anchor peptide signaling sequence (GPI-PSS) was not cleaved due to GPI anchor synthesis deficiency. To identify the mechanism for the production of pro-PrP in most pancreatic cancer cell lines, we found a pancreatic cancer cell line AsPC-1 expressing GPI-anchored PrP^C; profiling of GPI anchor synthesis machinery that identified several enzymes showed a downregulated expression in BxPC-3 cells. To prove that the lower expression of PIG F and PGAP1 in BxPC-3 cells is responsible for the production of pro-PrP, we transfected PGAP1 alone into BxPC-3 cells, and the pro-PrP isoform was not affected. However, when PIG F but not PIG P, another downregulated protein in BxPC-3 cells, was co-expressed with PGAP1, the pro-PrP isoform was successfully converted into GPI-anchored PrP and thus cleavable by phosphor-specific lipase C (PI-PLC) [47]. The generation of pro-PrP obviously enhances the mobility of the cancer cells due to pro-PrP retaining its GPI-PSS, which can interact with filamin A (FLNa), a multifunctional cytoskeletal linker [30, 46]. When comparing AsPC-1, a

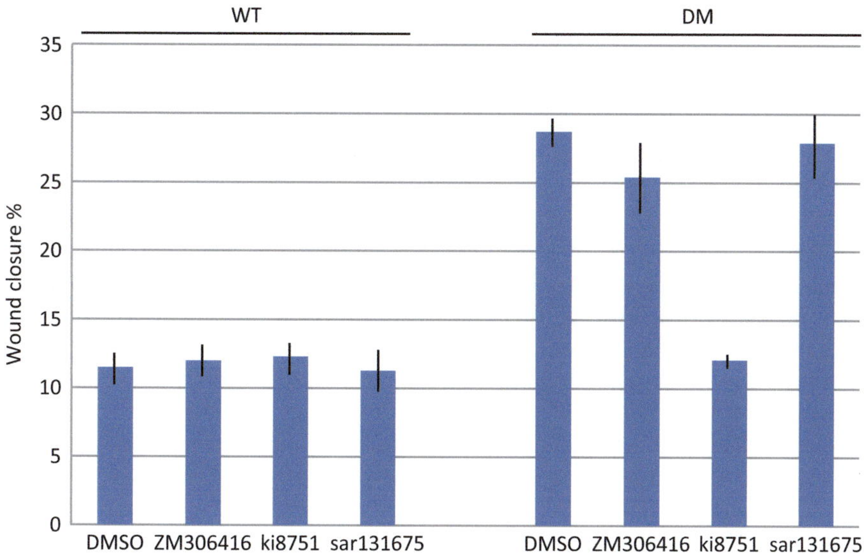

Fig. 13.2 CHO cells expressing unglycosylated PrP showed VEGFR2-dependent cell migration under serum-free culture condition. WT, WT human PrP expressed in CHO cells; DM, unglycosylated PrP expressed in CHO cells. DMSO, vehicle control; ZM306416, VEGFR1-specific inhibitor; ki8751, VEGFR2-specific inhibitor; sar131675, VEGFR3-specific inhibitor. Cell wound was inflicted, and closure was measured 18 h later. Wound closure was quantified based on image J software)

pancreatic cancer cell expressing GPI-anchored PrP to BxPC-3 in wound-healing assays, BxPC-3 cells migrate much faster. More importantly, when pro-PrP was converted into GPI-anchored PrP in PIG F and PGA1 expressing BxPC-3 cells, these cells showed a significant reduction in mobility [47]. To further confirm that interactions between FLNa and pro-PrP contribute to cancer cell migration, we compared the mobility of two melanoma cell lines, M2 and A7. M2 is the sole melanoma cell line that does not express FLNa, whereas A7 is an FLNa-expressing cell line derived from M2. A7 showed a much greater mobility compared to M2. However, when PrP was downregulated in A7 cells, those A7 cells with reduced PrP expression migrate significantly slower, thus proving that pro-PrP increases cancer cell migration via interacting with FLNa [30]. Moreover, when pro-PrP and FLNa binding was attenuated in A7 cells by a peptide corresponding to the sequence of PrP-interacting FLNa domain, A7 cell migration was also greatly inhibited [30]. This result not only points out the interaction between pro-PrP and FLNa indeed existed in vivo but also indicates a potential intervention method to interrupt melanoma metastasis. Overall, due to posttranslational differences, PrP bestows cell adhesion and migration through different pathways in interaction with distinct partner.

13.3 PrP Is Pro-survival for Cells

One of the important characteristics of cancer cell is its ability against cell death. Many proteins against apoptosis were upregulated in cancer cells including PrP. The first study to show that PrP is probably participating against apoptosis was shown by Kuwahara et al. [48]. Under serum-depletion condition, hippocampal neuron cells from PrP WT mice are more sustainable than their PrP null counterpart cells, thus implying that PrP is a pro-survival protein under this condition. However, it remains to be investigated why PrP null cells die faster than WT neurons, as serum-free condition may simply imply that nutrients are important for the survival of PrP null neuron. After that, many studies indicated that PrP protects cells against oxidative stress and ischemia. The effects of PrP on oxidative stress were revealed by showing that PrP affects Cu/Zn superoxide dismutase (SOD) activity. Increased level of Cu/Zn SOD activity is positively correlated with augmented levels of PrP expression, most likely due to copper incorporation into these enzymes, and were affected by the expression levels of the PrP which binds copper with relatively high affinity [49, 50] showing that recombinant mouse and chicken PrP or mouse brain tissue-purified PrP possess SOD activity. This SOD activity is acquired by PrP binding copper, and since octapeptide repeat is required to bind copper, deletion of octapeptide results in loss of SOD activity [51]. To further dissect which domain at the N-terminus of PrP is important for the protective role against oxidative stress, [52] found that the amino acids 23–50 from N2 fragment of PrP-ß cleavage are able to initiate the protection. In addition, this protective reaction requires cell surface proteins, such as glycosaminoglycans and intact lipid – raft domains – thus pointing out a copper-independent pathway for PrP to protect against oxidative stress. In considering the low affinity between PrP and copper compared to other copper-bound proteins, Wong et al. [53] set out to investigate if PrP in vivo can contribute to the SOD activity. They found that the level of PrP expression is positively correlated with the level of total SOD activity, thus supporting the idea that PrP differentially contributes to the total SOD activity in vivo. Even if PrP does not transport copper at physiological concentrations in the rabbit kidney epithelial cell (RK13) model system, murine PrP expression indeed increases antioxidant enzyme activity and glutathione levels [54]. This observation further supports a role for PrP in antioxidant reaction. In addition, increased oxidative stress markers such as oxidation of lipid and protein were detected in the brains of PrP null mice; thus, it is possible that the antioxidant activity requires constitutive expression of PrP [55, 56]. Protection of neuron cells by PrP against ischemia probably is the most studied function of PrP against oxidative stress in vivo. Maintained under normoxia condition, C57Bl/6J mice showed elevated PrP expression in cerebral microvessels and in microvessel-depleted brain homogenate at age 6, 18, and 24 months [57], thus implicating PrP may compensate for a loss of antioxidant activity by increasing expressing levels as mice aging. At the same year, McLennan found that mRNA and protein of PRNP were upregulated during hypoxia damage in neuronal processes in the penumbra [58]. This result suggests that upregulation of PrP may be a stress response of the cells to prevent cell death. To determine if PrP is a cellular component response to neuronal injury in vivo, Weise et al. [57] used the thread occlusion stroke mouse model and detected a significant upregulation of PrP in the ischemic hemisphere at 4 and 8 h after onset of permanent focal ischemia when compared to control animals. However, significant upregulation of PrP was not detected if the focal ischemia is transient. Thus, the extent of upregulation of PrP which is positively correlated with the severity of ischemia may reflect the protection level required for neuron cell survival. To further examine if overexpression of PrP may contribute to transient cerebral ischemia, [59] subject PrP overexpressing TG35 mice and WT mice to a 90 min focal cerebral ischemia and found a reduction of early activation of Erk1/Erk2 in TG35 mice compared to WT mice along with reduced infarct volume, thus supporting PrP indeed is a pro-survival for neuron cells and implies a neuroprotection role for PrP-Erk signaling axis. To determine if PrP plays a role in a short period of ischemia, [60] performs a 30 min of intraluminal middle cerebral artery occlusion in PrP null and PrP WT mice and found that increased infarct size by about 200% in PrP null mice than PrP WT mice, along with increment in activity of Erk1/Erk2, STAT-1, and caspase-3, thus imitating lack of PrP, may subject neuron cells to apoptosis via Erk1/Erk2-STAT-1-caspase-3 axis. Stress-inducible phosphoprotein 1 (STI-1) is a ligand for PrP, in a STI-1 mutation ischemia mouse model, transgenic expression of STI-1-rescued embryonic lethality in a prion protein-dependent way [61]. This result further supports the notion that normal physiological function of PrP is neuroprotective, maybe by interacting with STI-1. Besides interaction with its potential ligand, the octapeptide repeat region at the N-terminus of PrP together with the PI3K/Akt signaling pathway is important for neuroprotection when subject to ischemia. Comparing with WT PrP mice, infarct size of PrP null mice or octapeptide repeat truncation PrP mice is significantly increased [62]. It is likely that the octapeptide repeat region by modifying SOD activity thus preventing neuron cell death. The aforementioned ischemia models are focusing on acute injury with maximal 3-day observation. To investigate if PrP participates in poststroke neurogenesis and angiogenesis in an extended period, Doeppner et al. induced focal cerebral ischemia by intraluminal middle cerebral artery occlusion in a post-ischemia mouse model; increased neurogenesis and angiogenesis were observed in PrP overexpression mice compared with PrP WT mice, whereas exacerbated neurological deficits and brain injury were detected in PrP null mice with increased proteasome activity and oxidative stress [63]. This result suggests that PrP is not only neuroprotective but also angiogenetic. Similarly, in a stroke rat model, PrP expression that mainly occurred in neuron, glia, and vascular endothelial cells was upregulated significantly in the penumbra of stroke brain compared with the untreated brain in a time-dependent manner. In addition, the rat showed improved neurological behavior and reduced cerebral infarction volume when PrP was forced to express in the ischemic brain [64], thus further supporting a role for PrP in the neuroprotection not only in the mouse model but also in the rat model. Most recently, Guitart showed that it is the exosomal PrP from astrocytes but not neuronal PrP that improves the neuron cell survival under oxidative stress [65]. Although PrP has been shown to protect neuron cells from ischemia, it is not clear that the protection is due to PrP that prevents necrosis, autophagy, or apoptosis induced by ischemia. The underlying mechanism for PrP protection of neural cell survival under Bax challenge was first identified by Bounhar et al. [66]. Working with human primary neuron and breast carcinoma MCF-7, Bounhar et al. showed that PrP potently inhibits Bax-induced cell death, and the protection role requires the octapeptide repeat regions but not the GPI anchor, thus implying that lipid raft location of PrP is not necessary for this anti-apoptotic role of PrP. To further investigate if the protection by PrP against Bax-induced cell death requires other Bcl-2 family members, Bounhar et al. [67] challenged *Saccharomyces cerevisiae* lacking Bcl-2 genes with exogenous expression of Bax and found that PrP overcomes Bax-mediated S phase growth and cell death. This result suggests that PrP does not require other Bcl-2 family proteins to protect against Bax-mediated mitochondrion-dependent apoptosis. In addition to protect breast carcinoma MCF-7 cells from Bax-induced cell death, PrP also prevents breast cancer cell death induced by Adriamycin (ADR) and TRAIL. In contrast to Bcl-2-independent protection of *S. cerevisiae*, downregulation of PrP is accompanied with reduced Bcl-2 expression, thus suggesting Bcl-2 is the target downstream of PrP to prevent cancer cell death [24]. Drug resistance by PrP was also observed in gastric cancer cell line and colon cancer cells [25, 68], and such drug resistance in gastric cancer is mediated by the first four octapeptide repeats, but not the fifth octapeptide repeat at the N-terminus of PrP [69]. Although suffered a limited sample size, probably the most intriguing observation is a retrospective analysis of breast cancer patients that showed patients with estrogen receptor-negative and PrP-positive staining are resistant but patients with estrogen receptor negative and PrP negative are sensitive to adjuvant chemotherapy [70]. This result not only points out that cancer cells expressing PrP are drug resistant but also identifies PrP as a potential biomarker for molecular typing of cancer cells.

13.4 PrP Plays a Role in Inflammation Response

It is widely accepted that chronic inflammation may result in cancer. Due to its high expression in lymphatic tissues, PrP have long been implicated in immune response [71]. It was later shown that T-cell activation may be modulated by PrP expression [72]. However, the role of PrP in inflammation was not well established. In normal skin, there is little expression of PrP mainly confined to keratinocytes. But expression of PrP was strongly upregulated in both keratinocytes and infiltrating mononuclear cells in psoriasis and contact dermatitis, two inflammatory skin diseases. Strong expression of PrP can also be detected in viral warts. Cytokines such as transforming growth factor alpha (TGFα) or interferon-γ resulted in an increase of PrP in primary cultured keratinocytes [73]. Similarly, in patients infected with *Helicobacter pylori*, higher expression of PrP at mRNA and protein levels was detected, which was correlated with gastric inflammation. More importantly, when *H. pylori* are eradicated, expression of PrP is no longer detectable [74]. In peripheral immune system, lack of PrP expression affects peripheral inflammation induced by ligands of either Toll-like receptors or Fas, as well as phagocytosis by macrophages [75, 76], whereas the content of PrP in monocytes is modulated by IFN-γ [77]. Additionally, PrP is expressed on the surface of human and mouse mast cells and is constantly released. A rapid release by mast cells upon activation can be detected. More importantly, PrP is also released in vivo in responding to mast cell-dependent allergic inflammation. Although PrP release is activated upon activation, PrP expression is not required for mast cell differentiation [78]. When treated with dextran sodium sulfate, PrP null mice showed elevated BAD protein level, IL-1β, IL-6, TNF-α, IL-4, IFN-γ, and other cytokine profile to favor inflammation [79, 80]. All these evidences suggest that PrP plays a role in modulating inflammation response. However, whether to suppress or to activate inflammation reactions probably depends on the agents or the time of the challenge. Neutrophils play critical roles in both acute and chronic inflammatory responses. To investigate the role PrP plays in neutrophil, Mariante et al. injected LPS intraperitoneally into mice and detected a dose- and time-dependent upregulation of PrP expression. This elevated PrP expression depends on the serum content of TGFß and glucocorticoids, which, in turn, are contingent on the activation of the hypothalamic-pituitary-adrenal axis in response to systemic inflammation [81]. These results suggest that PrP participates in inflammation, but its physiological functions remain unanswered. Besides participation in peripheral immune system, PrP also played a role in immune reaction in central nervous system (CNS). When infected with *Mycobacterium bovis*, BV2 microglia cells showed a gradual increase in *PRNP* mRNA level along with an upregulation of pro-inflammatory factors. When PrP expression level is silenced by siRNA in *M. bovis*-infected BV2 microglia, a reduction of those pro-inflammatory cytokines was observed. As a consequence, increased apoptosis occurred in infected microglia. This result implies that microglia PrP is pro-inflammatory when infected by *M. bovis* [82]. In addition to microglia cells, neuron cells expressing PrP also respond to cytokine treatment. When hippocampal neuron was treated with pro-inflammatory cytokines TNFα, IL-6, and IL-1ß, neuron cells expressing PrP were induced to form cofilin/actin (1:1) rods, whereas PrP null neuron cells do not respond to those pro-inflammatory cytokines. It is worth noting that overexpression of PrP by itself is sufficient to induce this type of rod in an NOX-dependent manner [83]. Since cofilin/actin (1:1) rods have been shown to impair transport and synaptic function of neuron, an adverse effect of PrP responding to pro-inflammatory reaction in CNS is anticipated. The function of PrP in vivo was additionally proved in an experimental autoimmune encephalomyelitis (EAE) mouse model in which EAE is induced by tail base subcutaneous injection with myelin oligodendrocyte glycoprotein; higher levels of leukocytic infiltrates and pro-inflammatory cytokine gene expression as well as increased spinal cord myelin basic protein and axonal loss were detected in the *prnp* null spinal cord, cerebellum, and forebrain examined during the acute phase. In addition, during the chronic phase, a remarkable persistence of leukocytic infiltrates was present in the forebrain and cerebellum, accompanied by an increase in interferon-γ and IL-17 transcripts [84]. Other than bacteria and cytokine treatments, PrP also responds to virus infection in vivo. When infected with 600 pfu encephalomyocarditis virus B variant (EMCV-B) via an intracranial route, *Prnp*

WT mice showed more severe infiltration of inflammatory cells in the brain, accompanied by higher activation of microglia cells around the hippocampus, than *prnp* null mice, thus indicating that PrP play some roles in induction of inflammation [85]. Due to PrP response to both TNFα, a pro-inflammatory cytokine, and TGFß, an anti-inflammatory cytokine, it was assumed that PrP may play dual roles in inflammation reaction. Working with the PrP-deficient mouse model, Liu et al. showed that PrP participates in pro-inflammatory responses by upregulating cytokines such as TNFα, IL-1ß, IL-6, and NOS2 during acute stage to eliminate LPS. However, PrP plays a role in anti-inflammatory reaction by increasing expression of cytokines such as IL-10, Arg1, and Mrc1 during tissue-repairing stage [86]. Previously, we reported that about 13% of patients in PanIN-3 stage showed positive staining for PrP [46], and Zhang et al. [87] reported that PrP immune-reaction-positive specimens increased from 15–42 to 95% when comparing oral normal mucosa, leukoplakia, and squamous cell carcinoma tissues, suggesting a positive correlation from normal, inflammation transition to cancer. Since in some cases PrP favors inflammation response and chronic inflammation may cause tumor, expression of PrP in precancerous stage may be a key factor initiating tumor.

13.5 Remarks and Perspectives

The notion that PrP contributes to tumorigenesis was an unanticipated one since PrP null mice develop apparent normally without obvious phenotype. Those PrP null mice strongly suggest that PrP is not an essential protein during mice development. However, the concept that PrP is a garbage protein present in mammals is not supported by its conservation in evolution, its location in lipid raft, and its complex posttranslational modifications during protein biogenesis. On the contrary, PrP is a multifunctional protein, and many of its functions can be substituted by other proteins. In considering that most cancer patients die from cancer cell metastasis and multidrug resistance, it is natural to target any proteins expressed in cancer cells to fulfill these two functions. Based on the evidences that PrP facilitates cell adhesion and migration and protects cells against apoptosis, it is not surprising that expression of PrP is a biomarker for poor prognosis in some cancer types. Besides metastasis and drug resistance, angiogenesis is another important factor for cancer cell development. In fact, PrP was implicated in developmental angiogenesis although no detailed mechanism was provided [88]. It is well known that VEGFR2 is important for tumor angiogenesis; by searching the physiological function of unglycosylated PrP, we identified an interaction between unglycosylated PrP and VEGFR2 in CHO cells. This result strongly suggests that PrP, maybe behaving as Dpl, participates in tumor angiogenesis [89]. By binding to VEGFR2 in tumor endothelial cells, Dpl stimulates VEGFR2 signaling and enhances tumor vascularization. Blocking the doppel/VEGFR2 axis by an orally active glycosaminoglycan (LHbisD4) specifically represses tumor angiogenesis, thus reducing tumor growth [89].

Although PrP has been shown to contribute to tumorigenesis at cellular level, a mouse model to prove PrP attributing to tumorigenesis has yet to be reported. Another important question that remains unanswered is whether aberrant expression of PrP can cause cancer and, if so, why. Since PrP is an endogenous expressing protein in many tissues, it is difficult to imagine that PrP expression can result in cancer. However, considering that the expression of PrP may favor pro-inflammation response, the possibility that PrP causes cancer cannot be totally excluded. Finally, since PrP is a protein prone to aggregation and conformational altered PrP may lead to neurodegeneration, overexpressed PrP in tumor cells may be a cause for future problem for patients.

Acknowledgments This work was supported by National Science Foundation of China (31670170), by Nature Science Foundation of Hubei Province (2015CFA087), and by State Key Laboratory of Virology (2015IOV005).

References

1. Harrison PM, Khachane A, Kumar M (2010) Genomic assessment of the evolution of the prion protein gene family in vertebrates. Genomics 95:268–277
2. Pimenta J, Domingos A, Santos P, Marques CC, Cantante C, Santos A, Barbas JP, Baptista MC, Horta AE, Viegas A, Mesquita P, Goncalves J, Fontes CA, Prates JA, Pereira RM (2012) Is prnt a pseudogene? Identification of ram Prt in testis and ejaculated spermatozoa. PLoS One 7:e42957
3. Rivera-Milla E, Oidtmann B, Panagiotidis CH, Baier M, Sklaviadis T, Hoffmann R, Zhou Y, Solis GP, Stuermer CAO, Malaga-Trillo E (2006) Disparate evolution of prion protein domains and the distinct origin of Doppel- and prion-related loci revealed by fish-to-mammal comparisons. FASEB J 20:317–319
4. Premzl M, Gamulin V (2007) Comparative genomic analysis of prion genes. BMC Genomics 8:1
5. Schmitt-Ulms G, Ehsani S, Watts JC, Westaway D, Wille H (2009) Evolutionary descent of prion genes from the ZIP family of metal ion transporters. PLoS One 4:e7208
6. Ehsani S, Tao R, Pocanschi CL, Ren H, Harrison PM, Schmitt-Ulms G (2011) Evidence for retrogene origins of the prion gene family. PLoS One 6:e26800
7. Baybutt H, Manson J (1997) Characterisation of two promoters for prion protein (PrP) gene expression in neuronal cells. Gene 184:125–131
8. Wright JA, McHugh PC, Stockbridge M, Lane S, Kralovicova S, Brown DR (2009) Activation and repression of prion protein expression by key regions of intron 1. Cell Mol Life Sci 66:3809–3820
9. Vincent B, Sunyach C, Orzechowski HD, St George-Hyslop P, Checler F (2009) p53-dependent transcriptional control of cellular prion by presenilins. J Neurosci 29:6752–6760
10. Dery MA, Jodoin J, Ursini-Siegel J, Aleynikova O, Ferrario C, Hassan S, Basik M, LeBlanc AC. 2013. Endoplasmic reticulum stress induces PRNP prion protein gene expression in breast cancer. Breast Cancer Res 15
11. Cichon AC, Brown DR (2014) Nrf-2 regulation of prion protein expression is independent of oxidative stress. Mol Cell Neurosci 63:31–37

12. Burgess STG, Shen C, Ferguson LA, O'Neill GT, Docherty K, Hunter N, Goldmann W (2009) Identification of adjacent binding sites for the YY1 and E4BP4 transcription factors in the ovine PrP (prion) gene promoter. J Biol Chem 284:6716–6724
13. Bellingham SA, Coleman LA, Masters CL, Camakaris J, Hill AF (2009) Regulation of prion gene expression by transcription factors SP1 and metal transcription factor-1. J Biol Chem 284:1291–1301
14. Denman R, Potempska A, Wolfe G, Ramakrishna N, Miller DL (1991) Distribution and activity of alternatively spliced Alzheimer amyloid peptide precursor and scrapie Prp messenger-Rnas on rat-brain polysomes. Arch Biochem Biophys 288:29–38
15. Rane NS, Chakrabarti O, Feigenbaum L, Hegde RS (2010) Signal sequence insufficiency contributes to neurodegeneration caused by transmembrane prion protein. J Cell Biol 188:515–526
16. Chen R, Knez JJ, Merrick WC, Medof ME (2001) Comparative efficiencies of C-terminal signals of native glycophosphatidylinositol (GPI)-anchored proproteins in conferring GPI-anchoring. J Cell Biochem 84:68–83
17. Zhang H, Stockel J, Mehlhorn I, Groth D, Baldwin MA, Prusiner SB, James TL, Cohen FE (1997) Physical studies of conformational plasticity in a recombinant prion protein. Biochemistry 36:3543–3553
18. Decker L, ffrench-Constant C. (2004) Lipid rafts and integrin activation regulate oligodendrocyte survival. J Neurosci 24:3816–3825
19. Sun X, Fu Y, Gu M, Zhang L, Li D, Li H, Chien S, Shyy JY, Zhu Y (2016) Activation of integrin alpha5 mediated by flow requires its translocation to membrane lipid rafts in vascular endothelial cells. Proc Natl Acad Sci U S A 113:769–774
20. Mouillet-Richard S, Ermonval M, Chebassier C, Laplanche JL, Lehmann S, Launay JM, Kellermann O (2000) Signal transduction through prion protein. Science 289:1925–1928
21. Roffe M, Beraldo FH, Bester R, Nunziante M, Bach C, Mancini G, Gilch S, Vorberg I, Castilho BA, Martins VR, Hajj GN (2010) Prion protein interaction with stress-inducible protein 1 enhances neuronal protein synthesis via mTOR. Proc Natl Acad Sci U S A 107:13147–13152
22. Roucou X, Giannopoulos PN, Zhang Y, Jodoin J, Goodyer CG, LeBlanc A (2005) Cellular prion protein inhibits proapoptotic Bax conformational change in human neurons and in breast carcinoma MCF-7 cells. Cell Death Differ 12:783–795
23. Liang J, Pan Y, Zhang D, Guo C, Shi Y, Wang J, Chen Y, Wang X, Liu J, Guo X, Chen Z, Qiao T, Fan D (2007) Cellular prion protein promotes proliferation and G1/S transition of human gastric cancer cells SGC7901 and AGS. FASEB J 21:2247–2256
24. Meslin F, Hamai A, Gao P, Jalil A, Cahuzac N, Chouaib S, Mehrpour M (2007) Silencing of prion protein sensitizes breast adriamycin-resistant carcinoma cells to TRAIL-mediated cell death. Cancer Res 67:10910–10919
25. Du J, Pan Y, Shi Y, Guo C, Jin X, Sun L, Liu N, Qiao T, Fan D (2005) Overexpression and significance of prion protein in gastric cancer and multidrug-resistant gastric carcinoma cell line SGC7901/ADR. Int J Cancer 113:213–220
26. Malaga-Trillo E, Solis GP, Schrock Y, Geiss C, Luncz L, Thomanetz V, Stuermer CAO (2009) Regulation of embryonic cell adhesion by the prion protein. PLoS Biol 7:576–590
27. Huc-Brandt S, Hieu N, Imberdis T, Cubedo N, Silhol M, Leighton PLA, Domaschke T, Allison WT, Perrier V, Rossel M. 2014. Zebrafish prion protein PrP2 controls collective migration process during lateral line sensory system development. PLoS One 9
28. Horton ER, Byron A, Askari JA, Ng DHJ, Millon-Fremillon A, Robertson J, Koper EJ, Paul NR, Warwood S, Knight D, Humphries JD, Humphries MJ (2015) Definition of a consensus integrin adhesome and its dynamics during adhesion complex assembly and disassembly. Nat Cell Biol 17:1577–1587
29. Watts JC, Huo H, Bai Y, Ehsani S, Jeon AH, Shi T, Daude N, Lau A, Young R, Xu L, Carlson GA, Williams D, Westaway D, Schmitt-Ulms G (2009) Interactome analyses identify ties of PrP and its mammalian paralogs to oligomannosidic N-glycans and endoplasmic reticulum-derived chaperones. PLoS Pathog 5:e1000608

30. Li C, Yu S, Nakamura F, Pentikainen OT, Singh N, Yin S, Xin W, Sy MS (2010) Pro-prion binds filamin A, facilitating its interaction with integrin beta1, and contributes to melanomagenesis. J Biol Chem 285:30328–30339
31. Mange A, Milhavet O, Umlauf D, Harris D, Lehmann S (2002) PrP-dependent cell adhesion in N2a neuroblastoma cells. FEBS Lett 514:159–162
32. Kaneider NC, Kaser A, Dunzendorfer S, Tilg H, Patsch JR, Wiedermann CJ (2005) Neurokinin-1 receptor interacts with PrP(106-126)-induced dendritic cell migration and maturation. J Neuroimmunol 158:153–158
33. Zhou H, Zhou X, Kouadir M, Zhang Z, Yin X, Yang L, Zhao D (2009) Induction of macrophage migration by neurotoxic prion protein fragment. J Neurosci Methods 181:1–5
34. Tu J, Yang L, Zhou X, Qi K, Wang J, Kouadir M, Xu L, Yin X, Zhao D (2014) PrP106-126 and Abeta 1-42 peptides induce BV-2 microglia chemotaxis and proliferation. J Mol Neurosci: MN 52:107–116
35. Viegas P, Chaverot N, Enslen H, Perriere N, Couraud PO, Cazaubon S (2006) Junctional expression of the prion protein PrPC by brain endothelial cells: a role in trans-endothelial migration of human monocytes. J Cell Sci 119:4634–4643
36. Azzalin A, Sbalchiero E, Barbieri G, Palumbo S, Muzzini C, Comincini S (2008) The doppel (Dpl) protein influences in vitro migration capability in astrocytoma-derived cells. Cell Oncol: Off J Int Soc Cell Oncol 30:491–501
37. Watanabe T, Yasutaka Y, Nishioku T, Kusakabe S, Futagami K, Yamauchi A, Kataoka Y (2011) Involvement of the cellular prion protein in the migration of brain microvascular endothelial cells. Neurosci Lett 496:121–124
38. Mosesson Y, Mills GB, Yarden Y (2008) Derailed endocytosis: an emerging feature of cancer. Nat Rev Cancer 8:835–850
39. Green KJ, Getsios S, Troyanovsky S, Godsel LM (2010) Intercellular junction assembly, dynamics, and homeostasis. Csh Perspect Biol 2
40. Solis GP, Schrock Y, Hulsbusch N, Wiechers M, Plattner H, Stuermer CAO (2012) Reggies/flotillins regulate E-cadherin-mediated cell contact formation by affecting EGFR trafficking. Mol Biol Cell 23:1812–1825
41. Du L, Rao G, Wang H, Li B, Tian W, Cui J, He L, Laffin B, Tian X, Hao C, Liu H, Sun X, Zhu Y, Tang DG, Mehrpour M, Lu Y, Chen Q (2013) CD44-positive cancer stem cells expressing cellular prion protein contribute to metastatic capacity in colorectal cancer. Cancer Res 73:2682–2694
42. Cheng Y, Tao L, Xu J, Li Q, Yu J, Jin Y, Chen Q, Xu Z, Zou Q, Liu X (2014) CD44/cellular prion protein interact in multidrug resistant breast cancer cells and correlate with responses to neoadjuvant chemotherapy in breast cancer patients. Mol Carcinog 53:686–697
43. Richardson DD, Tol S, Valle-Encinas E, Pleguezuelos C, Bierings R, Geerts D, Fernandez-Borja M (2015) The prion protein inhibits monocytic cell migration by stimulating beta 1 integrin adhesion and uropod formation. J Cell Sci 128:3018–3029
44. Caughey B, Brown K, Raymond GJ, Katzenstein GE, Thresher W (1994) Binding of the protease-sensitive form of PrP (prion protein) to sulfated glycosaminoglycan and congo red [corrected]. J Virol 68:2135–2141
45. Gao ZX, Zhang HX, Hu F, Yang LH, Yang XW, Zhu Y, Sy MS, Li CY (2016) Glycan-deficient PrP stimulates VEGFR2 signaling via glycosaminoglycan. Cell Signal 28:652–662
46. Li C, Yu S, Nakamura F, Yin S, Xu J, Petrolla AA, Singh N, Tartakoff A, Abbott DW, Xin W, Sy MS (2009) Binding of pro-prion to filamin A disrupts cytoskeleton and correlates with poor prognosis in pancreatic cancer. J Clin Invest 119:2725–2736
47. Yang LH, Gao ZX, Hu LP, Wu GR, Yang XW, Zhang LH, Zhu Y, Wong BS, Xin W, Sy MS, Li CY (2016) Glycosylphosphatidylinositol anchor modification machinery deficiency is responsible for the formation of pro-prion protein (PrP) in BxPC-3 protein and increases cancer cell motility. J Biol Chem 291:3905–3917

48. Kuwahara C, Takeuchi AM, Nishimura T, Haraguchi K, Kubosaki A, Matsumoto Y, Saeki K, Matsumoto Y, Yokoyama T, Itohara S, Onodera T (1999) Prions prevent neuronal cell-line death. Nature 400:225–226
49. Brown DR, Wong BS, Hafiz F, Clive C, Haswell SJ, Jones IM (1999) Normal prion protein has an activity like that of superoxide dismutase. Biochem J 344:1–5
50. Brown DR, Besinger A (1998) Prion protein expression and superoxide dismutase activity. Biochem J 334:423–429
51. Brown DR, Clive C, Haswell SJ (2001) Antioxidant activity related to copper binding of native prion protein. J Neurochem 76:69–76
52. Haigh CL, McGlade AR, Collins SJ (2015) MEK1 transduces the prion protein N2 fragment antioxidant effects. Cell Mol Life Sci 72:1613–1629
53. Wong BS, Pan T, Liu T, Li RL, Gambetti P, Sy MS (2000) Differential contribution of superoxide dismutase activity by prion protein in vivo. Biochem Bioph Res Co 273:136–139
54. Rachidi W, Vilette D, Guiraud P, Arlotto M, Riondel J, Laude H, Lehmann S, Favier A (2003) Expression of prion protein increases cellular copper binding and antioxidant enzyme activities but not copper delivery. J Biol Chem 278:9064–9072
55. Wong BS, Liu T, Li R, Pan T, Petersen RB, Smith MA, Gambetti P, Perry G, Manson JC, Brown DR, Sy MS (2001) Increased levels of oxidative stress markers detected in the brains of mice devoid of prion protein. J Neurochem 76:565–572
56. Klamt F, Dal-Pizzol F, Conte da Frota ML, Walz R Jr, Andrades ME, da Silva EG, Brentani RR, Izquierdo I, Fonseca Moreira JC (2001) Imbalance of antioxidant defense in mice lacking cellular prion protein. Free Radic Biol Med 30:1137–1144
57. Williams WM, Stadtman ER, Moskovitz J (2004) Ageing and exposure to oxidative stress in vivo differentially affect cellular levels of PrP in mouse cerebral microvessels and brain parenchyma. Neuropathol Appl Neurobiol 30:161–168
58. McLennan NF, Brennan PM, McNeill A, Davies I, Fotheringham A, Rennison KA, Ritchie D, Brannan F, Head MW, Ironside JW, Williams A, Bell JE (2004) Prion protein accumulation and neuroprotection in hypoxic brain damage. Am J Pathol 165:227–235
59. Weise J, Doeppner TR, Muller T, Wrede A, Schulz-Schaeffer W, Zerr I, Witte OW, Bahr M (2008) Overexpression of cellular prion protein alters postischemic Erk1/2 phosphorylation but not Akt phosphorylation and protects against focal cerebral ischemia. Restor Neurol Neurosci 26:57–64
60. Spudich A, Frigg R, Kilic E, Kilic U, Oesch B, Raeber A, Bassetti CL, Hermann DM (2005) Aggravation of ischemic brain injury by prion protein deficiency: role of ERK-1/-2 and STAT-1. Neurobiol Dis 20:442–449
61. Beraldo FH, Soares IN, Goncalves DF, Fan J, Thomas AA, Santos TG, Mohammad AH, Roffe M, Calder MD, Nikolova S, Hajj GN, Guimaraes AL, Massensini AR, Welch I, Betts DH, Gros R, Drangova M, Watson AJ, Bartha R, Prado VF, Martins VR, Prado MA (2013) Stress-inducible phosphoprotein 1 has unique cochaperone activity during development and regulates cellular response to ischemia via the prion protein. FASEB J 27:3594–3607
62. Mitteregger G, Vosko M, Krebs B, Xiang W, Kohlmannsperger V, Nolting S, Hamann GF, Kretzschmar HA (2007) The role of the octarepeat region in neuroprotective function of the cellular prion protein. Brain Pathol 17:174–183
63. Doeppner TR, Kaltwasser B, Schlechter J, Jaschke J, Kilic E, Bahr M, Hermann DM, Weise J (2015) Cellular prion protein promotes post-ischemic neuronal survival, angioneurogenesis and enhances neural progenitor cell homing via proteasome inhibition. Cell Death Dis 6:e2024
64. Shyu WC, Lin SZ, Chiang MF, Ding DC, Li KW, Chen SF, Yang HI, Li H (2005) Overexpression of PrPC by adenovirus-mediated gene targeting reduces ischemic injury in a stroke rat model. J Neurosci 25:8967–8977
65. Guitart K, Loers G, Buck F, Bork U, Schachner M, Kleene R (2016) Improvement of neuronal cell survival by astrocyte-derived exosomes under hypoxic and ischemic conditions depends on prion protein. Glia 64:896–910

66. Bounhar Y, Zhang Y, Goodyer CG, LeBlanc A (2001) Prion protein protects human neurons against Bax-mediated apoptosis. J Biol Chem 276:39145–39149
67. Bounhar Y, Mann KK, Roucou X, LeBlanc AC (2006) Prion protein prevents Bax-mediated cell death in the absence of other Bcl-2 family members in Saccharomyces Cerevisiae. FEMS Yeast Res 6:1204–1212
68. Chieng CK, Say YH (2015) Cellular prion protein contributes to LS 174T colon cancer cell carcinogenesis by increasing invasiveness and resistance against doxorubicin-induced apoptosis. Tumour Biol: J Int Soc Oncodevelopmental Biol Med 36:8107–8120
69. Wang JH, Du JP, Li SJ, Zhai LP, Yang XY, Wang ZH, Wu ZT, Han Y (2012) Octarepeat peptides of prion are essential for multidrug resistance in gastric cancer cells. J Dig Dis 13:143–152
70. Meslin F, Conforti R, Mazouni C, Morel N, Tomasic G, Drusch F, Yacoub M, Sabourin JC, Grassi J, Delaloge S, Mathieu MC, Chouaib S, Andre F, Mehrpour M (2007) Efficacy of adjuvant chemotherapy according to prion protein expression in patients with estrogen receptor-negative breast cancer. Annals Oncol: Off J Eur Soc Med Oncol/ESMO 18:1793–1798
71. Cashman NR, Loertscher R, Nalbantoglu J, Shaw I, Kascsak RJ, Bolton DC, Bendheim PE (1990) Cellular isoform of the scrapie agent protein participates in lymphocyte activation. Cell 61:185–192
72. Mabbott NA, Brown KL, Manson J, Bruce ME (1997) T-lymphocyte activation and the cellular form of the prion protein. Immunology 92:161–165
73. Pammer J, Weninger W, Tschachler E (1998) Human keratinocytes express cellular prion-related protein in vitro and during inflammatory skin diseases. Am J Pathol 153:1353–1358
74. Konturek PC, Bazela K, Kukharskyy V, Bauer M, Hahn EG, Schuppan D (2005) Helicobacter pylori upregulates prion protein expression in gastric mucosa: a possible link to prion disease. World J Gastroenterol 11:7651–7656
75. Linden R, Martins VR, Prado MAM, Cammarota M, Izquierdo I, Brentani RR (2008) Physiology of the prion protein. Physiol Rev 88:673–728
76. de Almeida CJG, Chiarini LB, da Silva JP, Silva PMRE, Martins MA, Linden R (2005) The cellular prion protein modulates phagocytosis and inflammatory response. J Leukoc Biol 77:238–246
77. Durig J, Giese A, Schulz-Schaeffer W, Rosenthal C, Schmucker U, Bieschke J, Duhrsen U, Kretzschmar HA (2000) Differential constitutive and activation-dependent expression of prion protein in human peripheral blood leucocytes. Brit J Haematol 108:488–496
78. Haddon DJ, Hughes MR, Antignano F, Westaway D, Cashman NR, McNagny KM (2009) Prion protein expression and release by mast cells after activation. J Infect Dis 200:827–831
79. Petit CSV, Barreau F, Besnier L, Gandille P, Riveau B, Chateau D, Roy M, Berrebi D, Svrcek M, Cardot P, Rousset M, Clair C, Thenet S (2012) Requirement of cellular prion protein for intestinal barrier function and mislocalization in patients with inflammatory bowel disease. Gastroenterology 143:122–U679
80. Martin GR, Keenan CM, Sharkey KA, Jirik FR (2011) Endogenous prion protein attenuates experimentally induced colitis. Am J Pathol 179:2290–2301
81. Mariante RM, Nobrega A, Martins RAP, Areal RB, Bellio M, Linden R (2012) Neuroimmunoendocrine regulation of the prion protein in neutrophils. J Biol Chem 287:35506–35515
82. Ding TJ, Zhou XM, Kouadir M, Shi FS, Yang Y, Liu J, Wang M, Yin XM, Yang LF, Zhao DM (2013) Cellular prion protein participates in the regulation of inflammatory response and apoptosis in BV2 microglia during infection with mycobacterium Bovis. J Mol Neurosci 51:118–126
83. Walsh KP, Minamide LS, Kane SJ, Shaw AE, Brown DR, Pulford B, Zabel MD, Lambeth JD, Kuhn TB, Bamburg JR. 2014. Amyloid-beta and proinflammatory cytokines utilize a prion protein-dependent pathway to activate NADPH oxidase and induce Cofilin-actin rods in hippocampal neurons. PLoS One 9

84. Tsutsui S, Hahn JN, Johnson TA, Ali Z, Jirik FR (2008) Absence of the cellular prion protein exacerbates and prolongs neuroinflammation in experimental autoimmune encephalomyelitis. Am J Pathol 173:1029–1041
85. Nasu-Nishimura Y, Taniuchi Y, Nishimura T, Sakudo A, Nakajima K, Ano Y, Sugiura K, Sakaguchi S, Itohara S, Onodera T (2008) Cellular prion protein prevents brain damage after encephalomyocarditis virus infection in mice. Arch Virol 153:1007–1012
86. Liu J, Zhao DM, Liu CF, Ding TJ, Yang LF, Yin XM, Zhou XM (2015) Prion protein participates in the protection of mice from lipopolysaccharide infection by regulating the inflammatory process. J Mol Neurosci 55:279–287
87. Zhang J, Zeng Y, Zheng J, Xu J (2013) Expression of prion protein and its clinical significance in oral squamous cells carcinoma and oral leukoplakia. Zhonghua kou qiang yi xue za zhi = Zhonghua kouqiang yixue zazhi = Chin J Stomatology 48:752–754
88. Alfaidy N, Chauvet S, Donadio-Andrei S, Salomon A, Saoudi Y, Richaud P, Aude-Garcia C, Hoffmann P, Andrieux A, Moulis JM, Feige JJ, Benharouga M (2013) Prion protein expression and functional importance in developmental angiogenesis: role in oxidative stress and copper homeostasis. Antioxid Redox Signal 18:400–411
89. Al-Hilal TA, Chung SW, Choi JU, Alam F, Park J, Kim SW, Kim SY, Ahsan F, Kim IS, Byun Y (2016) Targeting prion-like protein doppel selectively suppresses tumor angiogenesis. J Clin Invest 126:1251–1266
90. Zhou L, Shang YL, Liu CH, Li JG, Hu H, Liang C, Han YA, Zhang W, Liang J, Wu KC (2014) Overexpression of PrPc, combined with MGr1-Ag/37LRP, is predictive of poor prognosis in gastric cancer. Int J Cancer 135:2329–2337
91. Wang JH, Du JP, Zhang YH, Zhao XJ, Fan RY, Wang ZH, Wu ZT, Han Y (2011) Dynamic changes and surveillance function of prion protein expression in gastric cancer drug resistance. World J Gastroenterol 17:3986–3993
92. Tang Z, Ma J, Zhang W, Gong C, He J, Wang Y, Yu G, Yuan C, Wang X, Sun Y, Ma J, Liu F, Zhao Y (2016) The role of prion protein expression in predicting gastric cancer prognosis. J Cancer 7:984–990
93. Liang J, Wang JB, Pan YL, Wang J, Liu LL, Guo XY, Sun L, Lin T, Han S, Xie HH, Yin F, Guo XG, Fan D (2006) High frequency occurrence of 1-OPRD variant of PRNP gene in gastric cancer cell lines and Chinese population with gastric cancer. Cell Biol Int 30:920–923
94. Liang J, Pan YL, Ning XX, Sun LJ, Lan M, Hong L, Du JP, Liu N, Liu CJ, Qiao TD, Fan DM (2006) Overexpression of PrPC and its antiapoptosis function in gastric cancer. Tumour Biol: J Int Soc Oncodevelopmental Biol Med 27:84–91
95. Lopes MH, Santos TG, Rodrigues BR, Queiroz-Hazarbassanov N, Cunha IW, Wasilewska-Sampaio AP, Costa-Silva B, Marchi FA, Bleggi-Torres LF, Sanematsu PI, Suzuki SH, Oba-Shinjo SM, Marie SK, Toulmin E, Hill AF, Martins VR (2015) Disruption of prion protein-HOP engagement impairs glioblastoma growth and cognitive decline and improves overall survival. Oncogene 34:3305–3314
96. Corsaro A, Bajetto A, Thellung S, Begani G, Villa V, Nizzari M, Pattarozzi A, Solari A, Gatti M, Pagano A, Wurth R, Daga A, Barbieri F, Florio T. 2016. Cellular prion protein controls stem cell-like properties of human glioblastoma tumor-initiating cells. Oncotarget
97. Sollazzo V, Galasso M, Volinia S, Carinci F (2012) Prion proteins (PRNP and PRND) are over-expressed in osteosarcoma. J Orthop Res: Off Publ Orthop Res Soc 30:1004–1012
98. de Wit M, Jimenez CR, Carvalho B, Belien JA, Delis-van Diemen PM, Mongera S, Piersma SR, Vikas M, Navani S, Ponten F, Meijer GA, Fijneman RJ (2012) Cell surface proteomics identifies glucose transporter type 1 and prion protein as candidate biomarkers for colorectal adenoma-to-carcinoma progression. Gut 61:855–864
99. Antonacopoulou AG, Grivas PD, Skarlas L, Kalofonos M, Scopa CD, Kalofonos HP (2008) POLR2F, ATP6V0A1 and PRNP expression in colorectal cancer: new molecules with prognostic significance? Anticancer Res 28:1221–1227
100. Wei W, Shi Q, Zhang NS, Xiao K, Chen LN, Yang XD, Ji JF, Dong XP (2016) Expression of prion protein is closely associated with pathological and clinical progression and abnormalities of p53 in head and neck squamous cell carcinomas. Oncol Rep 35:817–824

Chapter 14
Murine Gammaherpesvirus 68: A Small Animal Model for Gammaherpesvirus-Associated Diseases

Sihan Dong, J. Craig Forrest, and Xiaozhen Liang

Abstract Murine gammaherpesvirus 68 (MHV68) is a naturally occurring pathogen of murid rodents that is genetically related to the human gammaherpesviruses Epstein-Barr virus (EBV) and Kaposi sarcoma-associated herpesvirus (KSHV). Viral, immunologic, and disease parameters following experimental infection of laboratory mice with MHV68 closely resemble what occurs during primary EBV infection of humans, which suggests that MHV68 infection of mice offers a small animal model to study in general the pathogenesis of gammaherpesvirus infections. Diseases elicited by MHV68 infection include lymphoproliferative diseases, idiopathic pulmonary fibrosis, and autoimmune diseases, ailments also associated with EBV infection of humans. Furthermore, MHV68 infection also is linked to the development of vasculitis, encephalomyelitis, and other disorders that resemble pathologies with viral and nonviral etiologies in humans. This review aims to provide an overview of MHV68-associated diseases in infected mice that may provide a model for understanding basic mechanisms by which similar diseases in humans occur and can be treated.

Keywords Animal model • Murine gammaherpesvirus 68 • Epstein-Barr virus • Kaposi sarcoma-associated herpesvirus

S. Dong • X. Liang (✉)
Key Laboratory of Molecular Virology & Immunology, Institut Pasteur of Shanghai, Chinese Academy of Sciences, Shanghai 200031, People's Republic of China
e-mail: xzliang@ips.ac.cn

J.C. Forrest
Department of Microbiology and Immunology and Center for Microbial Pathogenesis and Host Inflammatory Responses, University of Arkansas for Medical Sciences, Little Rock, AR, USA

Q. Cai et al. (eds.), *Infectious Agents Associated Cancers: Epidemiology and Molecular Biology*, Advances in Experimental Medicine and Biology 1018,
DOI 10.1007/978-981-10-5765-6_14

14.1 Introduction

Murine gammaherpesvirus 68 (MHV68) is a naturally occurring virus that was originally isolated from Slovakian bank voles and is endemic in European wood mice [1–3]. MHV68 is genetically related to the human gammaherpesviruses (GHVs), Epstein-Barr virus (EBV), and Kaposi sarcoma-associated herpesvirus (KSHV), possessing a genome that is collinear with those of EBV and KSHV and contains large blocks of conserved genes with interspersed unique genes as well [1, 4, 5]. Following intranasal infection of mice, MHV68 undergoes acute infection in the lungs and nasal epithelium and establishes latency, a nonproductive, quiescent infection characterized by minimal viral gene expression and maintenance of the viral genome, in cells of the spleen and blood [6–9]. Productive replication mainly involves epithelial and mononuclear cells in the lungs, with acute infection resolving by approximately 2 weeks post-infection [10]. Acute replication precedes and is thought to be necessary for lifelong latent infection in lymphoid tissues, where B cells serve as the major latent reservoir for MHV68 in the spleen [6]. Peritoneal macrophages and lung epithelial cells also harbor latent MHV68 genomes [10, 11].

MHV68 infection of mice results in a variety of pathologies that resemble EBV-associated diseases and other human disorders. Mice chronically infected with MHV68 develop a marked splenomegaly and lymphoproliferative diseases (LPDs), similar to what is observed in patients infected with EBV [12–14]. MHV68 infection induces multi-organ fibrosis and vasculitis in interferon gamma receptor knockout (IFNγR-/-) mice [15–17]. In other disease models, MHV68 infection promotes systemic inflammation, exacerbates autoimmune encephalomyelitis, and influences development of other pathologies [18, 19]. Here, we will discuss MHV68-related diseases and the potential value of this small animal model for the study of similar diseases associated with infections by human GHVs.

14.2 MHV68-Associated Diseases

14.2.1 MHV68-Associated Lymphoproliferative Diseases

Infection of laboratory mice with MHV68 leads to a variety of pathological changes that mirror EBV-associated LPDs and other malignancies. Following intranasal infection of wild-type mice with MHV68, acute infection in the lung develops and is subsequently cleared, followed by the establishment of latency in the spleen [20, 21]. Latency establishment is accompanied by splenomegaly, which is characterized by a two- to threefold increase in the number of spleen cells, with the largest increase occurring in the CD8+ T cell population [13]. Polyclonal B-cell activation and autoantibody production also occur [13]. This is similar to what occurs during EBV-induced infectious mononucleosis in humans [22, 23]. Development of splenomegaly in MHV68 infection requires CD4+ T cells and organized secondary lymphoid tissue [13, 24, 25]. CD25-mediated IL-2 signaling also is necessary for the CD8+ T cell mononucleosis that occurs [26].

BALB/c mice chronically infected with MHV68 develop LPDs, including high-grade lymphomas that resemble centroblastic or plasmablastic non-Hodgkin lymphomas seen in humans [12]. MHV68-associated lymphomas primarily occur in older mice (0.75–3 years of age), and lymphoma incidence is greatly increased when infected mice are treated with the immunosuppressive drug cyclosporin A [12]. Since cyclosporin A functions chiefly through inhibition of T cell function, this finding strongly suggests that T cells are important for limiting tumor growth in MHV68-infected mice. Indeed, adoptive transfer of CD4+ T cells from infected mice promotes regression of lymphomas that developed following subcutaneous injection of an MHV68-positive B-cell lymphoma line, S11, isolated from a tumor-bearing BALB/c mouse [27]. Although MHV68 does not appear to transform primary murine B cells in culture, murine fetal liver-derived B cells are transformed by MHV68 into plasmablast-like B cells in vitro [28]. Similar to S11 cells, when these plasmablast-like B cells are injected into immunodeficient mice, the transformed B cells induce lymphomas that can be controlled by both CD4+ and CD8+ T cells [29]. Together, these findings illustrate (i) that MHV68 infection can cause lymphomas and (ii) that T cells are important for controlling infection-associated lymphomas.

MHV68 infection of BALB/c_2 microglobulin (B2M)-deficient mice (BALB B2M-/-) also results in B-cell lymphoma and an atypical lymphoid hyperplasia (ALH) [14]. ALH pathologically is differentially regulated by MHV68 genes and resembles posttransplant lymphoproliferative disease observed in some EBV-infected individuals that are immune suppressed for solid organ transplants [30, 31]. B2M is a critical component of the major histocompatibility I (MHCI) complex, a cell surface receptor necessary for CD8+ T cells to engage target cells [32]. This further illustrates the importance of T cells in preventing MHV68-associated LPDs.

Lymphomatoid granulomatosis (LYG) is a rare systemic angiodestructive LPD caused by the combination of EBV infection and immunosuppression [33, 34]. LYG mostly affects the lungs and is recently characterized as B-cell lymphomas with prominent pulmonary involvement [33]. MHV68-infected IFNγR-/- mice also develop pulmonary B-cell lymphomas which closely mimic EBV-associated LYG in human [35].

Nevertheless, there are differences between EBV-associated LPDs in humans and MHV68-associated LPDs in mice. For example, CD8+ T cell lymphocytosis associated with EBV-induced mononucleosis is predominantly an outgrowth of T cells responding to viral lytic epitopes [36, 37]. In contrast MHV68-induced mononucleosis in C57BL/6 mice represents expansion of CD8+ T cells that encode a Vβ4 T cell receptor that is not reactive to viral epitopes and appears to be stimulated by latently infected B cells [38, 39]. However, the striking pathological similarities between EBV-associated LPDs and the corresponding syndrome in MHV68-infected mice highlight MHV68 as a valuable small animal model for studying fundamental issues in gammaherpesvirus-associated LPD pathogenesis in a natural host.

14.2.2 MHV68-Associated Fibrosis

Several reports associate EBV infection with idiopathic pulmonary fibrosis (IPF), a chronic, progressive, fibrotic lung disorder of unknown etiology that is a risk factor for lung cancer development [40–43]. Although EBV is frequently detected in lung tissues of patients with IPF, an etiologic role for EBV in IPF is not established REF. Interestingly, MHV68 infection of IFNγR-/- mice leads to multi-organ fibrosis, which occurs in the lung, spleen, mediastinal lymph nodes, and liver of these mice [15, 16, 44, 45]. Lung fibrosis in MHV68-infected IFNγR-/- mice shares similar pathology to IPF in humans [45]. Mechanistic studies show that both viral and cellular factors are involved in MHV68-induced fibrosis in IFNγR-/- mice. Persistent MHV68 lytic replication apparently is essential for induction or exacerbation of IPF, because severe fibrosis is ameliorated in MHV68-infected IFNγR-/- mice that receive antiviral treatment and in IFNγR-/- mice infected with a reactivation-defective MHV68 mutant that fails to express v-cyclin [46]. Moreover, MHV68 superantigen-like M1 protein and activated Vβ4+ CD8+ T cells, which are driven to expand by M1, also are required for MHV68-induced inflammation and fibrosis in IFN-γR-/- mice [47, 48]. Additionally, inhibition of NF-κB signaling reduces virus persistence and pulmonary fibrosis in MHV68-infected IFNγR-/- mice, indicating that NF-κB signaling also is important for MHV68-induced pulmonary fibrosis [49]. Thus, MHV68 infection of IFNγR-/- mice could be used to model the association of gammaherpesvirus infection with IPF and define underlying molecular mechanisms of disease.

MHV68 infection of bleomycin-resistant BALB/c mice has also been used to study the association between GHV infection and IPF. Bleomycin-induced fibrosis is widely used experimental model for lung fibrosis occurring during chemotherapy [50]. BALB/c mice are inherently resistant to lung fibrosis due to bleomycin treatment and do not develop pulmonary fibrosis when infected with MHV68. However, when BALB/c mice are simultaneously infected with MHV68 and treated with bleomycin, lung fibrosis occurs [51], indicating that MHV68 functions as a cofactor in bleomycin-induced fibrosis. Another study demonstrated that TLR9 signaling protects against MHV68-induced exacerbation of lung fibrosis induced by bleomycin in BALB/c mice [52]. These findings support the role of GHV infection in human IPD, and future development of the MHV68/bleomycin model should further explore mechanisms by which GHV infection functions as a cofactor in the pathogenesis of pulmonary fibrosis.

Finally, MHV68 infection of IFNγ deficient (IFNγ-/-) mice on the BALB/c genetic background results in acute lethal pneumonia that is dependent on MHV68-encoded v-cyclin and v-bcl-2 [53]. However, whether MHV68-induced pneumonia in IFNγ-/- mice is pathologically similar to EBV-associated pneumonia, which mainly occurs in children and transplant patients, is not yet clear [54–57].

In addition to lung fibrosis, MHV68 infection of IFNγR-/- mice also induces fibrosis in the spleen. The prominent feature of splenic pathology in infected IFNγR-/- mice is a loss of B cells and CD4+ and CD8+ T cells, which correlates with significant changes in cytokines and chemokines in spleens. In contrast, a dramatic increase in T and B lymphocytes in peripheral blood occurs [15, 44]. CD8+ T cells are the major mediators of splenic damage, since depletion of CD8+ T cells completely reverses the pathological and histological changes in spleens of these mice. However, although removal of CD4+ T cells reverses the weight loss and reduces the number of infective centers, some pathological changes are still observed in CD4+ T cell-depleted mice. This suggests that CD4+ T cells are not the dominant mediators but still play an important role in splenic fibrosis [44].

Furthermore, MHV68 infection of IFNγR-/- mice leads to enhanced production of Th2 cytokines IL-5, IL-13, and IL-21 and increased expression of CCR4 in the spleens of infected mice [16]. This drives alternative activation of macrophages to produce arginase 1 (ARG1) and found in inflammatory zone 1 (FIZZ1)/resistin-like molecule-α (RELMα) to promote fibrotic disease in the spleen [16, 58]. Though EBV infection is not directly linked to fibrotic disease of the spleen in humans, the data from MHV68 infections suggest a role for GHV infection in such diseases. This model may therefore hold future relevance for understanding how viruses influence splenic fibrosis in general.

14.2.3 MHV68 Impact on Autoimmune Diseases

Multiple sclerosis (MS) is an autoimmune disorder in which the immune system attacks the central nervous system (CNS), damaging the myelin sheath of nerve cells in the brain and spinal cord. EBV is etiologically linked to MS [59–61]; however mechanisms by which EBV influences MS pathogenesis are not known. In mice induction of inflammatory immune responses in the brain triggers an MS-like syndrome known as experimental autoimmune encephalomyelitis (EAE) [62, 63]. Because MHV68 replicates in the mouse brain, infecting microglia and astrocytes [64, 65], and globally influences immune activation in infected animals [66], the impact of MHV68 infection on EAE pathogenesis was evaluated. Latent MHV68 infection enhances EAE pathogenesis and central nervous system pathology in a manner reminiscent of human MS [19, 67]. This observation demonstrates that GHV infection can influence the course of disease in CNS autoimmune disorders and highlights the potential of these small animal models in facilitating an understanding of mechanisms by which EBV influences MS.

EBV infection also is linked to development of lupus in humans, an autoimmune disease in which healthy tissues are attacked by the individual's immune system, leading to swelling and damage of various tissues of the body. In contrast to EAE models, MHV68 infection protects, rather than exacerbates, lupus-prone mice from the development and progression of autoimmunity [68]. Together, these findings demonstrate that GHV infection influences the course of CNS autoimmune disease. However, the data also demonstrate that pathogenesis is likely a multifactorial process in which GHV infections may have pleiotropic impacts on disease progression.

The impact of MHV68 infection in other mouse models of autoimmune disease also has been evaluated. For instance, $IL10^{-/-}$ mice are prone to developing inflammatory bowel disease (IBD), and infection with MHV68 promotes more rapid and severe disease in these mice [69]. This finding is similar to the observation that EBV infection correlates with disease severity in some IBD patients [70–74]. In contrast, MHV68 infection of nonobese diabetic (NOD) mice, a mouse model for evaluating type I diabetes (T1D), significantly delays diabetes onset [75], which supports the hypothesis that viruses are potential regulators of T1D [76–78]. Furthermore, transgenic mice expressing MHV68 chemokine decoy receptor M3 in beta cells are remarkably resistant to diabetes induced by multiple low doses of streptozotocin [79]. This suggests the importance of specific viral factors in regulating T1D. Together, these data highlight the manner in which MHV68 studies could be employed to define roles for GHVs in intestinal diseases and diabetes.

14.2.4 MHV68-Related Vascular and Ductal Disorders

In addition to lymphoma development and fibrosis, MHV68 causes severe large-vessel arteritis associated with lipid accumulation in the vessel wall and luminal thrombosis in IFNγR-/- mice. Lesions that develop are similar to those seen during the acute inflammatory phase of Takayasu arteritis, the nongranulomatous variant of temporal arteritis and Kawasaki diseases [17], suggesting possible GHV etiologies in these pathologies and demonstrating the utility of MHV68 infection of mice in dissecting GHV roles in human vasculitis. Furthermore, MHV68 induces chronic inflammation of intrahepatic bile ducts in infected IFNγR-/- mice, which is pathologically similar to the human fibrotic liver disorder primary sclerosing cholangitis [80]. Additionally, MHV68 reactivation from latency induces neointimal lesions in pulmonary arteries of S100A4/Mts1-overexpressing mice. These lesions are associated with elevated neutrophil elastase, which is produced by pulmonary artery smooth muscle cells and linked to experimental and clinical pulmonary vascular disease [81, 82]. Finally, MHV68 infection in mice also induces phenotypes that mimic rare diseases such as systemic lymphocytosis following gastric instillation and fatigue [83, 84]. MHV68 may therefore provide a useful model for the study of fatigue and other physiologic and behavioral perturbations that occur during acute and chronic infection with gammaherpesviruses.

14.3 Remarks and Perspectives

Human gammaherpesviruses are exquisitely species restricted, which limits possible approaches for defining precise mechanisms by which these viruses cause disease. The beauty of small animal models of viral pathogenesis is that they enable evaluations of both viral and host determinants of disease in experimentally controlled settings. In contrast, analogous studies of GHV infections in humans would require the presence of naturally occurring mutations in either virus or host, paired with the ability to identify individuals and viruses that possess such genetic variants. Even then, the studies would be necessarily associative, observational, and potentially influenced by numerous outside variables due to environment, lifestyle, additional genetic variations, age, coinfections, etc. The use of humanized mice, immunodeficient animals in which the immune system is reconstituted by human stem cells, allows an experimental system for evaluating certain aspects of GHV infection in human cells. But again the system is still genetically limited and may not faithfully recapitulate natural cellular development and cell-cell interactions, and not all tissues in the reconstituted mouse are of human origin. Humanized mice are also very expensive. Hence, the capacity to study a genetically related pathogen (MHV68) in a natural host (rodents) offers a powerful tool for understanding virus-host interactions in GHV infection-associated diseases. Here, infections of inbred mice with MHV68 provide a simplified and standardized analysis of immune responses against the virus and eliminate many potential experimental variables. Moreover, the ease of genetically manipulating both virus and host further highlights the tractability of the MHV68 system. Indeed, genetically modified mice and viruses enable many of disease models described above.

However, this is not to say that infection of mice with MHV68 is identical to human infections with EBV or KSHV. Clearly mice are not humans, and it is naïve to think that all aspects of the host response will be identical in two quite divergent species. Further, while genetic diversity is an experimental problem, genetic polymorphisms from human to human and population to population undoubtedly influence the pathogenesis and outcome of infection by GHVs.

Along these same lines, MHV68 is not EBV or KSHV. While all GHVs (and herpesviruses in general) possess blocks of conserved genes, each virus also encodes unique proteins that are not shared with their GHV relatives. These genes maintain no vestige of sequence homology and may have developed through convergent evolution to satisfy unique requirements of the virus-host relationship. It is however interesting to note that the products of these divergent genes likely perform conserved functions, for instance, both LMP2a of EBV and M2 of MHV68 manipulate B-cell survival and differentiation [85]. If these unique genes are under selection from the host, it is equally possible that they have simply diverged over millennia at a rate that made them nonhomologous at the sequence level by modern informatics techniques while maintaining critical functions. Perhaps the key take-home points are these: MHV68 provides a highly tractable experimental system for understanding how GHVs influence disease in a variety of experimental models. Though obvious differences exist between human and mouse infections, the data produced in the mouse models are real and may offer invaluable insights into factors that influence similar diseases in humans. As such, MHV68 infection of mice can serve as powerful tool in the arsenal for illuminating previously unappreciated factors and cofactors that influence human disease and allow for preclinical testing of novel hypotheses for treating related diseases.

Acknowledgments This work was supported by the grant (CA167065, J.C.F) from the National Cancer Institute of the National Institutes of Health and the grant (81371825, X.Z.L) from Natural Science Foundation of China.

References

1. Blaskovic D, Stancekova M, Svobodova J, Mistrikova J (1980) Isolation of five strains of herpesviruses from two species of free living small rodents. Acta Virol 24:468
2. Mistrikova J, Blaskovic D (1985) Ecology of the murine alphaherpesvirus and its isolation from lungs of rodents in cell culture. Acta Virol 29:312–317
3. Svobodova J, Blaskovic D, Mistrikova J (1982) Growth characteristics of herpesviruses isolated from free living small rodents. Acta Virol 26:256–263
4. Efstathiou S, Ho YM, Hall S, Styles CJ, Scott SD, Gompels UA (1990) Murine herpesvirus 68 is genetically related to the gammaherpesviruses Epstein-Barr virus and herpesvirus saimiri. J Gen Virol 71(Pt 6):1365–1372
5. Virgin HW, Latreille P, Wamsley P, Hallsworth K, Weck KE, Dal Canto AJ, Speck SH (1997) Complete sequence and genomic analysis of murine gammaherpesvirus 68. J Virol 71:5894–5904
6. Sunil-Chandra NP, Efstathiou S, Nash AA (1992) Murine gammaherpesvirus 68 establishes a latent infection in mouse B lymphocytes in vivo. J Gen Virol 73(Pt 12):3275–3279
7. Weck KE, Barkon ML, Yoo LI, Speck SH, Virgin HI (1996) Mature B cells are required for acute splenic infection, but not for establishment of latency, by murine gammaherpesvirus 68. J Virol 70:6775–6780
8. Weck KE, Kim SS, Virgin HI, Speck SH (1999) B cells regulate murine gammaherpesvirus 68 latency. J Virol 73:4651–4661
9. Milho R, Smith CM, Marques S, Alenquer M, May JS, Gillet L, Gaspar M, Efstathiou S, Simas JP, Stevenson PG (2009) In vivo imaging of murid herpesvirus-4 infection. J Gen Virol 90:21–32
10. Stewart JP, Usherwood EJ, Ross A, Dyson H, Nash T (1998) Lung epithelial cells are a major site of murine gammaherpesvirus persistence. J Exp Med 187:1941–1951
11. Weck KE, Kim SS, Virgin HI, Speck SH (1999) Macrophages are the major reservoir of latent murine gammaherpesvirus 68 in peritoneal cells. J Virol 73:3273–3283
12. Sunil-Chandra NP, Arno J, Fazakerley J, Nash AA (1994) Lymphoproliferative disease in mice infected with murine gammaherpesvirus 68. Am J Pathol 145:818–826
13. Usherwood EJ, Ross AJ, Allen DJ, Nash AA (1996) Murine gammaherpesvirus-induced splenomegaly: a critical role for CD4 T cells. J Gen Virol 77(Pt 4):627–630
14. Tarakanova VL, Suarez F, Tibbetts SA, Jacoby MA, Weck KE, Hess JL, Speck SH, Virgin HW (2005) Murine gammaherpesvirus 68 infection is associated with lymphoproliferative disease and lymphoma in BALB beta2 microglobulin-deficient mice. J Virol 79:14668–14679
15. Ebrahimi B, Dutia BM, Brownstein DG, Nash AA (2001) Murine gammaherpesvirus-68 infection causes multi-organ fibrosis and alters leukocyte trafficking in interferon-gamma receptor knockout mice. Am J Pathol 158:2117–2125
16. Gangadharan B, Hoeve MA, Allen JE, Ebrahimi B, Rhind SM, Dutia BM, Nash AA (2008) Murine gammaherpesvirus-induced fibrosis is associated with the development of alternatively activated macrophages. J Leukoc Biol 84:50–58
17. Weck KE, Dal Canto AJ, Gould JD, O'Guin AK, Roth KA, Saffitz JE, Speck SH, Virgin HW (1997) Murine gamma-herpesvirus 68 causes severe large-vessel arteritis in mice lacking interferon-gamma responsiveness: a new model for virus-induced vascular disease. Nat Med 3:1346–1353

18. Park S, Buck MD, Desai C, Zhang X, Loginicheva E, Martinez J, Freeman ML, Saitoh T, Akira S, Guan JL, He YW, Blackman MA, Handley SA, Levine B, Green DR, Reese TA, Artyomov MN, Virgin HW (2016) Autophagy genes enhance murine Gammaherpesvirus 68 reactivation from latency by preventing virus-induced systemic inflammation. Cell Host Microbe 19:91–101
19. Peacock JW, Elsawa SF, Petty CC, Hickey WF, Bost KL (2003) Exacerbation of experimental autoimmune encephalomyelitis in rodents infected with murine gammaherpesvirus-68. Eur J Immunol 33:1849–1858
20. Sunil-Chandra NP, Efstathiou S, Arno J, Nash AA (1992) Virological and pathological features of mice infected with murine gamma-herpesvirus 68. J Gen Virol 73(Pt 9):2347–2356
21. Cardin RD, Brooks JW, Sarawar SR, Doherty PC (1996) Progressive loss of CD8+ T cell-mediated control of a gamma-herpesvirus in the absence of CD4+ T cells. J Exp Med 184:863–871
22. Cantani A, Mastrantoni F (1989) Recent advances on Epstein-Barr virus infectious mononucleosis. Riv Eur Sci Med Farmacol 11:41–44
23. Flano E, Woodland DL, Blackman MA (2002) A mouse model for infectious mononucleosis. Immunol Res 25:201–217
24. Lee BJ, Santee S, Von Gesjen S, Ware CF, Sarawar SR (2000) Lymphotoxin-alpha-deficient mice can clear a productive infection with murine gammaherpesvirus 68 but fail to develop splenomegaly or lymphocytosis. J Virol 74:2786–2792
25. Ehtisham S, Sunil-Chandra NP, Nash AA (1993) Pathogenesis of murine gammaherpesvirus infection in mice deficient in CD4 and CD8 T cells. J Virol 67:5247–5252
26. Molloy M, Zhang W, Usherwood E (2010) Mononucleosis and antigen-driven T cell responses have different requirements for interleukin-2 signaling in murine gammaherpesvirus infection. J Virol 84:10923–10927
27. Robertson KA, Usherwood EJ, Nash AA (2001) Regression of a murine gammaherpesvirus 68-positive b-cell lymphoma mediated by CD4 T lymphocytes. J Virol 75:3480–3482
28. Liang X, Paden CR, Morales FM, Powers RP, Jacob J, Speck SH (2011) Murine gammaherpesvirus immortalization of fetal liver-derived B cells requires both the viral cyclin D homolog and latency-associated nuclear antigen. PLoS Pathog 7:e1002220
29. Liang X, Crepeau RL, Zhang W, Speck SH, Usherwood EJ (2013) CD4 and CD8 T cells directly recognize murine gammaherpesvirus 68-immortalized cells and prevent tumor outgrowth. J Virol 87:6051–6054
30. Good DJ, Gascoyne RD (2009) Atypical lymphoid hyperplasia mimicking lymphoma. Hematol Oncol Clin North Am 23:729–745
31. Tarakanova VL, Kreisel F, White DW, Virgin HW (2008) Murine gammaherpesvirus 68 genes both induce and suppress lymphoproliferative disease. J Virol 82:1034–1039
32. Peterson PA, Rask L, Ostberg L (1977) Beta2-microglobulin and the major histocompatibility complex. Adv Cancer Res 24:115–163
33. Dunleavy K, Roschewski M, Wilson WH (2012) Lymphomatoid granulomatosis and other Epstein-Barr virus associated lymphoproliferative processes. Curr Hematol Malig Rep 7:208–215
34. Wilson WH, Kingma DW, Raffeld M, Wittes RE, Jaffe ES (1996) Association of lymphomatoid granulomatosis with Epstein-Barr viral infection of B lymphocytes and response to interferon-alpha 2b. Blood 87:4531–4537
35. Lee KS, Groshong SD, Cool CD, Kleinschmidt-DeMasters BK, van Dyk LF (2009) Murine gammaherpesvirus 68 infection of IFNgamma unresponsive mice: a small animal model for gammaherpesvirus-associated B-cell lymphoproliferative disease. Cancer Res 69:5481–5489
36. Callan MF, Steven N, Krausa P, Wilson JD, Moss PA, Gillespie GM, Bell JI, Rickinson AB, McMichael AJ (1996) Large clonal expansions of CD8+ T cells in acute infectious mononucleosis. Nat Med 2:906–911

37. Callan MF, Tan L, Annels N, Ogg GS, Wilson JD, O'Callaghan CA, Steven N, McMichael AJ, Rickinson AB (1998) Direct visualization of antigen-specific CD8+ T cells during the primary immune response to Epstein-Barr virus in vivo. J Exp Med 187:1395–1402
38. Tripp RA, Hamilton-Easton AM, Cardin RD, Nguyen P, Behm FG, Woodland DL, Doherty PC, Blackman MA (1997) Pathogenesis of an infectious mononucleosis-like disease induced by a murine gamma-herpesvirus: role for a viral superantigen? J Exp Med 185:1641–1650
39. Doherty PC, Tripp RA, Hamilton-Easton AM, Cardin RD, Woodland DL, Blackman MA (1997) Tuning into immunological dissonance: an experimental model for infectious mononucleosis. Curr Opin Immunol 9:477–483
40. Vergnon JM, Vincent M, de The G, Mornex JF, Weynants P, Brune J (1984) Cryptogenic fibrosing alveolitis and Epstein-Barr virus: an association? Lancet 2:768–771
41. Egan JJ, Woodcock AA, Stewart JP (1997) Viruses and idiopathic pulmonary fibrosis. Eur Respir J 10:1433–1437
42. Stewart JP, Egan JJ, Ross AJ, Kelly BG, Lok SS, Hasleton PS, Woodcock AA (1999) The detection of Epstein-Barr virus DNA in lung tissue from patients with idiopathic pulmonary fibrosis. Am J Respir Crit Care Med 159:1336–1341
43. Tsukamoto K, Hayakawa H, Sato A, Chida K, Nakamura H, Miura K (2000) Involvement of Epstein-Barr virus latent membrane protein 1 in disease progression in patients with idiopathic pulmonary fibrosis. Thorax 55:958–961
44. Dutia BM, Clarke CJ, Allen DJ, Nash AA (1997) Pathological changes in the spleens of gamma interferon receptor-deficient mice infected with murine gammaherpesvirus: a role for CD8 T cells. J Virol 71:4278–4283
45. Vannella KM, Moore BB (2008) Viruses as co-factors for the initiation or exacerbation of lung fibrosis. Fibrogenesis Tissue Repair 1:2
46. Mora AL, Torres-Gonzalez E, Rojas M, Xu J, Ritzenthaler J, Speck SH, Roman J, Brigham K, Stecenko A (2007) Control of virus reactivation arrests pulmonary herpesvirus-induced fibrosis in IFN-gamma receptor-deficient mice. Am J Respir Crit Care Med 175:1139–1150
47. Evans AG, Moser JM, Krug LT, Pozharskaya V, Mora AL, Speck SH (2008) A gammaherpesvirus-secreted activator of Vbeta4+ CD8+ T cells regulates chronic infection and immunopathology. J Exp Med 205:669–684
48. O'Flaherty BM, Matar CG, Wakeman BS, Garcia A, Wilke CA, Courtney CL, Moore BB, Speck SH (2015) CD8+ T cell response to Gammaherpesvirus infection mediates inflammation and fibrosis in interferon gamma receptor-deficient mice. PLoS One 10:e0135719
49. Krug LT, Torres-Gonzalez E, Qin Q, Sorescu D, Rojas M, Stecenko A, Speck SH, Mora AL (2010) Inhibition of NF-kappaB signaling reduces virus load and gammaherpesvirus-induced pulmonary fibrosis. Am J Pathol 177:608–621
50. Izbicki G, Segel MJ, Christensen TG, Conner MW, Breuer R (2002) Time course of bleomycin-induced lung fibrosis. Int J Exp Pathol 83:111–119
51. Lok SS, Haider Y, Howell D, Stewart JP, Hasleton PS, Egan JJ (2002) Murine gammaherpes virus as a cofactor in the development of pulmonary fibrosis in bleomycin resistant mice. Eur Respir J 20:1228–1232
52. Luckhardt TR, Coomes SM, Trujillo G, Stoolman JS, Vannella KM, Bhan U, Wilke CA, Moore TA, Toews GB, Hogaboam C, Moore BB (2011) TLR9-induced interferon beta is associated with protection from gammaherpesvirus-induced exacerbation of lung fibrosis. Fibrogenesis Tissue Repair 4:18
53. Lee KS, Cool CD, van Dyk LF (2009) Murine gammaherpesvirus 68 infection of gamma interferon-deficient mice on a BALB/c background results in acute lethal pneumonia that is dependent on specific viral genes. J Virol 83:11397–11401
54. Andiman WA, McCarthy P, Markowitz RI, Cormier D, Horstmann DM (1981) Clinical, virologic, and serologic evidence of Epstein-Barr virus infection in association with childhood pneumonia. J Pediatr 99:880–886
55. Liu QF, Fan ZP, Luo XD, Sun J, Zhang Y, Ding YQ (2010) Epstein-Barr virus-associated pneumonia in patients with post-transplant lymphoproliferative disease after hematopoietic stem cell transplantation. Transpl Infect Dis 12:284–291

56. Krumbholz A, Sandhaus T, Gohlert A, Heim A, Zell R, Egerer R, Breuer M, Straube E, Wutzler P, Sauerbrei A (2010) Epstein-Barr virus-associated pneumonia and bronchiolitis obliterans syndrome in a lung transplant recipient. Med Microbiol Immunol 199:317–322
57. Abughali N, Khiyami A, Birnkrant DJ, Kumar ML (2002) Severe respiratory syncytial virus pneumonia associated with primary Epstein-Barr virus infection. Pediatr Pulmonol 33:395–398
58. Gordon S (2003) Alternative activation of macrophages. Nat Rev Immunol 3:23–35
59. Ascherio A, Munch M (2000) Epstein-Barr virus and multiple sclerosis. Epidemiology 11:220–224
60. Ascherio A, Munger KL, Lunemann JD (2012) The initiation and prevention of multiple sclerosis. Nat Rev Neurol 8:602–612
61. Tselis A (2012) Epstein-Barr virus cause of multiple sclerosis. Curr Opin Rheumatol 24:424–428
62. Mackay IR, Carnegie PR, Coates AS (1973) Immunopathological comparisons between experimental autoimmune encephalomyelitis and multiple sclerosis. Clin Exp Immunol 15:471–482
63. Swanborg RH (1995) Experimental autoimmune encephalomyelitis in rodents as a model for human demyelinating disease. Clin Immunol Immunopathol 77:4–13
64. Terry LA, Stewart JP, Nash AA, Fazakerley JK (2000) Murine gammaherpesvirus-68 infection of and persistence in the central nervous system. J Gen Virol 81:2635–2643
65. Taylor WR, Rasley A, Bost KL, Marriott I (2003) Murine gammaherpesvirus-68 infects microglia and induces high levels of pro-inflammatory cytokine production. J Neuroimmunol 136:75–83
66. Stevenson PG, Doherty PC (1998) Kinetic analysis of the specific host response to a murine gammaherpesvirus. J Virol 72:943–949
67. Casiraghi C, Shanina I, Cho S, Freeman ML, Blackman MA, Horwitz MS (2012) Gammaherpesvirus latency accentuates EAE pathogenesis: relevance to Epstein-Barr virus and multiple sclerosis. PLoS Pathog 8:e1002715
68. Larson JD, Thurman JM, Rubtsov AV, Claypool D, Marrack P, van Dyk LF, Torres RM, Pelanda R (2012) Murine gammaherpesvirus 68 infection protects lupus-prone mice from the development of autoimmunity. Proc Natl Acad Sci U S A 109:E1092–E1100
69. Nelson DA, Petty CC, Bost KL (2009) Infection with murine gammaherpesvirus 68 exacerbates inflammatory bowel disease in IL-10-deficient mice. Inflamm Res 58:881–889
70. Bertalot G, Villanacci V, Gramegna M, Orvieto E, Negrini R, Saleri A, Terraroli C, Ravelli P, Cestari R, Viale G (2001) Evidence of Epstein-Barr virus infection in ulcerative colitis. Dig Liver Dis 33:551–558
71. Kangro HO, Chong SK, Hardiman A, Heath RB, Walker-Smith JA (1990) A prospective study of viral and mycoplasma infections in chronic inflammatory bowel disease. Gastroenterology 98:549–553
72. Spieker T, Herbst H (2000) Distribution and phenotype of Epstein-Barr virus-infected cells in inflammatory bowel disease. Am J Pathol 157:51–57
73. Takeda Y, Takada K, Togashi H, Takeda H, Sakano M, Osada Y, Shinzawa H, Takahashi T (2000) Demonstration of Epstein-Barr virus localized in the colonic and ileal mucosa of a patient with ulcerative colitis. Gastrointest Endosc 51:205–209
74. Van Kruiningen HJ, Poulin M, Garmendia AE, Desreumaux P, Colombel JF, De Hertogh G, Geboes K, Vermeire S, Tsongalis GJ (2007) Search for evidence of recurring or persistent viruses in Crohn's disease. APMIS 115:962–968
75. Smith KA, Efstathiou S, Cooke A (2007) Murine gammaherpesvirus-68 infection alters self-antigen presentation and type 1 diabetes onset in NOD mice. J Immunol 179:7325–7333
76. Pane JA, Coulson BS (2015) Lessons from the mouse: potential contribution of bystander lymphocyte activation by viruses to human type 1 diabetes. Diabetologia 58:1149–1159
77. Gale EA (2012) Viruses and type 1 diabetes: ignorance acquires a better vocabulary. Clin Exp Immunol 168:1–4
78. Coppieters KT, Boettler T, von Herrath M (2012) Virus infections in type 1 diabetes. Cold Spring Harb Perspect Med 2:a007682

79. Martin AP, Alexander-Brett JM, Canasto-Chibuque C, Garin A, Bromberg JS, Fremont DH, Lira SA (2007) The chemokine binding protein M3 prevents diabetes induced by multiple low doses of streptozotocin. J Immunol 178:4623–4631
80. Gangadharan B, Dutia BM, Rhind SM, Nash AA (2009) Murid herpesvirus-4 induces chronic inflammation of intrahepatic bile ducts in mice deficient in gamma-interferon signalling. Hepatol Res 39:187–194
81. Spiekerkoetter E, Alvira CM, Kim YM, Bruneau A, Pricola KL, Wang L, Ambartsumian N, Rabinovitch M (2008) Reactivation of gammaHV68 induces neointimal lesions in pulmonary arteries of S100A4/Mts1-overexpressing mice in association with degradation of elastin. Am J Physiol Lung Cell Mol Physiol 294:L276–L289
82. Kim YM, Haghighat L, Spiekerkoetter E, Sawada H, Alvira CM, Wang L, Acharya S, Rodriguez-Colon G, Orton A, Zhao M, Rabinovitch M (2011) Neutrophil elastase is produced by pulmonary artery smooth muscle cells and is linked to neointimal lesions. Am J Pathol 179:1560–1572
83. Peacock JW, Bost KL (2000) Infection of intestinal epithelial cells and development of systemic disease following gastric instillation of murine gammaherpesvirus-68. J Gen Virol 81:421–429
84. Olivadoti MD, Weinberg JB, Toth LA, Opp MR (2011) Sleep and fatigue in mice infected with murine gammaherpesvirus 68. Brain Behav Immun 25:696–705
85. Speck SH, Ganem D (2010) Viral latency and its regulation: lessons from the gamma-herpesviruses. Cell Host Microbe 8:100–115

Chapter 15
Infection of KSHV and Interaction with HIV: The Bad Romance

Jie Qin and Chun Lu

Abstract Kaposi's sarcoma-associated herpesvirus (KSHV), namely, human herpesvirus 8 (HHV-8), is considered as the pathogen of Kaposi's sarcoma (KS), the most frequent cancer in untreated HIV-infected individuals. Patients infected with HIV have a much higher possibility developing KS than average individual. Researchers have found that HIV, which functions as a cofactor of KS, contributes a lot to the development of KS. In this article, we will give a brief introduction of KS and KSHV and how the interaction between KSHV and HIV contributes to the development of KS. Also we will take a glance at the development of treatment in KS, especially AIDS-KS.

Keywords Kaposi's sarcoma-associated herpesvirus (KSHV) • Human immunodeficiency virus (HIV) • Coinfection • HIV viral proteins • Treatment

15.1 Kaposi's Sarcoma-Associated Herpesvirus and Pathogenesis

15.1.1 Kaposi's Sarcoma and Kaposi's Sarcoma-Associated Herpesvirus

With complex histology feature, Kaposi's sarcoma (KS) shows abnormal vascular proliferation peculiarity. There are four types of KS, including classical KS, mainly affecting elderly men of Mediterranean or eastern European Jewish ancestry; AIDS-related KS, as its name shows, happens to AIDS patients; iatrogenic KS, usually

---

J. Qin • C. Lu (✉)
Key Laboratory of Pathogen Biology (Jiangsu Province), Nanjing Medical University, Nanjing, People's Republic of China

Department of Microbiology, Nanjing Medical University, Nanjing 210029, People's Republic of China
e-mail: clu@njmu.edu.cn

Q. Cai et al. (eds.), *Infectious Agents Associated Cancers: Epidemiology and Molecular Biology*, Advances in Experimental Medicine and Biology 1018,
DOI 10.1007/978-981-10-5765-6_15

happens to immunosuppressive patients after organ transplant; and African endemic KS, existing in parts of Central and Eastern Africa [1–4].

Investigation on KSHV seroprevalence shows that distribution of KSHV-positive individuals differs in regions and subpopulations. A report has been made that all forms of KS are more common in men than in woman, and further investigation showed that men from sub-Saharan Africa (50% KSHV prevalence) but not men from other district have a higher prevalence of KSHV than women [5, 6]. KSHV prevalence also shows distinct district differences. In endemic district, such as Uganda, KSHV prevalence of 50% has been reported, while in the USA, the report is 6% or even lowers [7–9]. Also in Xinjiang Uyghur Autonomous Region, China, a traditional endemic area, KSHV prevalence is much higher than other districts in China, with Han group showing a distinct lower rate [10]. Outside the endemic district, men who have sex with men (MSM) show a much higher KSHV prevalence than the average population. All around the world, KSHV prevalence are much high in MSM [11–14]. KSHV can also be found in saliva, and it is also reported as the highest shedding place; oral exposure to infectious saliva can be the transmission route of KSHV both sexually and nonsexually [15–17]. More researches have proved that in nonendemic districts, KS is more likely to happen to HIV-infected/AIDS population [18]. A recent study shows that HIV-1-infected children and adolescents in nonendemic districts have a higher possibility of KSHV seroprevalence [19].

Infectious saliva is the major route of KSHV transmission. However, increasing infection of KSHV among MSM strongly suggests that KSHV might transmit through sexual contact. More research has to be done to validate though [20].

Despite the fact that the discovery of KS is early in the late nineteenth century by Hungarian dermatologist Moritz Kaposi, it was not until the 1990s that KSHV, now considered as the pathogen of KS, was detected in KS tissues. In 1994, Chang and Moore identified KSHV genome in KS lesions [21]. They used representational difference analysis (a PCR-based technique) to identify and characterize alien DNA sequences in KS tissues. Sequences homologous to, but distinct from, capsid and tegument protein genes of the gammaherpesvirus *Saimiri* and Epstein-Barr virus were found in these tissues [21]. Now, 20 years has gone since this remarkable discovery. The characteristics of this virus have been mapped out by numerous scientists. Except KS, KSHV is also related to other two malignancies, primary effusion lymphoma (PEL) and multicentric Castleman disease (MCD) [21, 22].

15.1.2 KSHV Genome and Life Cycle

Soon after the discovery of KSHV, the genome of KSHV was mapped with cosmid and phage genomic libraries from the BC-1 cell line [23]. This group from New York found that the BC-1 KSHV genome consists of a 140.5-kb-long unique coding region flanked by multiple G+C-rich 801-bp terminal repeat sequences [23]. Now it has been found that KSHV encodes at least 86 open reading frames (ORFs), which are expressed during distinct phases of KSHV infection [23]. It is believed that among these ORFs, 22 of them have the capacity of modulating immune response, such as K3, K5, K7, and K11.1 [24].

As a member of the γ-herpesvirus, KSHV has two distinct phases of infection as well, the latent and lytic phases [24]. After primary infection, both latent and lytic genes are expressed. Expression of lytic genes is shut down after a few rounds of replication to avoid immune surveillance, and latent infection of KSHV is established. During latency, KSHV expresses a few viral genes, ORF73 (latency-associated nuclear antigen 1 [LANA-1]), ORF72 (viral cyclin [vCyclin]), ORF71 (K13/vFLIP), and ORFK12 (kaposins A, B, and C), along with at least 12 distinct microRNAs [24, 25]. These all together facilitate the establishment of KSHV latency in hosts for a lifetime, survival against the host innate, and adaptive immune surveillance mechanisms, contributing to KSHV-related malignancies [24]. These genes and miRNAs expressed during latency also aid malignant transformation and oncogenesis by coping with several signaling pathways [26]. Among them, KSHV LANA directly deregulates signaling pathways such as MAPK, JAK/STAT, MEK/ERK, PI3K/AKT, Notch, and Wnt signaling to help establish latent infection [24, 27, 28].

Multiple chemicals, including tetracycline [29], are able to trigger KSHV reactivation. Once lytic replication is activated, immediate early (IE), early and late genes are expressed [30]. Production of lytic genes switches infected cell into intense viral replication, contributing to KSHV-induced tumorigenesis [24, 30]. These proteins encoded by KSHV lytic genes are also involved in modulating immune system or pathogenesis. For instance, K2-encoded vIL-6 can regulate B-cell proliferation by activating JAK/STAT, MAPK, and PI3K/Akt signaling pathways [31].

15.1.3 Noncoding RNA Encoded by KSHV

KSHV also expresses noncoding genes during latent or lytic phase. During lytic production, a 1.1-kb-long long noncoding RNA, which is now known as polyadenylated nuclear RNA (PAN RNA), is produced to facilitate KSHV lytic production [32]. Recent study also shows that this particular noncoding RNA encodes three peptides [33]. And with chromatin isolation by RNA purification coupled with next-generation sequencing (ChIRP-seq), PAN is found binding to KSHV genome to initiate lytic phase [33].

MiRNAs are expressed in latent cells, helping establish lifetime infection in host cells. MiRNAs are a group of small, about 22 nt in length, noncoding RNAs that are capable of regulating gene expression posttranscriptionally [34, 35]. The mechanism of how these small RNAs works has been studied since its discovery. It is believed that miRs can regulate gene expression through inhibiting transcription or destabilizing target genes by targeting complementary sequences in the 3′ untranslated regions (3′UTR) [34–37].

Discovery of KSHV miRNAs went through a history of a half decade. In 2005, Pfeffer et al., Cai et al., and Samols et al. identified 11 precursor-miRNAs (pre-miRNA) coded by KSHV by cDNA cloning strategies [38–40]. Later, with the help of a combined computational and microarray-based approach, Grundhoff et al. uncovered a different hairpin that leads to the 12th pre-miRNA, miRNA-K12, as well as most of the miRNAs discovered before [40]. With more digging, in 2010, different groups ascertained that there were at least 25 mature miRNAs deriving from those previously found pre-miRNAs [41]. No more miRs have been found ever since.

In 2013, a group in the USA found out that KSHV miRNAs are essential for tumorigenesis of KS. In this particular research, they found that deletion of KSHV miRs fails to transform, and instead it caused cell cycle arrest and apoptosis [26]. These results show that KSHV miRs are of great significance in the tumorigenesis of KS. And in this same research, NF-κB pathway is found to be the critical pathway targeted by KSHV miRs [26].

Moreover, these miRNAs are capable of regulating viral life cycle and gene expression, facilitating the tumorigenesis of KS. In a project done by Lu et al., they discovered that KSHV miR-K3 regulates viral latency by targeting nuclear factor I/B (NFIB), which indicates that KSHV miRNAs play a significant role in KSHV life cycle [42]. Also, miR-K3, miR-K4, miR-K7-5p, and miR-K9 have been reported to be related with the KSHV lytic switch protein (RTA)-regulated KSHV life cycle [42–45]. Moreover, a recent study showed that KSHV miRNA miR-K12-6-5p (miR-K6-5) can directly target and suppress a human gene, breakpoint cluster region (Bcr), resulting the activation of Rac1-mediated angiogenesis [46]. MiR-K1 also target IκBα, leading to NF-κB-dependent viral latency [47].

Researches on miRs are developing rapidly. A lot of the target genes or pathways regulated by miRs have been confirmed. For example, by inhibiting SH3BGR, miR-K6-3p activates STAT3 pathway to aid the malignancy of KS [48]; and by targeting GRK2, miR-K3 activates the CXCR2/AKT pathway, which influences the angiogenesis, migration and invasion of KSHV-infected primary human umbilical vein endothelial cells (HUVECs) [49]. A lot more targets of KSHV miRs have been confirmed, parts of the targets are shown in Table 15.1.

15.2 Interaction Between KSHV and HIV Viral Proteins

Although KSHV is the pathogen of KS, KSHV alone is not sufficient for the tumorigenesis of KS. HIV infection is thought as the cofactor in tumorigenesis of KS [50]. Epidemiology research on KSHV showed that KS is of higher possibility developing in AIDS patients [51, 52]. The HIV-KSHV interaction must have a place in KS.

HIV genome encodes 16 viral proteins, which all play essential roles in HIV life cycle. In the coinfected hosts, more cytokines are induced by HIV-1 effecting KSHV life cycle. Experiment done on BCBL-1 cells found that cytokines, like OSM,

Table 15.1 KSHV miRNAs and confirmed target genes

miRNA	Related pathways or genes	Ref.
miRNA-K12-1	Nuclear factor-κb (NF-κB)	[109, 110]
	Signal transducer and activator of transcription 3 (STAT3)	
	Casp3	
miRNA-K12-3	G protein-coupled receptor (GPCR) kinase 2	[49, 110, 111]
	Casp3	
	Nuclear factor I/B (NFIB)	
miRNA-K12-4	Casp3	[110]
miRNA-K12-5	Tumor suppressor protein tropomyosin 1 (TPM1)	[112]
miRNA-K12-6	Breakpoint cluster region (Bcr) protein	[46, 48]
	SH3 domain-binding glutamate-rich protein (SH3BGR)	
miRNA-K12-7	Replication and transcription activator (RTA)	[44]
miRNA-K12-9	Interleukin-1 receptor (IL-1R)-associated kinase 1 (IRAK1)	[113]
MiRNA-K12-11	SMAD5	[114]

HGF/SF, and IFN-γ, can induce KSHV lytic reactivation [50]. Same results are shown in other two PEL cell lines, BC-1 and BC-3 [50].

Multiple signaling pathways are involved in HIV-1 induced KSHV reactivation. HIV-1-infected BCBL-1 activated several pathways, including phosphatidylinositol 3-kinase/AKT (also called protein kinase B, PKB), mitogen-activated protein kinase (MAPK), and nuclear factor-κB (NF-κB) signaling pathways [53]. All these three pathways are involved in KS pathogenesis. In HIV-1-infected BCBL-1 cells, phosphorylation of PI3K and AKT is dramatically increased, and meanwhile the negative regulator of PI3K and PTEN is decreased, all leading to the activation of the PI3K/AKT pathway [53]. Moreover, HIV-1-infected BCBL-1 cells showed increased expression of Ras and phosphorylation of c-Raf, MEK1/2, and extracellular signaling-regulated kinase (ERK), which all represent their activation [53]. Furthermore, activation of the Ras/c-Raf/MEK1/2 MAPK pathway leads to the activation of KSHV lytic production [53]. However, the role of NF-κB in the reactivation of KSHV remains a controversy.

15.2.1 HIV-1 Tat and Its Function in the Oncogenesis of KS

The HIV-1 Tat is a polypeptide with a length of 86–104 amino acids (aa) [54]. With its ability to transactivation, HIV-1 Tat is vital for HIV replication [54]. Extracellular Tat is capable of entering uninfected cells and transactivate endogenous genes, such as tumor necrosis factor, interleukin-2 (IL-2), and IL-6 [55]. Tat is positively charged, and with this feature, it is able to bind to negatively charged molecules, such as VEGFR-2, which significantly promotes angiogenesis in vivo [54]. With this effect, there is high possibility that Tat contributes to KSHV-inducing abnormal angiogenesis in KS formation.

Researches confirmed that Tat, regulatory protein encoded by HIV, is involved in several activities of KSHV. It has been proved that Tat, as a cofactor in pathogenesis of AIDS-KS, is a growth factor for KS spindle cells [56, 57]. Transgenic expression of Tat in mice helps in the formation of KS-like lesions [58]. Researches so far have found that HIV Tat can affect KSHV life cycle and facilitate AIDS-KS by inducing cellular proliferation and pro-inflammatory genes. In 2007, our group found that, by inducing human interleukin-6 (huIL-6) and its receptor (huIL-6Ra), Tat enhances KSHV lytic replication through modulation of the JAK/STAT pathway [59].

Far in the late 1990s, it is demonstrated that, for Tat being capable of inducing pro-inflammatory and proliferative genes in KS, it might contribute to the pathogen of KS [60]. Tat enhances the expression of IL-6, MCP-1, ICAM-1, and VCAM-1 in cultured KS cells [60]. Among these cytokines, IL-6 is a cytokine that activates leukocytes and induces the proliferation of KS cells [60]. The expression of MCP-1 and other cellular adhesion molecules could in return promote the expression of IL-6 [60].

In cooperation with a 13-amino-acid peptide corresponding to the basic region of Tat, HIV-1 Tat enhances KSHV infectivity by aiding KSHV entering into endothelial cells and other cells [61]. This might be the reason AIDS-KS is far more aggressive than KS in other immunodeficiency or immunocompromised states. In the pathogenesis of KS, HIV-1 Tat may cooperate with KSHV-encoded genes to facilitate KS tumorigenesis. Research found that HIV-1 Tat may enhance KSHV kaposin A-mediated tumorigenesis in vitro and in vivo through several signaling pathways, such as MEK/ERK, STAT3, and PI3K/Akt signals [58]. However, it is not only that kaposin A-mediated tumorigenesis is enhanced by Tat but also vIL-6. Through activating PI3K and AKT and inactivating PTEN and GSK-3β, Tat significantly promotes vIL-6-induced angiogenesis and tumorigenesis of fibroblasts and human endothelial cells in a chicken chorioallantoic membrane (CAM) model [62].

Most of the cells in KS are under latent infection; however, a few KSHV-infected cells are activated and express lytic genes, such as Orf-K1 and Orf-K2 [63]. Soluble Tat or ectopic expression of Tat enhanced K1-induced cell proliferation and angiogenesis in vitro and in vivo [63]. In synergy with K1, Tat induces the expression of miR-891a-5p of host cells, which activates NF-κB by targeting IκBα 3′ untranslated region [63]. Activation of NF-κB in turn contributes to the malignancy of KS.

Moreover, ectopic expression of HIV-1 Tat promotes HSV-2-induced KSHV reactivation, resulting in KSHV going into lytic phase [64].

15.2.2 HIV-1 Nef and Its Role in Tumorigenesis of KS

Nef, expressed during the early stage of infection, is encoded by the *nef* gene, which only exists in primate lentiviruses [65]. In 1991, Kestler et al. infected Rhesus macaques with a mutated strain of $SIVmac_{239}$ lacking the Nef ORF, which proved that the *nef* gene is vital in maintaining high viral load and viral infection [66]. Nef is structurally multifunction. Far in the 1990s, multiple groups confirmed that in HIV-1-infected cells, Nef assembles on the cell surface or in cytoplasm [67, 68]. Myristoylation of Nef and basic amino acids on its N-terminal helps the interaction between Nef and membrane [68, 69], which facilities its coping with host contents and helps the replication of HIV-1 [70]. Different groups confirmed that Nef is also able to enhance the infectivity of HIV-1 [70, 71]. Recent study shows that Nef is also involved in the localization of Gag, resulting in transferring viruses cell to cell [72]. With its ability to interact with multiple host factors, Nef displays remarkable ability in connecting with the cellular vesicular trafficking machinery and to perturb cell signaling [65].

Not only is HIV-1 Nef of great importance in HIV-1 infection, but also it plays significant roles in the oncogenesis of KSHV. Based on the fact that Nef localizes in the pulmonary arterial endothelial cells of AIDS patients, our group validated that in cooperation with KSHV viral interleukin-6 (vIL-6), HIV-1 Nef facilitates angiogenesis and oncogenesis of KSHV by manipulating AKT signaling pathway [73]. The experiment in vivo shows Nef boosts vIL-6-induced angiogenesis and tumorigenesis [73]. In this particular research, we found that exogenous Nef is able to penetrate endothelial cells, without impacting the apoptosis of endothelial cells [73]. That corresponds with Nef being able to get to cell membrane.

Despite vIL-6, HIV-1 Nef works in synergy with KSHV K1 to promote cell proliferation and tubulogenesis of human umbilical vessel endothelial cells (HUVEC) [74]. HIV-1 and KSHV K1 together induce cellular miR-718, which in turn regulates the PTEN/AKT/mTOR signaling pathway [74].

Moreover, Nef is capable of regulating KSHV life cycle. Our recent investigation shows that soluble and ectopic Nef can suppress KSHV lytic replication to promote latency in PEL cells [75]. Mechanism study revealed that cellular miR-1258 enhances Nef inhibition of KSHV reactivation [75].

Besides Tat and Nef, HIV-1 viral protein R (Vpr) is another viral protein that is involved in regulating KSHV life cycle. Researchers found that Vpr is able to activate KSHV transcription [76]. And with its ability of internalizing into PEL cells, Vpr can activate NF-κB signaling pathway, and cellular miR-942-5p directly target inhibitor of NF-κB, revealing the role of NF-κB in balancing KSHV latency and lytic production [77].

15.2.3 KSHV Affects HIV

HIV influences KSHV in multiple ways and plays important roles in the oncogenesis of KS; KSHV in turn influences host cell susceptibility of HIV-1 and replication in few ways as well [78]. The receptor for KSHV, DC-SIGN, is expressed on activated macrophages, B cells, and monocyte-derived dendritic cells (MDDCs). Also, isoform of DC-SIGN, DC-SIGNR, is also expressed on endothelial cells [79].

Among them, dendritic cells are of great significance in HIV-1 infection, which indicates the relationship of KSHV and HIV-1 in their coinfection.

KSHV plays a role in HIV viral transportation. Research found that dendritic cells stimulated by KSHV capture more HIV viral particles and enhance HIV-1 transport to CD4+ T cells, which is a key route of HIV-1 transfer between cells [80].

KSHV is also involved in regulating HIV-1 life cycle. KSHV ORF50 (encodes RTA) is an important gene in KSHV reactivation [81]. In KSHV and HIV coinfection case, KSHV ORF50 increases cell susceptibility of HIV-1 infection in vitro and is capable of transactivating the HIV-1 LTR in synergy with HIV-1 tat gene [81, 82]. In susceptible cell, like T cells and B cells, the expression of ORF50 activates HIV-1 replication, and in unsusceptible cells, HIV-1 alone is not able to launch reactivation, while transformed with ORF50, HIV-1 infection is more persist in parent cell and leads to low level of HIV-1 virus production, infecting susceptible cell by direct contact [83]. Meanwhile, KSHV ORF57 is found being able to activate HIV-1 replication by regulating ORF50 or other unidentified mechanism [82].

Despite OFR50, researchers found that KSHV-encoded ORF45 was the most robust in mediating transcriptional activation of HIV-1 TLR via the cellular p90 ribosomal S6 kinase (RSK2) as well [83].

In addition, KSHV-encoded viral FLIP (Fas-associated death domain-like IL-1 beta-converting enzyme inhibitory protein) K13 can activate the HIV-1 LTR in cooperation with HIV Tat [84]. The activation is done via K13 activating NF-κB pathway [84].

Moreover, KSHV latency-associated nuclear antigen (LANA) is constantly expressed in KSHV-infected cells. Research found that by functioning as a regulator of transcription, LANA is able to transactivate HIV-1 LTR in multiple cell lines, including human B-cell line BJAB, human monocytic cell line U937, and the human embryonic kidney fibroblast cell line 293 T [84]. And HIV-encoded Tat protein is in cooperation with LANA in the reactivation [84].

15.3 Effect of Antiviral Treatment on KS Development and New Treatment of KS

After the epidemic of HIV infection and outbreak of AIDS, till now, multiple anti-HIV drugs have been approved by US Food and Drug Administration (FDA). And in HIV-infected individuals, antiretroviral treatment (ART) is sufficient to prevent transmission [85–87]. At the same time, HAART has significantly reduced KS incidence in HIV-positive patients, while in Africa, where antiretroviral drugs are not easily accessible, KS remains a problem for HIV-infected patients [88, 89]. Effects of HAART on AIDS-KS are diverse, including inhibition of HIV replication, improved immune response, or direct inhibition of HIV-1 Tat [90]. However, no scant evidence or clinical evidence shows that HAART alone is sufficient to treat KS [89, 90]. KS several treatment methods have development in treating KS, while no standard methods have been made.

HAART, in combination with systemic and local therapy, is efficient in controlling KS, resulting in regression of KS both in size and number of KS lesions [88, 91, 92]. Such regimens include cytotoxic chemotherapy and protease inhibitor [90]. A trail involving chemotherapy and HAART elucidated that a combination of HAART and chemotherapy achieved higher overall KS response, resulting in higher overall survival and improved quality of life [93]. Chemotherapy is strongly recommended in treating KS, especially KS with pulmonary involvement [20]. HAART mainly controls HIV, while chemotherapy is specific to KS.

Together with HAART, FDA-approved chemotherapeutic drugs including pegylated liposomal doxorubicin (PLD), liposomal daunorubicin, and taxane paclitaxel are proved impactful in treating KS. PLD plus HAART showed better KS response after 48-week treatment than HAART alone, and it shows equal efficiency in advanced KS [20, 94, 95]. Later year, in 2005, researcher found that this combination can induce effective tumor remission and recovery of CD4+ cells [96]. The comparison between paclitaxel and PLD showed similar response toward KS (a rate of 50–60%), with paclitaxel showing hematologic toxicity and more alopecia and sensory neuropathy [97]. And liposomal daunorubicin was approved by US FDA as the first-line treatment of KS [98]. And KS patients benefit from higher cumulative chemotherapeutic doses without significant cardiotoxicity [99]. However, HAART in combination with chemotherapy is not as effective as expected. Still 51% of the patients have persistent KS 36 months after diagnosis.

New drugs targeting KSHV regulated pathways or factors are developed during recent decades. Rapamycin, an mTOR signaling pathway inhibitor, is proved effective in transplant-related KS, and in AIDS-KS, its effect still needs further investigation [100, 101]. And also there is a report on classic-KS regression after treatment with rapamycin [102]. Drugs or immune modulators like interferon-a, interleukin-12, thalidomide, and lenalidomide are effective either alone or in combination with other treatment [103–106]. Other drugs targeting KSHV-encoded genes regulated signaling or KSHV-induced angiogenesis; apoptosis is also under investigation [90].

15.4 Remarks and Perspectives

In this article, we made a discussion on KSHV and the heated topic of KSHV miRNAs during the last few years. These products of KSHV latency are of great significance in the angiogenesis, migration, and invasion of KS. This leads us one more step closer to the myths of KSHV and KS. However, cell origin of KS is still controversial and haunting around. The establishment of KSHV-infected MSCs is the first step in searching the secret behind KS [107, 108]. Besides that, the network of interaction between KSHV miRNAs and its target genes deserves more digging to clarify the underlying secrets of KSHV miRNAs in the tumorigenesis of KS. Researches on HIV and KSHV coinfection now mainly focus on the HAART treatment. Drugs and methods in treating AIDS-KS have been found and proved effective, saving thousands of lives. However, works on prevention of KS is still slow. No vaccines or other drugs have been found or developed in preventing KSHV infection or KS development. And works on non-AIDS-related KS still need more attention.

References

1. Martin JN et al (1998) Sexual transmission and the natural history of human herpesvirus 8 infection. N Engl J Med 338(14):948–954
2. Mendez JC et al (1999) Relationship of HHV8 replication and Kaposi's sarcoma after solid organ transplantation. Transplantation 67(8):1200–1201
3. Martin RR, Hood AF, Farmer ER (1993) Kaposi sarcoma. Medicine (Baltimore) 72(4):245–261
4. Sternbach G, Varon J (1995) Moritz Kaposi: idiopathic pigmented sarcoma of the skin. J Emerg Med 13(5):671–674
5. Begré L et al (2016) Is human herpesvirus 8 infection more common in men than in women? Systematic review and meta-analysis. Int J Cancer 139(4):776–783
6. Wu XJ et al (2014) One hundred and five Kaposi sarcoma patients: a clinical study in Xinjiang, Northwest of China. J Eur Acad Dermatol Venereol 28(11):1545–1552
7. Biryahwaho B et al (2010) Sex and geographic patterns of human herpesvirus 8 infection in a nationally representative population-based sample in Uganda. J Infect Dis 202(9):1347–1353
8. Butler LM et al (2011) Human herpesvirus 8 infection in children and adults in a population-based study in rural Uganda. J Infect Dis 203(5):625–634
9. Engels EA et al (2007) Risk factors for human herpesvirus 8 infection among adults in the United States and evidence for sexual transmission. J Infect Dis 196(2):199–207
10. Zhang T et al (2012) Human herpesvirus 8 seroprevalence. China Emerg Infect Dis 18(1):150–152
11. Bhutani M et al (2015) Kaposi sarcoma-associated herpesvirus-associated malignancies: epidemiology, pathogenesis, and advances in treatment. Semin Oncol 42(2):223–246
12. O'Brien TR et al (1999) Evidence for concurrent epidemics of human herpesvirus 8 and human immunodeficiency virus type 1 in US homosexual men: rates, risk factors, and relationship to Kaposi's sarcoma. J Infect Dis 180(4):1010–1017
13. Regamey, N., et al., (1998) High human herpesvirus 8 seroprevalence in the homosexual population in Switzerland. J Clin Microbiol 36(6):1784–1786
14. Zhang T et al (2013) Prevalence and correlates of Kaposi's sarcoma-associated herpesvirus infection in a sample of men who have sex with men in Eastern China. Epidemiol Infect 141(9):1823–1830
15. Casper C et al (2004) HIV infection and human herpesvirus-8 oral shedding among men who have sex with men. J Acquir Immune Defic Syndr 35(3):233–238
16. Pauk J et al (2000) Mucosal shedding of human herpesvirus 8 in men. N Engl J Med 343(19):1369–1377
17. Rohner E et al (2014) HHV-8 seroprevalence: a global view. Syst Rev 3:11
18. Rohner E et al (2016) HIV and human herpesvirus 8 co-infection across the globe: systematic review and meta-analysis. Int J Cancer 138(1):45–54
19. Feiterna-Sperling C et al (2016) High seroprevalence of antibodies against Kaposi's sarcoma-associated herpesvirus (KSHV) among HIV-1-infected children and adolescents in a non-endemic population. Med Microbiol Immunol 205:425–434
20. Uldrick TS, Whitby D (2011) Update on KSHV epidemiology, Kaposi sarcoma pathogenesis, and treatment of Kaposi sarcoma. Cancer Lett 305(2):150–162

21. Chang Y et al (1994) Identification of herpesvirus-like DNA sequences in AIDS-associated Kaposi's sarcoma. Science 266(5192):1865–1869
22. Cesarman E et al (1995) Kaposi's sarcoma-associated herpesvirus-like DNA sequences in AIDS-related body-cavity-based lymphomas. N Engl J Med 332(18):1186–1191
23. Russo JJ et al (1996) Nucleotide sequence of the Kaposi sarcoma-associated herpesvirus (HHV8). Proc Natl Acad Sci U S A 93(25):14862–14867
24. Purushothaman P et al (2016) KSHV genome replication and maintenance. Front Microbiol 7
25. Mesri EA, Cesarman E, Boshoff C (2010) Kaposi's sarcoma and its associated herpesvirus. Nat Rev Cancer 10(10):707–719
26. Moody R et al (2013) KSHV microRNAs mediate cellular transformation and tumorigenesis by redundantly targeting cell growth and survival pathways. PLoS Pathog 9(12):e1003857
27. Verma SC, Borah S, Robertson ES (2004) Latency-associated nuclear antigen of Kaposi's sarcoma-associated herpesvirus up-regulates transcription of human telomerase reverse transcriptase promoter through interaction with transcription factor Sp1. J Virol 78(19):10348–10359
28. Lan K, Kuppers DA, Robertson ES (2005) Kaposi's sarcoma-associated herpesvirus reactivation is regulated by interaction of latency-associated nuclear antigen with recombination signal sequence-binding protein Jkappa, the major downstream effector of the notch signaling pathway. J Virol 79(6):3468–3478
29. Nakamura H et al (2003) Global changes in Kaposi's sarcoma-associated virus gene expression patterns following expression of a tetracycline-inducible Rta transactivator. J Virol 77(7):4205–4220
30. Balistreri G et al (2016) Oncogenic herpesvirus utilizes stress-induced cell cycle checkpoints for efficient lytic replication. PLoS Pathog 12(2):e1005424
31. Purushothaman P, Uppal T, Verma SC (2015) Molecular biology of KSHV lytic reactivation. Virus 7(1):116–153
32. Sun R et al (1996) Polyadenylylated nuclear RNA encoded by Kaposi sarcoma-associated herpesvirus. Proc Natl Acad Sci U S A 93(21):11883–11888
33. Arias C et al (2014) KSHV 2.0: a comprehensive annotation of the Kaposi's sarcoma-associated herpesvirus genome using next-generation sequencing reveals novel genomic and functional features. PLoS Pathog 10(1):e1003847
34. Bartel DP (2004) MicroRNAs: genomics, biogenesis, mechanism, and function. Cell 116(2):281–297
35. Rana TM (2007) Illuminating the silence: understanding the structure and function of small RNAs. Nat Rev Mol Cell Biol 8(1):23–36
36. Ambros V (2004) The functions of animal microRNAs. Nature 431(7006):350–355
37. Lei, X., et al. (2010) Regulation of herpesvirus lifecycle by viral microRNAs. Virulence 1(5):433–435
38. Cai X et al (2005) Kaposi's sarcoma-associated herpesvirus expresses an array of viral microRNAs in latently infected cells. Proc Natl Acad Sci U S A 102(15):5570–5575
39. Pfeffer S et al (2005) Identification of microRNAs of the herpesvirus family. Nat Methods 2(4):269–276
40. Grundhoff A, Sullivan CS, Ganem D (2006) A combined computational and microarray-based approach identifies novel microRNAs encoded by human gamma-herpesviruses. RNA 12(5):733–750
41. Lin YT et al (2010) Small RNA profiling reveals antisense transcription throughout the KSHV genome and novel small RNAs. RNA 16(8):1540–1558
42. Lu CC et al (2010) MicroRNAs encoded by Kaposi's sarcoma-associated herpesvirus regulate viral life cycle. EMBO Rep 11(10):784–790
43. Lu F et al (2010) Epigenetic regulation of Kaposi's sarcoma-associated herpesvirus latency by virus-encoded microRNAs that target Rta and the cellular Rbl2-DNMT pathway. J Virol 84(6):2697–2706

44. Lin X et al (2011) miR-K12-7-5p encoded by Kaposi's sarcoma-associated herpesvirus stabilizes the latent state by targeting viral ORF50/RTA. PLoS One 6(1):–e16224
45. Bellare P, Ganem D (2009) Regulation of KSHV lytic switch protein expression by a virus-encoded microRNA: an evolutionary adaptation that fine-tunes lytic reactivation. Cell Host Microbe 6(6):570–575
46. Ramalingam D, Happel C, Ziegelbauer JM (2015) Kaposi's sarcoma-associated herpesvirus microRNAs repress breakpoint cluster region protein expression, enhance Rac1 activity, and increase in vitro angiogenesis. J Virol 89(8):4249–4261
47. Lei X et al (2012) A Kaposi's sarcoma-associated herpesvirus microRNA and its variants target the transforming growth factor beta pathway to promote cell survival. J Virol 86(21):11698–11711
48. Li W et al (2016) The SH3BGR/STAT3 pathway regulates cell migration and angiogenesis induced by a Gammaherpesvirus MicroRNA. PLoS Pathog 12(4):e1005605
49. Hu M et al (2015) A KSHV microRNA directly targets G protein-coupled receptor kinase 2 to promote the migration and invasion of endothelial cells by inducing CXCR2 and activating AKT signaling. PLoS Pathog 11(9):e1005171
50. Mercader M et al (2000) Induction of HHV-8 lytic cycle replication by inflammatory cytokines produced by HIV-1-infected T cells. Am J Pathol 156(6):1961–1971
51. Merat R et al (2002) HIV-1 infection of primary effusion lymphoma cell line triggers Kaposi's sarcoma-associated herpesvirus (KSHV) reactivation. Int J Cancer 97(6):791–795
52. Greene W et al (2007) Molecular biology of KSHV in relation to AIDS-associated oncogenesis. Cancer Treat Res 133:69–127
53. Zhu X et al (2011) Human immunodeficiency virus type 1 induces lytic cycle replication of Kaposi's-sarcoma-associated herpesvirus: role of Ras/c-Raf/MEK1/2, PI3K/AKT, and NF-κB signaling pathways. J Mol Biol 410(5):1035–1051
54. Aoki Y, Tosato G (2007) Interactions between HIV-1 Tat and KSHV. Curr Top Microbiol Immunol 312:309–326
55. Rusnati M et al (2000) Thrombospondin-1/HIV-1 tat protein interaction: modulation of the biological activity of extracellular Tat. FASEB J 14(13):1917–1930
56. Barillari G et al (1993) The tat protein of human immunodeficiency virus type 1, a growth factor for AIDS Kaposi sarcoma and cytokine-activated vascular cells, induces adhesion of the same cell types by using integrin receptors recognizing the RGD amino acid sequence. Proc Natl Acad Sci U S A 90(17):7941–7945
57. Huang SK et al (1993) Extravasation and transcytosis of liposomes in Kaposi's sarcoma-like dermal lesions of transgenic mice bearing the HIV tat gene. Am J Pathol 143(1):10–14
58. Chen X et al (2009) Human immunodeficiency virus type 1 tat accelerates Kaposi sarcoma-associated herpesvirus Kaposin A-mediated tumorigenesis of transformed fibroblasts in vitro as well as in nude and immunocompetent mice. Neoplasia 11(12):1272–1284
59. Zeng Y et al (2007) Intracellular tat of human immunodeficiency virus type 1 activates lytic cycle replication of Kaposi's sarcoma-associated herpesvirus: role of JAK/STAT signaling. J Virol 81(5):2401–2417
60. Kelly GD et al (1998) Purified tat induces inflammatory response genes in Kaposi's sarcoma cells. AIDS 12(14):1753–1761
61. Aoki Y (2004) HIV-1 Tat enhances Kaposi sarcoma-associated herpesvirus (KSHV) infectivity. Blood 104(3):810–814
62. Zhou F et al (2013) HIV-1 tat promotes Kaposi's sarcoma-associated herpesvirus (KSHV) vIL-6-induced angiogenesis and tumorigenesis by regulating PI3K/PTEN/AKT/GSK-3beta signaling pathway. PLoS One 8(1):e53145
63. Yao S et al (2015) MiRNA-891a-5p mediates HIV-1 tat and KSHV Orf-K1 synergistic induction of angiogenesis by activating NF-κB signaling. Nucleic Acids Res 43(19):9362–9378
64. Tang Q et al (2012) Herpes simplex virus type 2 triggers reactivation of Kaposi's sarcoma-associated herpesvirus from latency and collaborates with HIV-1 tat. PLoS One 7(2):e31652
65. Basmaciogullari S, Pizzato M (2014) The activity of Nef on HIV-1 infectivity. Front Microbiol 5:232

66. Kestler HR et al (1991) Importance of the nef gene for maintenance of high virus loads and for development of AIDS. Cell 65(4):651–662
67. Fujii Y et al (1996) Clustered localization of oligomeric Nef protein of human immunodeficiency virus type 1 on the cell surface. FEBS Lett 395(2–3):257–261
68. Greenberg ME et al (1997) Co-localization of HIV-1 Nef with the AP-2 adaptor protein complex correlates with Nef-induced CD4 down-regulation. EMBO J 16(23):6964–6976
69. Bentham M, Mazaleyrat S, Harris M (2006) Role of myristoylation and N-terminal basic residues in membrane association of the human immunodeficiency virus type 1 Nef protein. J Gen Virol 87(Pt 3):563–571
70. Kim S et al (1989) Lack of a negative influence on viral growth by the nef gene of human immunodeficiency virus type 1. Proc Natl Acad Sci U S A 86(23):9544–9548
71. Goldsmith MA et al (1995) Dissociation of the CD4 downregulation and viral infectivity enhancement functions of human immunodeficiency virus type 1 Nef. J Virol 69(7):4112–4121
72. Malbec M et al (2013) HIV-1 Nef promotes the localization of Gag to the cell membrane and facilitates viral cell-to-cell transfer. Retrovirology 10:80
73. Zhu X et al (2014) Synergy between Kaposi's sarcoma-associated herpesvirus (KSHV) vIL-6 and HIV-1 Nef protein in promotion of angiogenesis and oncogenesis: role of the AKT signaling pathway. Oncogene 33(15):1986–1996
74. Xue M et al (2014) HIV-1 Nef and KSHV oncogene K1 synergistically promote angiogenesis by inducing cellular miR-718 to regulate the PTEN/AKT/mTOR signaling pathway. Nucleic Acids Res 42(15):9862–9879
75. Yan Q et al (2014) Inhibition of Kaposi's sarcoma-associated herpesvirus lytic replication by HIV-1 Nef and cellular MicroRNA hsa-miR-1258. J Virol 88(9):4987–5000
76. Huang LM et al (2001) Reciprocal regulatory interaction between human herpesvirus 8 and human immunodeficiency virus type 1. J Biol Chem 276(16):13427–13432
77. Yan Q et al (2016) HIV-1 Vpr Inhibits Kaposi's Sarcoma-Associated Herpesvirus Lytic Replication by Inducing MiR-942-5p and Activating NF-kappaB Signaling. J Virol 90:8739–8753
78. Caselli E et al (2005) Human herpesvirus 8 enhances human immunodeficiency virus replication in acutely infected cells and induces reactivation in latently infected cells. Blood 106(8):2790–2797
79. Hensler HR et al (2014) Human herpesvirus 8 glycoprotein B binds the entry receptor DC-SIGN. Virus Res 190:97–103
80. Liu W et al (2013) Kaposi's-sarcoma-associated-herpesvirus-activated dendritic cells promote HIV-1 trans-infection and suppress CD4(+) T cell proliferation. Virology 440(2):150–159
81. Caselli E et al (2003) Transient expression of human herpesvirus-8 (Kaposi's sarcoma-associated herpesvirus) ORF50 enhances HIV-1 replication. Intervirology 46(3):141–149
82. Caselli E et al (2001) Human herpesvirus-8 (Kaposi's sarcoma-associated herpesvirus) ORF50 interacts synergistically with the tat gene product in transactivating the human immunodeficiency virus type 1 LTR. J Gen Virol 82(Pt 8):1965–1970
83. Karijolich J et al (2014) Kaposi's sarcoma-associated herpesvirus ORF45 mediates transcriptional activation of the HIV-1 long terminal repeat via RSK2. J Virol 88(12):7024–7035
84. Hyun TS et al (2001) Latency-associated nuclear antigen encoded by Kaposi's sarcoma-associated herpesvirus interacts with tat and activates the long terminal repeat of human immunodeficiency virus type 1 in human cells. J Virol 75(18):8761–8771
85. Gunthard HF et al (2014) Antiretroviral treatment of adult HIV infection: 2014 recommendations of the International Antiviral Society-USA Panel. JAMA 312(4):410–425
86. Cohen MS et al (2011) Prevention of HIV-1 infection with early antiretroviral therapy. N Engl J Med 365(6):493–505
87. Gunthard HF et al (2016) Antiretroviral drugs for treatment and prevention of HIV infection in adults: 2016 recommendations of the International Antiviral Society-USA Panel. JAMA 316(2):191–210
88. La Ferla L et al (2013) Kaposi's sarcoma in HIV-positive patients: the state of art in the HAART-era. Eur Rev Med Pharmacol Sci 17(17):2354–2365

89. Krown SE (2004) Highly active antiretroviral therapy in AIDS-associated Kaposi's sarcoma: implications for the design of therapeutic trials in patients with advanced, symptomatic Kaposi's sarcoma. J Clin Oncol 22(3):399–402
90. Sullivan RJ, Pantanowitz L (2010) New drug targets in Kaposi sarcoma. Expert Opin Ther Targets 14(12):1355–1366
91. Sparano JA et al (1999) Effect of highly active antiretroviral therapy on the incidence of HIV-associated malignancies at an urban medical center. J Acquir Immune Defic Syndr 21(Suppl 1):S18–S22
92. Aboulafia DM (1998) Regression of acquired immunodeficiency syndrome-related pulmonary Kaposi's sarcoma after highly active antiretroviral therapy. Mayo Clin Proc 73(5):439–443
93. Mosam A et al (2012) A randomized controlled trial of highly active antiretroviral therapy versus highly active antiretroviral therapy and chemotherapy in therapy-naive patients with HIV-associated Kaposi sarcoma in South Africa. J Acquir Immune Defic Syndr 60(2):150–157
94. Martin-Carbonero L et al (2004) Pegylated liposomal doxorubicin plus highly active antiretroviral therapy versus highly active antiretroviral therapy alone in HIV patients with Kaposi's sarcoma. AIDS 18(12):1737–1740
95. Nunez M et al (2001) Response to liposomal doxorubicin and clinical outcome of HIV-1-infected patients with Kaposi's sarcoma receiving highly active antiretroviral therapy. HIV Clin Trials 2(5):429–437
96. Lichterfeld M et al (2005) Treatment of HIV-1-associated Kaposi's sarcoma with pegylated liposomal doxorubicin and HAART simultaneously induces effective tumor remission and CD4+ T cell recovery. Infection 33(3):140–147
97. Cianfrocca M et al (2010) Randomized trial of paclitaxel versus pegylated liposomal doxorubicin for advanced human immunodeficiency virus-associated Kaposi sarcoma: evidence of symptom palliation from chemotherapy. Cancer 116(16):3969–3977
98. Johnson JR (2003) End points and United States Food and Drug Administration approval of oncology drugs. J Clin Oncol 21(7):1404–1411
99. Petre CE, Dittmer DP (2007) Liposomal daunorubicin as treatment for Kaposi's sarcoma. Int J Nanomedicine 2(3):277–288
100. Krown SE et al (2012) Rapamycin with antiretroviral therapy in AIDS-associated Kaposi sarcoma. JAIDS J Acquir Immune Defic Syndr 59(5):447–454
101. Hernandez-Sierra A et al (2016) Role of HHV-8 and mTOR pathway in post-transplant Kaposi sarcoma staging. Transpl Int 29:1008–1016
102. Diaz-Ley B et al (2015) Classic Kaposi's sarcoma treated with topical rapamycin. Dermatol Ther 28(1):40–43
103. Krown SE (1998) Interferon-alpha: evolving therapy for AIDS-associated Kaposi's sarcoma. J Interf Cytokine Res 18(4):209–214
104. Krown SE et al (1990) Interferon-alpha with zidovudine: safety, tolerance, and clinical and virologic effects in patients with Kaposi sarcoma associated with the acquired immunodeficiency syndrome (AIDS). Ann Intern Med 112(11):812–821
105. Yarchoan R et al (2007) Treatment of AIDS-related Kaposi's sarcoma with interleukin-12: rationale and preliminary evidence of clinical activity. Crit Rev Immunol 27(5):401–414
106. Little RF et al (2007) Phase 2 study of pegylated liposomal doxorubicin in combination with interleukin-12 for AIDS-related Kaposi sarcoma. Blood 110(13):4165–4171
107. Yoo S et al (2014) Kaposi's sarcoma-associated herpesvirus infection of human bone-marrow-derived mesenchymal stem cells and their angiogenic potential. Arch Virol 159(9):2377–2386
108. Lee MS et al (2016) Human mesenchymal stem cells of diverse origins support persistent infection with Kaposi's sarcoma-associated herpesvirus and manifest distinct angiogenic, invasive, and transforming phenotypes. MBio 7(1):e02109–e02115
109. Chen M et al (2016) Kaposi's sarcoma herpesvirus (KSHV) microRNA K12-1 functions as an oncogene by activating NF-kappaB/IL-6/STAT3 signaling. Oncotarget 7:33363–33373
110. Suffert G et al (2011) Kaposi's sarcoma herpesvirus microRNAs target caspase 3 and regulate apoptosis. PLoS Pathog 7(12):e1002405

111. Li W et al (2016) A KSHV microRNA enhances viral latency and induces angiogenesis by targeting GRK2 to activate the CXCR2/AKT pathway. Oncotarget 7:32286–32305
112. Kieffer-Kwon P et al (2015) KSHV MicroRNAs repress tropomyosin 1 and increase anchorage-independent growth and endothelial tube formation. PLoS One 10(8):e0135560
113. Abend JR et al (2012) Kaposi's sarcoma-associated herpesvirus microRNAs target IRAK1 and MYD88, two components of the toll-like receptor/interleukin-1R signaling cascade, to reduce inflammatory-cytokine expression. J Virol 86(21):11663–11674
114. Liu Y et al (2012) Kaposi's sarcoma-associated herpesvirus-encoded microRNA miR-K12-11 attenuates transforming growth factor beta signaling through suppression of SMAD5. J Virol 86(3):1372–1381

Chapter 16
Interplay Between Microenvironmental Abnormalities and Infectious Agents in Tumorigenesis

Qing Zhu, Feng Gu, Caixia Zhu, Yuyan Wang, Fang Wei, and Qiliang Cai

Abstract Emerging evidence has shown that the cell of microenvironmental abnormalities is a key factor that controls many cellular physiological processes including cellular communication, homing, proliferation, and survival. Given its central regulatory role, it is therefore not surprising that it is widely exploited by infectious agents for inducing pathogenesis. In the past decade, a number of oncogenic pathogens including viruses, bacteria, and parasites are demonstrated to take advantage of the tumor microenvironmental factors including hypoxia, oxidative stress, and cytokines, to create an extracellular environment more favorable for pathogen survival and propagation and escape from the host immune surveillance. Here we summarize and highlight the current understanding of the interplay between common tumor microenvironmental factors and oncogenic pathogens in promoting tumorigenesis.

Keywords Tumor microenvironment • Pathogen • Oncogenesis

Qing Zhu and Feng Gu equally contributed to this work.

Q. Zhu • F. Gu • C. Zhu • Y. Wang • Q. Cai (✉)
Key Laboratory of Medical Molecular Virology (Ministries of Education and Health), School of Basic Medical Sciences, Fudan University, Shanghai 200032, People's Republic of China
e-mail: qiliang@fudan.edu.cn

F. Wei (✉)
ShengYushou Center of Cell Biology and Immunology, School of Life Sciences and Biotechnology, Shanghai Jiao Tong University, Shanghai 200240, People's Republic of China
e-mail: fangwei@sjtu.edu.cn

Q. Cai et al. (eds.), *Infectious Agents Associated Cancers: Epidemiology and Molecular Biology*, Advances in Experimental Medicine and Biology 1018,
DOI 10.1007/978-981-10-5765-6_16

16.1 Introduction

Oxygen and glucose are not only essential nutrients but also key microenvironment factors to maintain cell survival. Imbalance between nutrients supply and demand can lead to nutrient stress within regions of tumor tissues. The growth of solid tumors, which are significantly different from the normal tissues, possesses the characteristic of rapid expansion of tumor mass and chaotic growth of tumor vasculature [1, 2]. Thus, excessive metabolism rate of tumor cells and insufficient blood supply could profoundly influence the tumor microenvironment where forms hypoxia and glucose starvation. To survive in hypoxia and glucose starvation stress, tumor cells have evolved strategies of adaptive cellular response by acting on various signaling pathways that are responsible for angiogenesis, glucose metabolism, cell proliferation, and apoptosis [3, 4]. Importantly, increasing evidence suggests that these adaptive strategies in cancer cells profoundly drive tumor growth and aggressive progression [1, 4, 5]. In addition, the consequence of limitation on the uptake of oxygen and glucose has also been shown to associate with the physiochemistry change within tumor microenvironment such as increase of acidic (H^+) concentration and ROS production [6]. Conversely, these physiochemistry changes acting as a selective stress influence cellular signaling pathways and can be exploited in tumorigenesis. Together, nutrient stress (hypoxia or glucose starvation) in synergizing with the accompanied production of metabolites constitutes a unique tumor microenvironment where it produces a potent selective stress in driving carcinogenesis.

Distinct from noninfectious agent-associated cancer, pathogen–host interaction has been causally demonstrated in the carcinogenesis of pathogen-associated cancer [7]. The hemostasis of both extracellular and intracellular metabolic environment is equally essential for oncogenic pathogen survival, especially for virus that absolutely relied on cells for living. Whether these oncogenic pathogens are directly capable of sensing changes in extracellular or intracellular microenvironment remains to be exploited. However, the factors including low oxygen and ROS generation have been indicated to influence virus replication and virions production [8]. On the other hand, emerging evidence has also suggested that many oncogenic pathogens participate in modulating key signaling pathways and gene expression that triggered cellular response to metabolic stress. The adaptive genetic alteration of signaling pathways by oncogenic pathogens may reflect the interaction between pathogen-associated cancer cells and tumor microenvironment. Therefore, it is highly possible that some of these oncogenic pathogens have evolved their own unique adaptive mechanisms. The pathogen-specific subversion response of signaling pathway not only facilitates the survival of infected cells under stress but also promotes pathogen-mediated oncogenesis. Hence, the understandings of these pathogen-associated critical signaling pathways in adaption to hypoxia and glucose starvation stress will not only expand the oncogenesis mechanism induced by pathogen in a microenvironment base but will also favor the identification of both

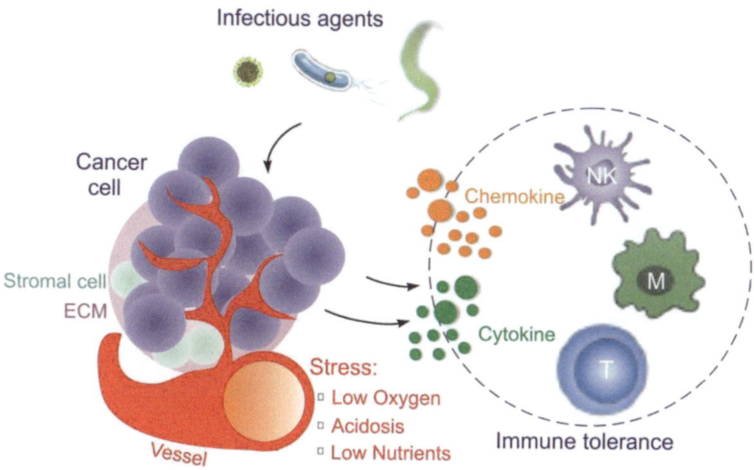

Fig. 16.1 Schematic representation of microenvironmental abnormalities including immune (immune cells, cytokines, and chemokines) and nonimmune (extracellular matrix, stromal cells, blood vessels) components associated with infectious agents (virus, bacterium, and parasite)

pathogen and microenvironment based on potential therapeutic targets for the treatment of pathogen-associated cancer (Fig. 16.1).

In this review, we summarize the key cellular adaptive signaling pathways that are modified by oncogenic pathogens and highlight the common or unique mechanisms utilized by these oncogenic pathogens for oncogenesis.

16.2 Pathogen-Mediated Alteration of Hypoxic Signaling and Response to Hypoxic Stress

Cellular oxygen homeostasis is highly dependent on the regulation of oxygen-sensitive signaling pathway. Accumulated evidence has strongly shown the activated oxygen-sensitive signaling is the first line to respond to hypoxic stress within tumor microenvironment [9–11]. Activations of hypoxia-inducible factor 1 (HIF-1) and HIF-dependent downstream gene are the master regulatory pathway during hypoxia. In addition, mTOR kinase signaling pathway and unfolded protein response (UPR) are another two oxygen-sensitive signaling that are individually activated under the condition of severe and durative hypoxia stress [11]. Therefore, it is not surprising that most oncogenic pathogens are involved in the deregulation of key molecules in controlling these hypoxic signaling pathways.

16.2.1 Deregulation of HIF-Dependent Hypoxic Signaling

HIF-1, which consists of a constitutively expressed β-subunit and an inducible α-subunit, is a central transcriptional factor of HIF-dependent signaling in response to hypoxia stress [12]. The modulation of HIF-1 is mainly through the stability and availability of the inducible subunit HIF-1α. The stabilization of HIF-1α is oxygen-dependent and is tightly regulated in the presence of oxygen [13]. More recently, it has been proven that many oncogenic viruses can directly enhance the accumulation of HIF-1α and promote its transcriptional activity through various mechanisms even in normoxia [14]. Given the role of HIF-1α in inducing the expression of proangiogenic factors, the subversion of HIF-1-dependnent angiogenesis has been deeply involved in oncoprotein-stimulated tumor angiogenesis.

16.2.1.1 Synthesis of HIF-1α Protein

Activation of growth factor signaling pathways including MAPK signaling, PI3K/Akt signaling, and TSC/mTOR signaling has been indicated to be involved in the synthesis of HIF-1α protein [15]. Several oncogenic viruses have been found to hijack these signaling pathways to enhance the synthesis of HIF-1α protein in hypoxia or normoxia. For instance, KSHV vGPCR-mediated paracrine secretion can activate TSC/mTOR signaling and mTOR-dependent upregulation of HIF-1α/HIF-2α [16]. Similarly, EBV-encoded latent membrane protein LMP-1 is also shown to induce the activation of p42/p44 MAPK signaling pathway to promote the synthesis of HIF-1α proteins [17], and HPV16-encoded E6 associate with ERK1/2 signaling pathway to enhance HIF-1α accumulation [18]. In addition, some viral oncoproteins are shown to regulate HIF-1α at a transcription level. For example, HTLV encodes Tax to promote the expression and DNA-binding activity of HIF-1α by means of activating PI3K/Akt signaling [19]. EBV-encoded LMP-1 is also shown to enhance the stability of HIF-1α RNA transcripts through ERK1/2 and STAT3 signaling targeting the expression of RNA-destabilizing proteins TTP and PUM2 [20].

16.2.1.2 Stability of HIF-1α Protein

The accumulation of HIF-1α protein not only depends on the constitutive synthesis of HIF-1α but also requires the modulation of HIF-1α degradation. The degradation of HIF-1α is primarily induced by PHD/HIF/VHL pathway in an oxygen-dependent manner [21, 22]. The tumor suppressor VHL acts as an E3 ubiquitin ligase to induce prolyl-hydroxylated HIF-1α for ubiquitylation and in turn proteasomal degradation. The hydroxylation of HIF-1α in the specific proline residue is mediated by oxygen-sensor prolyl hydroxylase (PHD) enzymes. Interestingly, increasing evidences have shown that oncogenic viruses have exploited diverse strategies to interfere PHD/

HIF/VHL pathway. For instance, both KSHV-encoded LANA and EBV-encoded LMP-1 have been demonstrated to induce the proteasome-mediated degradation of HIF-1α suppressor. LANA can stimulate the degradation of HIF-suppressor VHL and p53, which is dependent on the recruitment of Cul5-Elongin BC complex by the cytokine signaling-box motif within LANA [23]. In contrast, LMP-1 can induce the degradation of oxygen-sensor PHD1 and PHD3 via recruitment of Siah1 E3 ubiquitin ligase [24]. Distinct from LANA and LMP-1, KSHV-encoded IFN-regulatory factor 3 (vIRF3), a viral homologue of cellular IRF gene, can stabilize HIF-1α protein through forming a complex with HIF-1α, although the machinery of the inhibition of HIF-1α degradation remains unclear [25]. The EBV oncoproteins EBNA3 and EBNA5 are shown to bind to PHD1 and PHD2 for blocking the hydroxylation of HIF-1α [26]. Interestingly, in order to stabilize HIF-1α, the HBV-encoded HBx not only blocks the formation of VHL-HIF complex but also induces interaction between MTA1/HDAC and HIF-1α to promote the deacetylation of HIF-1α within the oxygen-sensitive domain [27, 28].

16.2.1.3 Transcriptional Activity of HIF-1α

In addition to the accumulation of HIF-1α protein, the regulators of HIF-1α transcriptional activity including nuclear translocation, and interaction with coactivators, DNA-binding capacity also plays a critical role in activating HIF signaling, which is targeted by different viral proteins [29]. For example, KSHV-encoded LANA and vIRF3 have been reported to promote nuclear accumulation of HIF-1α [23]. EBV oncoprotein LMP-1 enhances DNA-binding ability of HIF-1α to hypoxia-responsive DNA elements within the VEGF promoter [17], while HBx enhances the transcriptional activity of HIF-1α through the activation of p42/p44 MAPK signaling, leading to the interaction between HIF-1α and coactivator CREB-binding protein [30]. In addition, some viral oncoproteins are also involved in stimulating HIF-1α activity through posttranslational modification. For instance, the p38/MAPK signaling activated by KSHV vGPCR can phosphorylate HIF-1α and enhance its transcriptional activity [31], and HPV E7 prevents deacetylation of HIF-1α through dissociation with histone deacetylases HDAC1, HDAC4, and HDAC7 [32].

16.2.2 Deregulation of HIF-Independent mTOR Signaling

It has been demonstrated that the adaptive response to hypoxia stress involves not only stimulation of angiogenesis but also inhibition of protein synthesis [33]. mTOR kinase signaling pathway, as a central regulator of protein synthesis that integrates various physiological signals [34], has been shown to respond to hypoxia and restrain the growth of tumor [33]. mTOR-mediated protein synthesis is a process involving the phosphorylation of the eukaryotic initiation factor 4E binding protein

1 (4E-BP1) and the p70 ribosomal S6 kinase 1(S6K1) [34]. These two are critical effectors of the downstream of mTOR signaling and responsible for the initiation process of translation. It has been shown that inhibition of mTOR by hypoxia involves three hypoxia-inducible proteins REDD1, BNIP3, and PML [35–37]. Both REDD1 and BNIP3 can directly suppress mTORC1 activity by disrupting Rheb-mTOR interaction, whereas REDD-mediated downregulation of mTORC1 by hypoxia is dependent on TSC1/TSC2 complex (a negative regulator of mTOR). In view of the fact that the deregulation of mTOR signaling appears in many advanced cancers [38, 39], the constitutive activation of mTOR could be an adaptive strategy in response to hypoxia. Intriguingly, a growing number of evidence has shown the positive regulation of mTOR activity by several oncogenic viruses. For instance, HPV16-encoded E6 and HBV-encoded HBx are shown to target TSC1/TSC2 complex for stimulating protein synthesis. Moreover, HPV16 E6 not only induces the activation of mTOR/SK61 signaling, which is dependent on the disruption of TSC2 by E6-tuberin interaction and the proteasomal degradation of tuberin [40], but also enhances Atk/mTOR activity to initiate cap-dependent translation [41]. For HBV, the overexpression of HBx activates TSC1/mTOR/SK61 signaling by means of IKKβ [42]. Meanwhile, HCV NS5A-mediated activation of mTOR presents a positive effect on two key translation initiation-associated proteins S6 K1 and 4EBP1, by which NS5A promotes the dissociation of FKBP38 from mTOR by competitive binding to mTOR [43, 44]. This indicates to some extent that activation of mTORC1 and protein synthesis could be potent strategies targeted by oncogenic viruses in response to hypoxia. Nonetheless, the increasing severity and duration of hypoxia will conversely cause the suppression of protein synthesis in most cells. Therefore, the mTOR signaling is also a critical regulator in hypoxia toleration. However, whether the subversion of mTOR signaling by oncogenic virus for carcinogenesis will still benefit to the survival of tumor cells during severe hypoxia remains elusive. It is likely that the oncogenic virus will shift the regulatory mechanism of mTOR signaling or constitutively activate mTOR-dependent protein synthesis to promote viral replication in response to sever hypoxia.

16.3 Pathogen-Mediated Alteration of ROS Signaling and Response to Oxidative Stress

Mounting evidence has indicated the excess generation of intrinsic or extrinsic ROS in cancer cells. It has been proven that several factors including mitochondrial dysfunction and oncoprotein activity contribute to the accumulation of ROS [45]. In tumor microenvironment, hypoxia stress and glucose starvation have been clearly linked to the induction of intracellular ROS production [46, 47]. The constitutive production of ROS (i.e., hydroperoxides) and the consequence of oxidative stress will cause DNA damage and genomic instability and trigger the normal cell death signaling. To date it is well known that oxidative DNA damage caused by ROS will activate p53 signaling through the enhancement of p53 stability and DNA-binding activity [48], which is tightly modulated by negative regulator MDM2 (an ubiquitin E3 ligase) and ATM (an important sensor of DNA damage) [49, 50]. Different from the effect on normal cells, it has been well established that the oxidative stress in microenvironment profoundly contributes to tumor progression by affecting cell proliferation, apoptosis sensitivity, and genome stability [45]. Therefore, adaptive genetic change to subvert the death signaling induced by oxidative stress has evolved in tumorigenesis.

Since the tumor suppressor p53 is demonstrated as a central regulator in both cell cycle arrest and apoptosis and is potently activated in response to oxidative stress [51], several viruses have been found to evolve an adaptive mechanism to directly block p53 function. For example, the expression of vIRF1 encoded by KSHV can attenuate ATM/p53-mediated DNA damage response through directly blocking ATM-mediated phosphorylation of p53 on serine 15 which in turn increases the degradation of p53 by MDM2. In addition, vIRF1 can also reduce the transcriptional activation of p53 [52]. In contrast, EBV-encoded lytic protein BZLF-1 can induce the degradation of p53 in ATM-dependent DNA damage response which is independent of MDM2 [53]. The deregulation of p53-dependent oxidative stress response is also found in HCV infection. The overexpression of DHCR24 induced by HCV infection can suppress the activation of p53 through the accumulation of p53-MDM2 complex, although the specific viral protein involved in this process remains unknown [54]. In addition, KSHV-encoded LANA and some structural proteins expressed during the late stage of lytic replication have been found to inhibit p53-mediated apoptosis [55, 56]. Both EBV nuclear antigen 3C and viral oncoprotein LMP-1 have also been shown to be involved in repressing p53-induced apoptosis and transcriptional activity [57–59]. However, despite that the activation of p53 signaling has been linked to multiple types of DNA damage, how p53 is regulated by viral oncogene and in turn responsible for oxidative DNA damage is still elusive and requires to be further investigated.

16.4 Cross Talk Between Pathogens and Cytokines in Tumor Microenvironment

Cytokines and chemokines, existing in tumor microenvironment, are a series of small proteins that exert great effects on host response to pathogen infection. Despite antiviral activity induced by cytokines and chemokines, extensive evidence demonstrates that some pathogens, especially oncogenic viruses and bacteria, utilize cytokines and chemokines to promote tumor progression [60, 61]. Here, we summarized and highlighted several cytokines and chemokines that play a vital role in tumorigenesis during infection of oncogenic pathogens (Fig. 16.1).

16.4.1 IL-6

Interleukin 6 (IL-6), secreted by a variety of host cells such as T cells, macrophages, fibroblasts, and malignant cells, is a multifunctional inflammatory cytokine, inducing various biological effects including tumorigenesis [62]. Increasing evidence indicates that IL-6 has a strong link with pathogen-mediated carcinomas. For example, it has been found that IL-6 acts as an autocrine growth factor targeted by EBV to promote immortalization of B cells and tumor growth [63–65]. In contrast, KSHV encodes viral IL-6 (vIL-6), sharing about 25% homology with human IL-6 (hIL-6). Different from hIL-6, vIL-6 stimulates almost each type of cells through directly binding to gp130 without hIL-6 receptor [66]. vIL-6 is able to promote the growth and survival of PEL cells and tumorigenesis of nude mice [67, 68]. Blockading vIL-6 expression or neutralizing antibody against gp130 could efficiently inhibit the growth of PEL cells [69, 70]. Further studies revealed that vIL-6 blocks IFN signaling, which contributes to tumor cell proliferation [71]. In addition, miRNA K12-1, a viral miRNA encoded by KSHV, was found to activate NF-κB/IL-6/STAT3 pathway to promote tumorigenesis [72]. In the HPV-associated cervical cancer, recent studies reported that IL-6/STAT3 is activated by the E6 oncoprotein encoded by high-risk HPV for tumorigenesis [73, 74], while HBV-encoded X protein modulates IL-6 to promote the progression of liver cancer [75]. In the HTLV-1-associated T-cell malignancy, the viral protein Tax is shown to enhance the expression of IL-6 receptor and leads to the malignant growth of T cells [76]. In the bacterium-associated cancers, *Helicobacter pylori*, a gram-negative microaerophilic bacteria, is found to parasitize in the stomach and results in chronic gastritis that is intensely associated with gastric neoplasm [77]. Several reports indicated that the interplay between *H. pylori* and TLR2 induces the expression of IL-6 and subsequently activates IL-6/STAT3 signaling pathway, which strongly contributes to immortality of gastric cancer cells. Interestingly, TLR2 is also directly upregulated by STAT3 in gastric tumors [78–81]. Therefore, TLR2/IL-6/STAT3 pathway may form a positive loop to promote gastric tumorigenesis [77].

16.4.2 IL-10

Interleukin 10 (IL-10), initially identified as an inhibitor of cytokine synthesis, has been shown to play a vital role in regulating cell differentiation and immune response, including limiting inflammatory response to pathogens and thereby reducing damage to host [82, 83]. However, it is also reported that IL-10 is utilized by various viruses to favor viral survival and pathogenesis, among which some even encode IL-10 homologs. For instance, EBV encodes vIL-10, imitating biological activities of cellular IL-10, to inhibit cytokine synthesis and regulate immune response [83]. In addition, vIL-10 prevents EBV-infected B cells from being eliminated by NK cell and protects antigen-specific T-cell proliferation by downregulating MHCII antigen on monocytes and ultimately maintains EBV latent infection [84, 85]. On the other hand, human IL-10 (hIL-10) expression is also induced in the EBV-infected B cells. Evidence shows that hIL-10 is upregulated by latent membrane protein 1(LMP-1) via p38/SAPK2 pathway [86]. Another mechanism study revealed that EBV transcription factor Zta, previously recognized as a master regulator of EBV productive cycle, is also involved in the expression of hIL-10 [87]. Furthermore, EBV-encoded small RNA, EBER, was found to induce hIL-10 through RIG-I-mediated IRF3 pathway [88]. For another herpesvirus, KSHV was found to force PEL cells to release hIL-10 into culture supernatant. Moreover, neutralizing antibodies against IL-10 and IL-10 receptor shows that IL-10 is critical for the progression of PEL [67]. In the HPV-associated malignant cervical cancer, HPV drives immune cells to produce IL-10 to facilitate viral persistence and tumorigenesis [89]. During chronic HBV infection, high production of IL-10 suppresses the biological activity of $CD8^+$ and $CD4^+$ T cells, which favors the progression of tumorigenesis [90, 91]. Recently, it has been found that HTLV-1 bZIP factor (HBZ) upregulates T-cell immunoglobulin and ITIM domain (TIGIT) and enhances expression of IL-10 for evading host immune response [92].

16.4.3 IL-13

Interleukin 13 (IL-13) is known as inflammation regulatory factor, mainly generated by B cells, T cells, and NK cells [93]. The main function of IL-13 is to induce IgE switching and CD23 expression in B cells, promoting antigen presentation ability of MHCII, inhibiting inflammation in human monocytes, and suppressing apoptosis [94–97]. Increasing evidence has shown that IL-13 collaborates with various viruses including EBV, KSHV, and HTLV-1 to promote tumorigenesis [61, 98–101]. In the EBV-associated Hodgkin lymphoma, the expression of IL-13 is upregulated, and the underlying molecular mechanism is that Zta serving as a EBV lytic protein elicits IL-13 production via directly binding to IL-13 promoter. Furthermore, neutralizing antibody against IL-13 suggests that IL-13 is vital for proliferation and latency of EBV-immortalized lymphoblastoid cell lines [98]. STAT6, a key downstream effector of IL-13, is a remarkable transcriptional factor whose constitutive phosphorylation has been indicated in controlling tumorigenesis [102]. Our group found that the constitutive phosphorylation of STAT6 is due to autocrine/paracrine of IL-13 and downregulation of SHP1 mediated by KSHV, which is closely associated with oncogenesis. Strikingly, neutralizing antibody against IL-13 suppresses the proliferation and survival of PEL, suggesting IL-13 plays a significant role in KSHV-associated latency and subsequent tumorigenesis [61]. Though previous studies reveal that IL-4, sharing the same receptor IL-4Rα/IL-13Rα1 with IL-13, also leads to the phosphorylation of STAT6, we found that IL-4/STAT6 pathway is negatively regulated in the KSHV-infected cells through dephosphorylation of STAT6 by latency-associated nuclear antigen (LANA), an important viral oncoprotein for maintaining viral latency [103, 104]. Similar to the effect of EBV and

KSHV, HTLV-1 Tax protein also induces production of IL-13, which is capable of promoting cell proliferation and anti-apoptosis of infected cells in an autocrine manner [96, 97, 99, 105]. This indicates that IL-13 is often targeted by oncogenic virus for host immune escape and cell survival.

16.4.4 IL-8

Chemokines are a large family of small proteins that regulates inflammation. The main property of chemokines is to attract immune cells to the site of inflammation, resulting from various causes including infection, autoimmune disease, and carcinomas [106]. Interleukin 8 (IL-8), a member of CXC chemokine subfamily, is responsible for recruiting neutrophils and T lymphocytes to the site of inflammation. Extensive evidence demonstrates that IL-8 is closely implicated in tumorigenesis such as breast cancer, gastric cancer, and pancreatic cancer [107–109]. Thus, it is not surprising that IL-8 was also found to participate in pathogen-related tumorigenesis. For example, EBV-encoded Zta protein activates IL-8 through binding to two elements within IL-8 promoter and subsequently upregulates IL-8 production, which is crucial for NPC development by recruiting infiltrates around infected cells [110]. In contrast to EBV, KSHV adopts different mechanisms to regulate IL-8 expression. It has been demonstrated that LANA-1 boosts IL-8 production to assist KSHV-infected cells in evading host immune response [111]. Similarly, in the context of lung adenocarcinomas, HPV16 infection upregulates IL-8 expression and in turn promotes angiogenesis and metastasis through inducing MMP2 and MMP9 [112], while the downregulation of IL-8 in the HPV-immortalized exocervical cells or primary keratinocytes could create a favorable microenvironment for HPV infection and subsequent tumorigenesis [113, 114]. In addition, previous studies showed that HCV infection could upregulate IL-8 expression and contribute to host immune tolerance and viral pathogenesis [115]. Interestingly, the similar phenomenon occurs to HBV, which was found to increase viral tolerance to IFN-α by inducing IL-8 production [116]. HTLV-1 also encodes Tax oncoprotein to activate IL-8 production and in turn contribute to HTLV-1-associated pathogenesis [117]. In bacteria-associated cancer, the discoveries from gastric epithelial cells exposed to *Helicobacter pylori* indicate that bacterial infection could also upregulate the expression of IL-8. The fact that high production of IL-8 is tightly associated with tumor cell proliferation, angiogenesis, and metastasis suggests that IL-8 plays a key role in *H. pylori*-associated gastric cancer [118–121].

16.4.5 CCL20

CCL20 is a member of CC chemokine subfamily and acts as a potent chemotaxin of immature dendritic cells, B lymphocytes, and T lymphocytes [122]. The main characteristic of CCL20 is to recruit immune cells to the site of inflammation, and in turn it is also involved in host immune response and tumorigenesis, such as breast adenocarcinoma, hepatocellular carcinoma, and pancreatic cancer [123–125]. Many reports have indicated that CCL20 is connected with pathogen-related tumorigenesis. For instance, in the EBV-positive Burkitt lymphoma (BL) cells or EBV-negative cells overexpressing LMP-1, the production of CCL20 is highly upregulated, indicating viral oncoprotein LMP-1 is involved in inducing chemokine CCL20 [126]. In addition, CCL20 is also upregulated by another EBV latent antigen called EBNA1 [127]. Further studies showed that high-level CCL20 could recruit Treg and is capable of inhibiting $CD4^+$ and $CD8^+$ T cells [127, 128]. By which, EBV-infected cells can inhibit host immune response and may promote tumorigenesis. Similarly, overexpressions of CCL20 and its receptor CCR6 are also observed in both KSHV-positive cells and HTLV-1-infected cells, which may drive virus-infected cells to migrate in an autocrine or paracrine manner. In contrast, high-risk HPVs were also found to escape immune response by downregulating CCL20 through E6 and E7 proteins [129].

16.5 Remarks and Perspectives

It is well known that metabolic stress within solid tumors is characterized by hypoxia, nutrient deprivation, oxidative stress, and lactic acidosis as a hostile microenvironment for the survival of cancer cells. Nonetheless, these adverse microenvironments have been successfully exploited by cancer cells and have been converted as driving force in the initiation and progression of cancer. The same cases have been extended to the mechanism by which oncogenic pathogen utilized to involve in carcinogenesis. Here, we have summarized the potential roles of metabolic stress like hypoxia, glucose starvation, and ROS accumulation in promoting viral oncoprotein-induced adaptive signaling change and oncogenesis. Among these, extracellular lactic acidosis has also been confirmed as a potent metabolic stress that plays a multiple role in promoting cancer progression. However, rare information was reported about the interaction between oncogenic pathogens and lactic acidosis stress. In addition, the consequence of the interplay between oncogenic pathogen and metabolic stress microenvironment is complicated and elusive. For examples, although the metabolic stress environment could drive tolerance change of cancer cells, the adaptive response strategies exploited by virus may not stable but adjust to the severity and duration of metabolic stress. On ther other hand, the metabolic stress factors are not isolated but cross talked, which may imply a cooperative or opposed effect of these selective stress on the same viral-mediated adaptive signaling pathway. Thus, a dynamic and comprehensive perspective in understanding oncogenesis mechanism induced by microenvironment abnormalities and oncogenic pathogen interaction will facilitate the development of a precise pathogen-specific therapeutic strategy.

Cytokines and chemokines are crucial factors that benefit not only hosts but also viruses. For hosts, cytokines and chemokines play a key role in regulating immune system, in order to limit and eliminate harmful pathogens. In contrast, for pathogens, many of them would adopt various mechanisms to evade host immune response by manipulating cytokines and chemokines. Moreover, many pathogens, in particular oncogenic viruses, even utilize cytokines and chemokines to promote persistent infection, even tumorigenesis [60]. In recent years, increasing evidence demonstrates the strong relationship between pathogens and carcinoma, and more and more cytokines and chemokines are proved to participate in pathogen-associated tumorigenesis, whereas the potential related to underlying mechanisms remains to be fully understood. As research exploited, more knowledge about virus-mediated tumorigenesis by manipulating cytokines and chemokines will be unveiled, and this knowledge could potentially be utilized to design therapies to defeat pathogen-related malignancies.

Acknowledgments The authors would like to apologize to the many researchers who have contributed to this area of research but have not been cited in this review due to space limitations. This work is supported by the National Key Research and Development Program of China (2016YFC1200400) and the National Natural Science Foundation of China (81471930, 81402542, 81672015). FW is a scholar of Pujiang Talents in Shanghai. QC is a scholar of New Century Excellent Talents in University of China.

References

1. Eales KL, Hollinshead KE, Tennant DA (2016) Hypoxia and metabolic adaptation of cancer cells. Oncogene 25:50
2. Pouyssegur J, Dayan F, Mazure NM (2006) Hypoxia signalling in cancer and approaches to enforce tumour regression. Nature 441:437–443
3. Semenza GL (2012) Hypoxia-inducible factors in physiology and medicine. Cell 148:399–408
4. Rankin EB, Giaccia AJ (2016) Hypoxic control of metastasis. Science 352:175–180
5. Fang JS, Gillies RD, Gatenby RA (2008) Adaptation to hypoxia and acidosis in carcinogenesis and tumor progression. Semin Cancer Biol 18:330–337
6. Patel A, Sant S (2016) Hypoxic tumor microenvironment: opportunities to develop targeted therapies. Biotechnol Adv 34:803–812
7. McLaughlin-Drubin ME, Munger K (2008) Viruses associated with human cancer. Biochim Biophys Acta 3:127–150
8. Purushothaman P, Uppal T, Verma SC (2015) Molecular biology of KSHV lytic reactivation. Virus 7:116–153
9. Bruick RK (2003) Oxygen sensing in the hypoxic response pathway: regulation of the hypoxia-inducible transcription factor. Genes Dev 17:2614–2623
10. Ratcliffe PJ (2013) Oxygen sensing and hypoxia signalling pathways in animals: the implications of physiology for cancer. J Physiol 591:2027–2042

11. Wouters BG, Koritzinsky M (2008) Hypoxia signalling through mTOR and the unfolded protein response in cancer. Nat Rev Cancer 8:851–864
12. Wang GL, Jiang BH, Rue EA, Semenza GL (1995) Hypoxia-inducible factor 1 is a basic-helix-loop-helix-PAS heterodimer regulated by cellular O2 tension. Proc Natl Acad Sci U S A 92:5510–5514
13. Lee JW, Bae SH, Jeong JW, Kim SH, Kim KW (2004) Hypoxia-inducible factor (HIF-1) alpha: its protein stability and biological functions. Exp Mol Med 36:1–12
14. Zhu C, Zhu Q, Wang C, Zhang L, Wei F, Cai Q (2016) Hostile takeover: manipulation of HIF-1 signaling in pathogen-associated cancers (review). Int J Oncol 49:1269–1276
15. Masoud GN, Li W (2015) HIF-1alpha pathway: role, regulation and intervention for cancer therapy. Acta Pharm Sin B 5:378–389
16. Jham BC, Ma T, Hu J, Chaisuparat R, Friedman ER, Pandolfi PP, Schneider A, Sodhi A, Montaner S (2011) Amplification of the angiogenic signal through the activation of the TSC/mTOR/HIF axis by the KSHV vGPCR in Kaposi's sarcoma. PLoS One 6:0019103
17. Wakisaka N, Kondo S, Yoshizaki T, Murono S, Furukawa M, Pagano JS (2004) Epstein-Barr virus latent membrane protein 1 induces synthesis of hypoxia-inducible factor 1 alpha. Mol Cell Biol 24:5223–5234
18. Liu F, Lin B, Liu X, Zhang W, Zhang E, Hu L, Ma Y, Li X, Tang X (2016) ERK signaling pathway is involved in HPV-16 E6 but not E7 Oncoprotein-induced HIF-1alpha protein accumulation in NSCLC cells. Oncol Res 23:109–118
19. Tomita M, Semenza GL, Michiels C, Matsuda T, Uchihara JN, Okudaira T, Tanaka Y, Taira N, Ohshiro K, Mori N (2007) Activation of hypoxia-inducible factor 1 in human T-cell leukaemia virus type 1-infected cell lines and primary adult T-cell leukaemia cells. Biochem J 406:317–323
20. Sung WW, Chu YC, Chen PR, Liao MH, Lee JW (2016) Positive regulation of HIF-1A expression by EBV oncoprotein LMP1 in nasopharyngeal carcinoma cells. Cancer Lett 382:21–31
21. Maxwell PH, Wiesener MS, Chang GW, Clifford SC, Vaux EC, Cockman ME, Wykoff CC, Pugh CW, Maher ER, Ratcliffe PJ (1999) The tumour suppressor protein VHL targets hypoxia-inducible factors for oxygen-dependent proteolysis. Nature 399:271–275
22. Semenza GL (2001) HIF-1, O(2), and the 3 PHDs: how animal cells signal hypoxia to the nucleus. Cell 107:1–3
23. Cai Q, Murakami M, Si H, Robertson ES (2007) A potential alpha-helix motif in the amino terminus of LANA encoded by Kaposi's sarcoma-associated herpesvirus is critical for nuclear accumulation of HIF-1alpha in normoxia. J Virol 81:10413–10423
24. Kondo S, Seo SY, Yoshizaki T, Wakisaka N, Furukawa M, Joab I, Jang KL, Pagano JS (2006) EBV latent membrane protein 1 up-regulates hypoxia-inducible factor 1alpha through Siah1-mediated down-regulation of prolyl hydroxylases 1 and 3 in nasopharyngeal epithelial cells. Cancer Res 66:9870–9877
25. Shin YC, Joo CH, Gack MU, Lee HR, Jung JU (2008) Kaposi's sarcoma-associated herpesvirus viral IFN regulatory factor 3 stabilizes hypoxia-inducible factor-1 alpha to induce vascular endothelial growth factor expression. Cancer Res 68:1751–1759
26. Darekar S, Georgiou K, Yurchenko M, Yenamandra SP, Chachami G, Simos G, Klein G, Kashuba E (2012) Epstein-Barr virus immortalization of human B-cells leads to stabilization of hypoxia-induced factor 1 alpha, congruent with the Warburg effect. PLoS One 7:27
27. Yoo YG, Cho S, Park S, Lee MO (2004) The carboxy-terminus of the hepatitis B virus X protein is necessary and sufficient for the activation of hypoxia-inducible factor-1alpha. FEBS Lett 577:121–126
28. Yoo YG, Na TY, Seo HW, Seong JK, Park CK, Shin YK, Lee MO (2008) Hepatitis B virus X protein induces the expression of MTA1 and HDAC1, which enhances hypoxia signaling in hepatocellular carcinoma cells. Oncogene 27:3405–3413

29. Kallio PJ, Okamoto K, O'Brien S, Carrero P, Makino Y, Tanaka H, Poellinger L (1998) Signal transduction in hypoxic cells: inducible nuclear translocation and recruitment of the CBP/p300 coactivator by the hypoxia-inducible factor-1alpha. EMBO J 17:6573–6586
30. Yoo YG, Oh SH, Park ES, Cho H, Lee N, Park H, Kim DK, Yu DY, Seong JK, Lee MO (2003) Hepatitis B virus X protein enhances transcriptional activity of hypoxia-inducible factor-1alpha through activation of mitogen-activated protein kinase pathway. J Biol Chem 278:39076–39084
31. Sodhi A, Montaner S, Patel V, Zohar M, Bais C, Mesri EA, Gutkind JS (2000) The Kaposi's sarcoma-associated herpes virus G protein-coupled receptor up-regulates vascular endothelial growth factor expression and secretion through mitogen-activated protein kinase and p38 pathways acting on hypoxia-inducible factor 1alpha. Cancer Res 60:4873–4880
32. Bodily JM, Mehta KP, Laimins LA (2011) Human papillomavirus E7 enhances hypoxia-inducible factor 1-mediated transcription by inhibiting binding of histone deacetylases. Cancer Res 71:1187–1195
33. Liu L, Simon MC (2004) Regulation of transcription and translation by hypoxia. Cancer Biol Ther 3:492–497
34. Laplante M, Sabatini DM (2009) mTOR signaling at a glance. J Cell Sci 122:3589–3594
35. Brugarolas J, Lei K, Hurley RL, Manning BD, Reiling JH, Hafen E, Witters LA, Ellisen LW, Kaelin WG Jr (2004) Regulation of mTOR function in response to hypoxia by REDD1 and the TSC1/TSC2 tumor suppressor complex. Genes Dev 18:2893–2904
36. Bernardi R, Guernah I, Jin D, Grisendi S, Alimonti A, Teruya-Feldstein J, Cordon-Cardo C, Simon MC, Rafii S, Pandolfi PP (2006) PML inhibits HIF-1alpha translation and neoangiogenesis through repression of mTOR. Nature 442:779–785
37. Li Y, Wang Y, Kim E, Beemiller P, Wang CY, Swanson J, You M, Guan KL (2007) Bnip3 mediates the hypoxia-induced inhibition on mammalian target of rapamycin by interacting with Rheb. J Biol Chem 282:35803–35813
38. Connolly E, Braunstein S, Formenti S, Schneider RJ (2006) Hypoxia inhibits protein synthesis through a 4E-BP1 and elongation factor 2 kinase pathway controlled by mTOR and uncoupled in breast cancer cells. Mol Cell Biol 26:3955–3965
39. Populo H, Lopes JM, Soares P (2012) The mTOR signalling pathway in human cancer. Int J Mol Sci 13:1886–1918
40. Lu Z, Hu X, Li Y, Zheng L, Zhou Y, Jiang H, Ning T, Basang Z, Zhang C, Ke Y (2004) Human papillomavirus 16 E6 oncoprotein interferences with insulin signaling pathway by binding to tuberin. J Biol Chem 279:35664–35670
41. Spangle JM, Munger K (2010) The human papillomavirus type 16 E6 oncoprotein activates mTORC1 signaling and increases protein synthesis. J Virol 84:9398–9407
42. Yen CJ, Lin YJ, Yen CS, Tsai HW, Tsai TF, Chang KY, Huang WC, Lin PW, Chiang CW, Chang TT (2012) Hepatitis B virus X protein upregulates mTOR signaling through IKKbeta to increase cell proliferation and VEGF production in hepatocellular carcinoma. PLoS One 7:27
43. George A, Panda S, Kudmulwar D, Chhatbar SP, Nayak SC, Krishnan HH (2012) Hepatitis C virus NS5A binds to the mRNA cap-binding eukaryotic translation initiation 4F (eIF4F) complex and up-regulates host translation initiation machinery through eIF4E-binding protein 1 inactivation. J Biol Chem 287:5042–5058
44. Panda S, Vedagiri D, Viveka TS, Harshan KH (2014) A unique phosphorylation-dependent eIF4E assembly on 40S ribosomes co-ordinated by hepatitis C virus protein NS5A that activates internal ribosome entry site translation. Biochem J 462:291–302
45. Fiaschi T, Chiarugi P (2012) Oxidative stress, tumor microenvironment, and metabolic reprogramming: a diabolic liaison. Int J Cell Biol 762825:13
46. Marambio P, Toro B, Sanhueza C, Troncoso R, Parra V, Verdejo H, Garcia L, Quiroga C, Munafo D, Diaz-Elizondo J, Bravo R, Gonzalez MJ, Diaz-Araya G, Pedrozo Z, Chiong M, Colombo MI, Lavandero S (2010) Glucose deprivation causes oxidative stress and stimulates

aggresome formation and autophagy in cultured cardiac myocytes. Biochim Biophys Acta 6:509–518
47. Clanton TL (1985) Hypoxia-induced reactive oxygen species formation in skeletal muscle. J Appl Physiol 102:2379–2388
48. Martindale JL, Holbrook NJ (2002) Cellular response to oxidative stress: signaling for suicide and survival. J Cell Physiol 192:1–15
49. Haupt Y, Maya R, Kazaz A, Oren M (1997) Mdm2 promotes the rapid degradation of p53. Nature 387:296–299
50. Meek DW (2004) The p53 response to DNA damage. DNA Repair 3:1049–1056
51. Chen X, Ko LJ, Jayaraman L, Prives C (1996) p53 levels, functional domains, and DNA damage determine the extent of the apoptotic response of tumor cells. Genes Dev 10:2438–2451
52. Shin YC, Nakamura H, Liang X, Feng P, Chang H, Kowalik TF, Jung JU (2006) Inhibition of the ATM/p53 signal transduction pathway by Kaposi's sarcoma-associated herpesvirus interferon regulatory factor 1. J Virol 80:2257–2266
53. Sato Y, Shirata N, Kudoh A, Iwahori S, Nakayama S, Murata T, Isomura H, Nishiyama Y, Tsurumi T (2009) Expression of Epstein-Barr virus BZLF1 immediate-early protein induces p53 degradation independent of MDM2, leading to repression of p53-mediated transcription. Virology 388:204–211
54. Nishimura T, Kohara M, Izumi K, Kasama Y, Hirata Y, Huang Y, Shuda M, Mukaidani C, Takano T, Tokunaga Y, Nuriya H, Satoh M, Saito M, Kai C, Tsukiyama-Kohara K (2009) Hepatitis C virus impairs p53 via persistent overexpression of 3beta-hydroxysterol Delta24-reductase. J Biol Chem 284:36442–36452
55. Friborg J Jr, Kong W, Hottiger MO, Nabel GJ (1999) p53 inhibition by the LANA protein of KSHV protects against cell death. Nature 402:889–894
56. Chudasama P, Konrad A, Jochmann R, Lausen B, Holz P, Naschberger E, Neipel F, Britzen-Laurent N, Sturzl M (2015) Structural proteins of Kaposi's sarcoma-associated herpesvirus antagonize p53-mediated apoptosis. Oncogene 34:639–649
57. Yi F, Saha A, Murakami M, Kumar P, Knight JS, Cai Q, Choudhuri T, Robertson ES (2009) Epstein-Barr virus nuclear antigen 3C targets p53 and modulates its transcriptional and apoptotic activities. Virology 388:236–247
58. Cai Q, Guo Y, Xiao B, Banerjee S, Saha A, Lu J, Glisovic T, Robertson ES (2011) Epstein-Barr virus nuclear antigen 3C stabilizes Gemin3 to block p53-mediated apoptosis. PLoS Pathog 7:8
59. Li L, Li W, Xiao L, Xu J, Chen X, Tang M, Dong Z, Tao Q, Cao Y (2012) Viral oncoprotein LMP1 disrupts p53-induced cell cycle arrest and apoptosis through modulating K63-linked ubiquitination of p53. Cell Cycle 11:2327–2336
60. Mogensen TH, Paludan SR (2001) Molecular pathways in virus-induced cytokine production. Microbiol Mol Biol Rev 65:131–150
61. Wang C, Zhu C, Wei F, Zhang L, Mo X, Feng Y, Xu J, Yuan Z, Robertson E, Cai Q (2015) Constitutive activation of interleukin-13/STAT6 contributes to Kaposi's sarcoma-associated Herpesvirus-related primary effusion lymphoma cell proliferation and survival. J Virol 89:10416–10426
62. Sin S-H, Dittmer DP (2012) Cytokine homologs of human gammaherpesviruses. J Interf Cytokine Res 32:53–59
63. Yokoi T, Miyawaki T, Yachie A, Kato K, Kasahara Y, Taniguchi N (1990) Epstein-Barr virus-immortalized B cells produce IL-6 as an autocrine growth factor. Immunology 70:100
64. Tosato G, Tanner J, Jones K, Revel M, Pike S (1990) Identification of interleukin-6 as an autocrine growth factor for Epstein-Barr virus-immortalized B cells. J Virol 64:3033–3041
65. Cordano P, Lake A, Shield L, Taylor G, Alexander FE, Taylor PR, White J, Jarrett RF (2005) Effect of IL-6 promoter polymorphism on incidence and outcome in Hodgkin's lymphoma. Br J Haematol 128:493–495
66. Suthaus J, Adam N, Grötzinger J, Scheller J, Rose-John S (2011) Viral interleukin-6: structure, pathophysiology and strategies of neutralization. Eur J Cell Biol 90:495–504

67. Jones KD, Aoki Y, Chang Y, Moore PS, Yarchoan R, Tosato G (1999) Involvement of interleukin-10 (IL-10) and viral IL-6 in the spontaneous growth of Kaposi's sarcoma herpesvirus-associated infected primary effusion lymphoma cells. Blood 94:2871–2879
68. Aoki Y, Tosato G (1999) Role of vascular endothelial growth factor/vascular permeability factor in the pathogenesis of Kaposi's sarcoma-associated herpesvirus-infected primary effusion lymphomas. Blood 94:4247–4254
69. Zhang Y-J, Bonaparte RS, Patel D, Stein DA, Iversen PL (2008) Blockade of viral interleukin-6 expression of Kaposi's sarcoma–associated herpesvirus. Mol Cancer Ther 7:712–720
70. Drexler H, Meyer C, Gaidano G, Carbone A (1999) Constitutive cytokine production by primary effusion (body cavity-based) lymphoma-derived cell lines. Leukemia 08876924:13
71. Chatterjee M, Osborne J, Bestetti G, Chang Y, Moore PS (2002) Viral IL-6-induced cell proliferation and immune evasion of interferon activity. Science (New York, NY) 298:1432–1435
72. Chen M, Sun F, Han L, Qu Z (2016) Kaposi's sarcoma herpesvirus (KSHV) microRNA K12-1 functions as an oncogene by activating NF-κB/IL-6/STAT3 signaling. Oncotarget 7:33363–33373
73. Walboomer JM, Acos MV, Manos MM, Xavier Bosch F, Kummer JA (1999) Human papillomavirus is a necessary cause of invasive cervical cancer. Worldwide J pathol 189:12–19
74. Ren C, Cheng X, Lu B, Yang G (2013) Activation of interleukin-6/signal transducer and activator of transcription 3 by human papillomavirus early proteins 6 induces fibroblast senescence to promote cervical tumourigenesis through autocrine and paracrine pathways in tumour microenvironment. Eur J Cancer (Oxford, England : 1990) 49:3889–3899
75. Tang Y, Kitisin K, Jogunoori W, Li C, Deng C-X, Mueller SC, Ressom HW, Rashid A, He AR, Mendelson JS (2008) Progenitor/stem cells give rise to liver cancer due to aberrant TGF-β and IL-6 signaling. Proc Natl Acad Sci 105:2445–2450
76. Horiuchi S, Yamamoto N, Dewan M, Takahashi Y, Yamashita A, Yoshida T, Nowell MA, Richards PJ, Jones SA, Yamamoto N (2006) Human T-cell leukemia virus type-I tax induces expression of interleukin-6 receptor (IL-6R): shedding of soluble IL-6R and activation of STAT3 signaling. Int J Cancer 119:823–830
77. Uno K, Kato K, Shimosegawa T (2014) Novel role of toll-like receptors in helicobacter pylori-induced gastric malignancy. World J Gastroenterol 20:5244–5251
78. Tye H, Jenkins BJ (2013) Tying the knot between cytokine and toll-like receptor signaling in gastrointestinal tract cancers. Cancer Sci 104:1139–1145
79. Grivennikov SI, Karin M (2010) Dangerous liaisons: STAT3 and NF-κB collaboration and crosstalk in cancer. Cytokine Growth Factor Rev 21:11–19
80. Deng J-Y, Sun D, Liu X-Y, Pan Y, Liang H (2010) STAT-3 correlates with lymph node metastasis and cell survival in gastric cancer. World J Gastroenterol 16:5380–5387
81. Tye H, Kennedy CL, Najdovska M, McLeod L, McCormack W, Hughes N, Dev A, Sievert W, Ooi CH, T-o I (2012) STAT3-driven upregulation of TLR2 promotes gastric tumorigenesis independent of tumor inflammation. Cancer Cell 22:466–478
82. Fiorentino DF, Bond MW, Mosmann T (1989) Two types of mouse T helper cell. IV. Th2 clones secrete a factor that inhibits cytokine production by Th1 clones. J Exp Med 170:2081–2095
83. Moore KW, de Waal MR, Coffman RL, O'Garra A (2001) Interleukin-10 and the interleukin-10 receptor. Annu Rev Immunol 19:683–765
84. Jochum S, Moosmann A, Lang S, Hammerschmidt W, Zeidler R (2012) The EBV immunoevasins vIL-10 and BNLF2a protect newly infected B cells from immune recognition and elimination. PLoS Pathog 8:e1002704
85. de Waal MR, Haanen J, Spits H, Roncarolo M-G, Te Velde A, Figdor C, Johnson K, Kastelein R, Yssel H, De Vries JE (1991) Interleukin 10 (IL-10) and viral IL-10 strongly reduce antigen-specific human T cell proliferation by diminishing the antigen-presenting capacity of monocytes via downregulation of class II major histocompatibility complex expression. J Exp Med 174:915–924

86. Vockerodt M, Haier B, Buttgereit P, Tesch H, Kube D (2001) The Epstein-Barr virus latent membrane protein 1 induces interleukin-10 in Burkitt's lymphoma cells but not in Hodgkin's cells involving the p38/SAPK2 pathway. Virology 280:183–198
87. Mahot S, Sergeant A, Drouet E, Gruffat H (2003) A novel function for the Epstein–Barr virus transcription factor EB1/Zta: induction of transcription of the hIL-10 gene. J Gen Virol 84:965–974
88. Samanta M, Iwakiri D, Takada K (2008) Epstein-Barr virus-encoded small RNA induces IL-10 through RIG-I-mediated IRF-3 signaling. Oncogene 27:4150–4160
89. Prata TT, Bonin CM, Ferreira AM, Padovani CT, Fernandes CE, Machado AP, Tozetti IA (2015) Local immunosuppression induced by high viral load of human papillomavirus: characterization of cellular phenotypes producing interleukin-10 in cervical neoplastic lesions. Immunology 146:113–121
90. Das A, Ellis G, Pallant C, Lopes AR, Khanna P, Peppa D, Chen A, Blair P, Dusheiko G, Gill U (2012) IL-10–producing regulatory B cells in the pathogenesis of chronic hepatitis B virus infection. J Immunol 189:3925–3935
91. Xue H, Lin F, Tan H, Zhu Z-Q, Zhang Z-Y, Zhao L (2016) Overrepresentation of IL-10-expressing B cells suppresses cytotoxic CD4+ T cell activity in HBV-induced hepatocellular carcinoma. PLoS One 11:e0154815
92. Yasuma K, J-i Y, Takemoto K, Sugata K, Mitobe Y, Takenouchi N, Nakagawa M, Suzuki Y, Matsuoka M (2016) HTLV-1 bZIP factor impairs anti-viral immunity by inducing co-inhibitory molecule, T cell immunoglobulin and ITIM domain (TIGIT). PLoS Pathog 12:e1005372
93. Wynn TA (2003) IL-13 effector functions*. Annu Rev Immunol 21:425–456
94. de Waal MR, Figdor CG, Huijbens R, Mohan-Peterson S, Bennett B, Culpepper J, Dang W, Zurawski G, de Vries JE (1993) Effects of IL-13 on phenotype, cytokine production, and cytotoxic function of human monocytes. Comparison with IL-4 and modulation by IFN-gamma or IL-10. J Immunol 151:6370–6381
95. Punnonen J, Aversa G, Cocks BG, McKenzie A, Menon S, Zurawski G, de Waal MR, de Vries JE (1993) Interleukin 13 induces interleukin 4-independent IgG4 and IgE synthesis and CD23 expression by human B cells. Proc Natl Acad Sci 90:3730–3734
96. Manna SK, Aggarwal BB (1998) IL-13 suppresses TNF-induced activation of nuclear factor-κB, activation protein-1, and apoptosis. J Immunol 161:2863–2872
97. Relić B, Guicheux J, Mezin F, Lubberts E, Togninalli D, Garcia I, van den Berg WB, Guerne P-A (2001) IL-4 and IL-13, but not IL-10, protect human synoviocytes from apoptosis. J Immunol 166:2775–2782
98. Tsai SC, Lin SJ, Chen PW, Luo WY, Yeh TH, Wang HW, Chen CJ, Tsai CH (2009) EBV Zta protein induces the expression of interleukin-13, promoting the proliferation of EBV-infected B cells and lymphoblastoid cell lines. Blood 114:109–118
99. Silbermann K, Schneider G, Grassmann R (2008) Stimulation of interleukin-13 expression by human T-cell leukemia virus type 1 oncoprotein tax via a dually active promoter element responsive to NF-κB and NFAT. J Gen Virol 89:2788–2798
100. Wäldele K, Schneider G, Ruckes T, Grassmann R (2004) Interleukin-13 overexpression by tax transactivation: a potential autocrine stimulus in human T-cell leukemia virus-infected lymphocytes. J Virol 78:6081–6090
101. Chung H-K, Young HA, Goon PK, Heidecker G, Princler GL, Shimozato O, Taylor GP, Bangham CR, Derse D (2003) Activation of interleukin-13 expression in T cells from HTLV-1-infected individuals and in chronically infected cell lines. Blood 102:4130–4136
102. Bruns HA, Kaplan MH (2006) The role of constitutively active Stat6 in leukemia and lymphoma. Crit Rev Oncol Hematol 57:245–253
103. Cai Q, Verma SC, Choi J-Y, Ma M, Robertson ES (2010) Kaposi's sarcoma-associated herpesvirus inhibits interleukin-4-mediated STAT6 phosphorylation to regulate apoptosis and maintain latency. J Virol 84:11134–11144

104. Wang C, Wei F, Cai Q (2016) Deregulation of IL-4/IL-13-induced STAT6 signaling in viral oncogenesis. Oncol Commun 1:e1131
105. Lømo J, Blomhoff HK, Jacobsen SE, Krajewski S, Reed JC, Smeland EB (1997) Interleukin-13 in combination with CD40 ligand potently inhibits apoptosis in human B lymphocytes: upregulation of Bcl-xL and Mcl-1. Blood 89:4415–4424
106. Jundi K, Greene CM (2015) Transcription of interleukin-8: how altered regulation can affect cystic fibrosis lung disease. Biomol Ther 5:1386–1398
107. Matsuo Y, Ochi N, Sawai H, Yasuda A, Takahashi H, Funahashi H, Takeyama H, Tong Z, Guha S (2009) CXCL8/IL-8 and CXCL12/SDF-1α co-operatively promote invasiveness and angiogenesis in pancreatic cancer. Int J Cancer 124:853–861
108. Lu W, Pan K, Zhang L, Lin D, Miao X, You W (2005) Genetic polymorphisms of interleukin (IL)-1B, IL-1RN, IL-8, IL-10 and tumor necrosis factor α and risk of gastric cancer in a Chinese population. Carcinogenesis 26:631–636
109. Freund A, Chauveau C, Brouillet J-P, Lucas A, Lacroix M, Licznar A, Vignon F, Lazennec G (2003) IL-8 expression and its possible relationship with estrogen-receptor-negative status of breast cancer cells. Oncogene 22:256–265
110. Hsu M, Wu SY, Chang SS, Su IJ, Tsai CH, Lai SJ, Shiau AL, Takada K, Chang Y (2008) Epstein-Barr virus lytic transactivator Zta enhances chemotactic activity through induction of interleukin-8 in nasopharyngeal carcinoma cells. J Virol 82:3679–3688
111. Li X, Liang D, Lin X, Robertson ES, Lan K (2011) Kaposi's sarcoma-associated herpesvirus-encoded latency-associated nuclear antigen reduces interleukin-8 expression in endothelial cells and impairs neutrophil chemotaxis by degrading nuclear p65. J Virol 85:8606–8615
112. Shiau M-Y, Fan L-C, Yang S-C, Tsao C-H, Lee H, Cheng Y-W, Lai L-C, Chang Y-H (2013) Human papillomavirus up-regulates MMP-2 and MMP-9 expression and activity by inducing interleukin-8 in lung adenocarcinomas. PLoS One 8:e54423
113. Woodworth C, Simpson S (1993) Comparative lymphokine secretion by cultured normal human cervical keratinocytes, papillomavirus-immortalized, and carcinoma cell lines. Am J Pathol 142:1544
114. Huang S-M, McCance D (2002) Down regulation of the interleukin-8 promoter by human papillomavirus type 16 E6 and E7 through effects on CREB binding protein/p300 and P/CAF. J Virol 76:8710–8721
115. Polyak SJ, Khabar KS, Paschal DM, Ezelle HJ, Duverlie G, Barber GN, Levy DE, Mukaida N, Gretch DR (2001) Hepatitis C virus nonstructural 5A protein induces interleukin-8, leading to partial inhibition of the interferon-induced antiviral response. J Virol 75:6095–6106
116. Pollicino T, Bellinghieri L, Restuccia A, Raffa G, Musolino C, Alibrandi A, Teti D, Raimondo G (2013) Hepatitis B virus (HBV) induces the expression of interleukin-8 that in turn reduces HBV sensitivity to interferon-alpha. Virology 444:317–328
117. Mori N, Murakami S, Oda S, Prager D, Eto S (1995) Production of interleukin 8 in adult T-cell leukemia cells: possible transactivation of the interleukin 8 gene by human T-cell leukemia virus type I tax. Cancer Res 55:3592–3597
118. Asfaha S, Dubeykovskiy AN, Tomita H, Yang X, Stokes S, Shibata W, Friedman RA, Ariyama H, Dubeykovskaya ZA, Muthupalani S (2013) Mice that express human interleukin-8 have increased mobilization of immature myeloid cells, which exacerbates inflammation and accelerates colon carcinogenesis. Gastroenterology 144:155–166
119. Beales IL, Calam J (1997) Stimulation of IL-8 production in human gastric epithelial cells byhelicobacter pylori, IL-1β and TNF-α requires tyrosine kinase activity, but not protein kinase C. Cytokine 9:514–520
120. Kitadai Y, Haruma K, Sumii K, Yamamoto S, Ue T, Yokozaki H, Yasui W, Ohmoto Y, Kajiyama G, Fidler IJ (1998) Expression of interleukin-8 correlates with vascularity in human gastric carcinomas. Am J Pathol 152:93
121. Lee KE, Khoi PN, Xia Y, Park JS, Joo YE, Kim KK, Choi SY, Jung YD (2013) Helicobacter pylori and interleukin-8 in gastric cancer. World J Gastroenterol 19:8192–8202

122. Schutyser E, Struyf S, Van Damme J (2003) The CC chemokine CCL20 and its receptor CCR6. Cytokine Growth Factor Rev 14:409–426
123. Bell D, Chomarat P, Broyles D, Netto G, Harb GM, Lebecque S, Valladeau J, Davoust J, Palucka KA, Banchereau J (1999) In breast carcinoma tissue, immature dendritic cells reside within the tumor, whereas mature dendritic cells are located in peritumoral areas. J Exp Med 190:1417–1426
124. Shimizu Y, Murata H, Kashii Y, Hirano K, Kunitani H, Higuchi K, Watanabe A (2001) CC-chemokine receptor 6 and its ligand macrophage inflammatory protein 3α might be involved in the amplification of local necroinflammatory response in the liver. Hepatology 34:311–319
125. Kleeff J, Kusama T, Rossi DL, Ishiwata T, Maruyama H, Friess H, Büchler MW, Zlotnik A, Korc M (1999) Detection and localization of MIP-3α/LARC/exodus, a macrophage proinflammatory chemokine, and its CCR6 receptor in human pancreatic cancer. Int J Cancer 81:650–657
126. Okudaira T, Yamamoto K, Kawakami H, Uchihara JN, Tomita M, Masuda M, Matsuda T, Sairenji T, Iha H, Jeang KT (2006) Retracted: transactivation of CCL20 gene by Epstein–Barr virus latent membrane protein 1. Br J Haematol 132:293–302
127. Baumforth KR, Birgersdotter A, Reynolds GM, Wei W, Kapatai G, Flavell JR, Kalk E, Piper K, Lee S, Machado L (2008) Expression of the Epstein-Barr virus-encoded Epstein-Barr virus nuclear antigen 1 in Hodgkin's lymphoma cells mediates up-regulation of CCL20 and the migration of regulatory T cells. Am J Pathol 173:195–204
128. Satoh T, Wada R, Yajima N, Imaizumi T, Yagihashi S (2014) Tumor microenvironment and RIG-I signaling molecules in Epstein Barr virus-positive and -negative classical Hodgkin lymphoma of the elderly. J Clin Exp Hematopathol: JCEH 54:75–84
129. Jiang B, Xue M (2015) Correlation of E6 and E7 levels in high-risk HPV16 type cervical lesions with CCL20 and Langerhans cells. Genet Mol Res 14:10473–10481

MIX
Papier aus verantwortungsvollen Quellen
Paper from responsible sources
FSC® C105338

If you have any concerns about our products,
you can contact us on
ProductSafety@springernature.com

In case Publisher is established outside the EU,
the EU authorized representative is:
Springer Nature Customer Service Center GmbH
Europaplatz 3, 69115 Heidelberg, Germany

Printed by Libri Plureos GmbH
in Hamburg, Germany